Lockhart & Wiseman's

Crop Husbandry
including grassland

SEVENTH EDITION

by

A. J. L. WISEMAN
H. J. S. FINCH

and

A. M. SAMUEL

PERGAMON PRESS

OXFORD · NEW YORK · SEOUL · TOKYO

UK	Pergamon Press Ltd, Headington Hill Hall, Oxford OX3 0BW, England
USA	Pergamon Press Inc., 660 White Plains Road, Tarrytown, New York 10591-5153, USA
KOREA	Pergamon Press Korea, KPO Box 315, Seoul 110-603, Korea
JAPAN	Pergamon Press Japan, Tsunashima Building Annex, 3-20-12 Yushima, Bunkyo-ku, Tokyo 113, Japan

First edition 1966
Second edition 1970
Third edition 1975
Reprinted 1976
Fourth edition 1978
Reprinted (with additions) 1980
Fifth edition 1983
Reprinted 1984
Sixth edition 1988
Seventh edition 1993

Library of Congress Cataloging in Publication Data
Wiseman, A. J. L.
[crop husbandry and grassland]
Lockhart & Wiseman's crop husbandry and grassland —
7th ed. / by A. J. L. Wiseman, S. J. Finch, and
A. M. Samuel.
p.   cm.
Rev. ed. of: crop husbandry and grassland / by J. A. R.
Lockhart and A. J. L. Wiseman. 6th ed. c1988.
Includes bibliographical references and index.
1. Crops. 2. Agriculture. 3. Crops—Great Britain
4. Agriculture—Great Britain. I. Finch, S. J. (Steve J.)
II. Samuel, A. M. (Alison M.) III. Lockhart, J. A. R.
crop husbandry and grassland. IV. Title. V. Title: Lockhart
and Wiseman's crop husbandry and grassland. VI. Title:
crop husbandry and grassland.
SB98.W58 1993
630'.941—dc20   93-8466

ISBN 0 08 042002 8 hardcover
ISBN 0 08 042003 6 flexicover

Printed in Great Britain by B.P.P.C. Wheatons Ltd, Exeter

PREFACE

MY co-author, Jim Lockhart, has now retired, and without his experience and expertise I was faced with the task of compiling a new edition of what we understand is generally recognized as the standard textbook on crop husbandry. So specialized has the subject now become that it is almost impossible for one person alone to do it adequately. It is fortunate, therefore, that Alison Samuel (now Lecturer at the Lincolnshire College of Agriculture) and Steve Finch (Senior Lecturer at the Royal Agricultural College) agreed to join me as co-authors for this edition.

New authors have new ideas, and it was inevitable that my colleagues would, in some instances, wish to present a somewhat different approach to a particular aspect of a subject. This they have done and, at the same time, we have split into six single chapters the previously very long chapter on Cropping; these new chapters also include a number of the previous Appendices. All this should make for easier reference.

As well as the alterations which are inevitable in any new edition, we have enlarged on the changes that were beginning to take place in UK cropping (and which we discussed) when the sixth edition was being prepared. In spite of increasing restrictions on production, there is still a food surplus in the western half of the world. CAP changes, and yet more restrictions, will slow down production even further and, in not just a few cases, make the crop producer's task that much more untenable.

Set-Aside, the various Codes of Good Agricultural Practice, the Control of Substances Hazardous to Health (COSHH) Regulations and other legislation concerning the use of pesticides, are all discussed in this edition. Additionally, and as in the past, we have endeavoured to update on new husbandry techniques, weed, pest and disease control.

The COSHH legislation now has to be seriously adhered to by UK growers. The vast majority of them accept this as a necessary, almost welcome, burden and particular attention is drawn to these regulations where relevant.

The Grassland chapter shows some changes from the previous edition. More emphasis has been given to site class and how this can assist the farmer in a perhaps more judicious use of nitrogen fertilizer on grassland. It is appreciated, of course, that with our fickle climate and varying soil types, grass farming can only be an approximate science.

Today's thinking and pressures on the countryside inevitably bring farming, wildlife and countryside conservation closer together. Julian Hosking, one of our colleagues, has kindly updated the previous Appendix on this subject which has now been moved to the chapter on Cropping in the United Kingdom.

We are also glad to acknowledge the help given to us by Barbara Hart and Ray Churchill in the revision of the chapter on Pests and Diseases. Other colleagues have also given us helpful advice.

Three of us are, in effect, responsible for this edition and we are all very grateful to Diana Wiseman who has spent many hours assisting me in putting the text together. Hopefully, it now has a reasonably uniform style without prejudice to the technical contributions from the respective authors.

Cirencester   A. J. L. WISEMAN
May 1993

CONTENTS

1

PLANTS

PLANTS are living organisms consisting of innumerable tiny cells. They differ from animals in many ways and a very important difference is that they can build up valuable organic substances from simple materials. The most important part of this building process, called *photosynthesis*, is the production of *carbohydrates* such as *sugars*, *starches* and *cellulose*.

Photosynthesis

In photosynthesis a special green substance called chlorophyll uses light energy (normally sunlight) to change carbon dioxide and water into sugars (carbohydrates) in the green parts of the plant. The daily amount of photosynthesis is limited by the duration and intensity of sunlight. The amount of carbon dioxide available is a limiting factor. Shortage of water and low temperatures can also reduce photosynthesis.

The cells which contain chlorophyll also have yellow pigments such as carotene. Crop plants can only build up chlorophyll in the light and so any leaves which develop in the dark are yellow and cannot produce carbohydrates.

Oxygen is released during photosynthesis and the process may be set out as follows:

This process not only provides the basis for all food production but it also supplies the oxygen which animals and plants need for respiration.

The simple carbohydrates, such as glucose, may build up to form starch for storage purposes or to cellulose for building cell walls. Fats and oils are formed from carbohydrates. Protein material, which is an essential part of all living cells, is made from carbohydrates and nitrogen compounds.

Most plants consist of roots, stems, leaves and reproductive parts and need soil in which to grow.

The roots spread through the spaces between the particles in the soil and anchor the plant. In a plant such as wheat the root system may total many miles.

The leaves, with their broad surfaces, are the main parts of the plant where photosynthesis occurs (Fig. 1).

A very important feature of the leaf structure is the presence of large numbers of tiny pores (stomata) on the surface of the leaf (Fig. 2). There are usually thousands of stomata per square centimetre of leaf surface. Each pore (stoma) is oval-shaped and surrounded by two guard cells. When the guard cells are turgid (full of water) the stoma is open and when they lose water the stoma closes.

$$\text{Carbon dioxide} + \text{water} + \text{energy} \xrightarrow{\text{chlorophyll}} \text{carbohydrates} + \text{oxygen}$$
$$nCO_2 \qquad nH_2O \quad \text{(light)} \quad = \quad (CH_2O)n \qquad nO_2$$

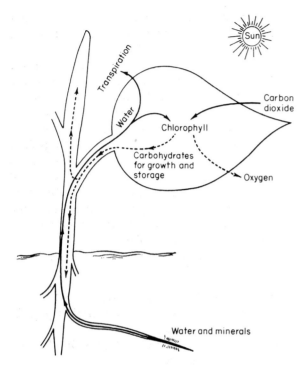

FIG. 1. Photosynthesis.

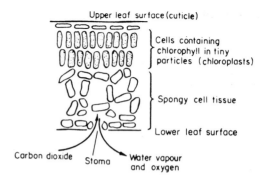

FIG. 3 Cross-section of green leaf showing gaseous movements during daylight.

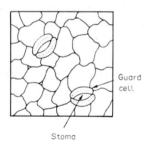

FIG. 2. Stomata on leaf surface.

The carbon dioxide used in photosynthesis diffuses into the leaf through the stomata. Most of the water vapour leaving the plant, as well as the oxygen from photosynthesis, diffuses out through the stomata.

Transpiration

The evaporation of water from plants is called transpiration. It mainly occurs through the sto-mata and has a cooling effect on the leaf cells. Water in the cells of the leaf can pass into the pore spaces in the leaf and then out through the stomata as water vapour (Fig. 3).

The rate of transpiration varies considerably. It is greatest when the plant is well supplied with water and the air outside the leaf is warm and dry. In very hot or windy weather water evaporates from the guard cells and so the stomata close and reduce the rate of transpiration. The stomata also close in very cold weather, e.g. 0°C.

The rate of loss is reduced if the plant is short of water because the guard cells then lose water and close the stomata. It is also retarded if the humidity of the atmosphere is high.

The stomata guard cells close (and so trans-piration ceases) during darkness. This is because photosynthesis ceases and water is lost from the guard cells when some of the sugars present change to starch.

Respiration

Plants breathe, like animals, i.e. they take in oxygen which combines with organic foodstuffs and this releases energy, carbon dioxide and water. Plants are likely to be checked in growth if the roots are deprived of oxygen for respiration, which might occur in a waterlogged soil.

Translocation

The movement of materials through the plant is known as translocation.

The xylem or wood vessels which carry the water and mineral salts from the roots to the leaves are tubes made from dead cells. The cross walls of the cells are no longer present and the longitudinal walls are thickened with lignum to form wood. These tubes help to strengthen the stem.

The phloem tubes (bast) carry organic material through the plant, e.g. sugars and amino acids from the leaves to storage parts or growing points. These vessels are chains of living cells, not lignified, and with cross walls which are perforated. They are sometimes referred to as sieve tubes.

In the stem, the xylem and phloem tubes are usually found in a ring near the outside of the stem.

In the root, the xylem and phloem tubes form separate bundles and are found near the centre of the root.

Uptake of water

Water is taken into the plant from the soil. This occurs mainly through the root hairs near the root tip. There are thousands of root tips on a single healthy crop plant (Fig. 4).

The absorption of water into the plant in this way is due to suction pull which starts in the leaves. As water transpires (evaporates) from the cells in the leaf, more water is drawn from the xylem tubes which extend from the leaves to the root tips. In these tubes the water is stretched like a taut wire. This is possible because the tiny particles of water hold together very firmly when in narrow tubes. The pull of this water in the xylem tubes of the root is transferred through the root cells to the root hairs and so water is absorbed into the roots and up to the leaves. In general, the greater the rate of transpiration, the greater is the amount of water taken into the plant.

The rate of absorption is slowed down by:

(1) Shortage of water in the soil.
(2) Lack of oxygen for root respiration (e.g. in waterlogged soils).
(3) A high concentration of salts in the soil water near the roots.

Normally, the concentration of the soil solution does not interfere with water absorption. High soil water concentration can occur in salty soils and near bands of fertilizer. Too much fertilizer near developing seedlings may damage germination by restricting the uptake of water.

Osmosis

Much of the water movements into and from cell to cell in plants is due to osmosis. This is a process in which a solvent, such as water, will flow through a semipermeable membrane (e.g. a cell wall) from a weak solution to a more concentrated one. The cell wall only allows the water to pass through. The force exerted by such a flow is called the osmotic pressure. In plants the normal movement of the water is into the cell. However, if the concentration of a solution outside the cell is greater than that inside, there is a loss of water from the cell, and its contents contract; this is called plasmolysis.

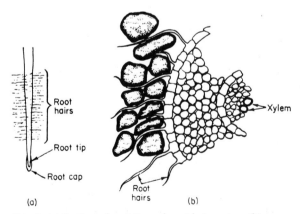

FIG. 4. (a) Section of root tip and root hair region, (b) cross-section of root showing the root hairs as tube-like elongations of the surface cells in contact with soil particles.

Uptake of nutrients

The absorption of chemical substances (nutrients) into the root cells is partly due to a diffusion process but it is mainly due to the ability of the cells near the root tips to accumulate such nutrients. The process is complicated and not fully understood. It is slowed up if root respiration is checked by a shortage of oxygen.

PLANT GROUPS

Plants can be divided into annuals, biennials and perennials according to their total length of life.

Annuals

Typical examples are wheat, barley and oats which complete their life history in one growing season, i.e. starting from the seed, in one year they develop roots, stem and leaves and then produce flowers and seed before dying.

Biennials

These plants grow for two years. They spend the first year in producing roots, stem and leaves, and the following year in producing the flowering stem and seeds, after which they die.

Sugar beet, swedes and turnips are typical biennials, although the grower treats these crops as annuals, harvesting them at the end of the first year when all the foodstuff is stored up in the root.

Perennials

They live for more than two years and, once fully developed, they usually produce seeds each year. Many of the grasses and legumes are perennials.

STRUCTURE OF THE SEED

Plants are also classified as dicotyledons and monocotyledons according to the structure of the seed.

Dicotyledon

A good example of a dicotyledon seed is the field bean. If its pod is opened when nearly ripe it will be seen that each seed is attached to the inside of the pod by a short stalk called the funicle. All the nourishment which the developing seed requires passes through the funicle from the bean plant.

When the seed is ripe and has separated from the pod, a black scar, known as the hilum, can be seen where the funicle was attached.

Near one end of the hilum is a minute hole called the micropyle (Fig. 5).

If a bean is soaked in water the seed coat can be removed easily and all that is left is largely made up of the embryo (germ). This consists of two seed leaves (cotyledons) which contain the food for the young seedling.

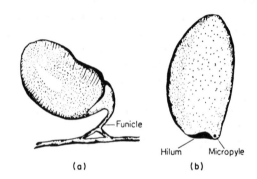

FIG. 5. (a) Bean seed attached to the inside of the pod by the funicle, (b) bean seed showing the hilum and micropyle.

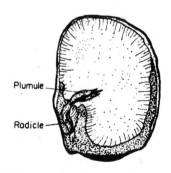

FIG. 6. Bean seed with one cotyledon removed.

Lying between the two cotyledons is the radicle which eventually forms the primary root, and a continuation of the radicle the other end, the plumule (Fig. 6). This develops into the young shoot and is the first bud of the plant.

Monocotyledon

This important class includes all the cereals and grasses.

The wheat grain is a typical example. It is not a true seed (it should be called a single-seeded fruit). The seed completely fills the whole grain, being practically united with the inside wall of the grain or fruit.

This fruit wall is made up of many different layers which are separated on milling into varying degrees of fineness, e.g. bran and pollards which are valuable livestock feed.

Most of the interior of the grain is taken up by the floury endosperm. The embryo occupies the small raised area at the base. The scutellum, a shield-like structure, separates the embryo from the endosperm. Attached to the base of the scutellum are the five roots of the embryo, one primary and two pairs of secondary rootlets. The roots are enclosed by a sheath called the coleorhiza. The position of the radicle and the plumule can be seen in the diagram (Fig. 7).

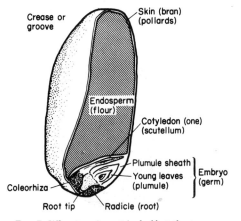

FIG. 7. Wheat grain cut in half at the crease.

The scutellum can be regarded as the cotyledon of the seed. There is only **one** cotyledon present and so wheat is a monocotyledon.

Germination of the bean—the dicotyledon

Given suitable conditions for germination, i.e. water, heat and air, the seed coat of the dormant but living seed splits near the micropyle and the radicle begins to grow downwards through this split to form the main, or primary root, from which lateral branches will soon develop (Fig. 8).

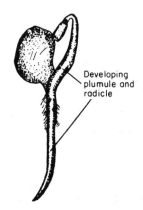

FIG. 8. Germination of the bean, one cotyledon removed.

When the root is firmly held in the soil, the plumule starts to grow by pushing its way out of the same opening in the seed coat. As it grows upwards its tip is bent to protect it from injury in passing through the soil, but it straightens out on reaching the surface, and leaves develop very quickly from the plumular shoot.

With the field bean the cotyledons remain underground, gradually giving up their stored food materials to the developing plant. However, with the French bean and many other dicotyledon seeds, the cotyledons are brought above ground with the plumule.

Germination of wheat—the monocotyledon

When the grain germinates the coleorhiza expands and splits open the seed coat and,

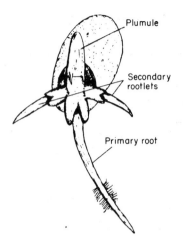

FIG. 9. Germination of the wheat grain.

at the same time, the roots break through the coleorhiza (Fig. 9).

The primary root is soon formed, supported by the two pairs of secondary rootlets, but this root system (the seminal roots) is only temporary and is soon replaced by adventitious roots (Figs 10 and 14). As the first root system is being formed at the base of the stem, so the

plumule starts to grow upwards, and its first leaf, the coleoptile, appears above the ground as a single pale tube-like structure.

From a slit in the top of the leaf there appears the first true leaf which is quickly followed by others, the younger leaves growing from the older leaves (Fig. 11).

As the wheat embryo grows, so the floury endosperm is used up by the developing roots and plumule, and the scutellum has the important function of changing the endosperm into digestible food for the growing parts.

With the field bean, the cotyledons provide the food for the early nutrition of the plant, whilst the wheat grain is dependent upon the endosperm and scutellum. In both cases it is not until the plumule has reached the light and turned green that the plant can begin to be independent.

This point is important in relation to the depth at which seeds should be sown. Small seeds such as the clovers and many of the grasses must, as far as possible, be sown shallow. Their food reserves will be exhausted before the shoot reaches the surface if sown too deep. Larger seeds such as the beans and peas can and should be sown deeper.

When the leaves of the plant begin to manufacture food by photosynthesis, and when the primary root has established itself sufficiently well to absorb nutrients from the soil, the plant can develop independently, provided there is sufficient moisture and air present.

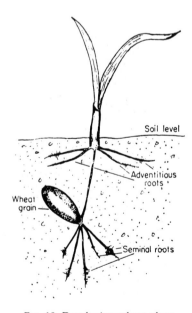

FIG. 10. Developing wheat plant.

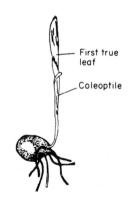

FIG. 11. Seedling wheat plant.

The main differences between the two groups of plants can be summarized as follows:

Dicotyledons	Monocotyledons
The embryo has two seed leaves.	The embryo has one seed leaf.
A primary root system is developed and persists.	A primary root system is developed, but is replaced by an adventitious root system.
Usually broad-leaved plants, e.g. legumes and sugar beet.	Usually narrow-leaved plants, e.g. cereals and grasses.

These two great groups of flowering plants can be further divided in the following way:

Families or orders, e.g.   Legumes
Genus   Clovers of the legume family
Species   Red clover
Cultivar or Variety   Late red clover

PLANT STRUCTURE

The plant can be divided into two parts:

The root system

The root system is concerned with the parts of the plant growing in the soil; there are two main types:

(1) The tap root or primary system

This is made up of the primary root called the tap root with lateral secondary roots branching out from it and, from these, tertiary roots may develop obliquely to form in some cases a very extensive system of roots (Fig. 12).

The root of the bean plant is a good example of a tap root system. If this is split it will be seen that there is a slightly darker central woody core, the skeleton of the root, which helps to anchor the plant and transport foodstuffs. The lateral secondary roots arise from this central core (Fig. 13).

Carrots and other true root crops such as sugar beet and mangels have very well developed tap roots. These biennials store food in their roots

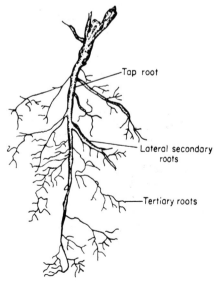

FIG. 12. Tap root or primary root system.

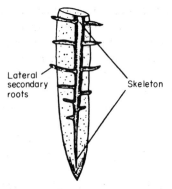

FIG. 13. Tap root of the bean plant.

during the first year of growth to be used in the following year for the production of the flowering shoot and seeds. However, they are normally harvested after one season and the roots are used as food for man and stock.

(2) *The adventitious root system*

This is found on all grasses and cereals, and it is the main root system of most monocotyledons. The primary root is quickly replaced by adventitious roots which arise from the base of the stem (Fig. 14). These roots can, in fact, develop from any part of the stem, and they are found on some dicotyledons as well, but not as the main root system, e.g. underground stems of the potato.

Root hairs (Fig. 4) are very small white hair-like structures which are found near the tips of all roots. As the root grows, the hairs on the older parts die off and others develop on the younger parts of the root. They play a very important part in the life of a plant.

The stem

The second part of the flowering plant is the shoot which normally grows upright above the ground. It is made up of a main stem, branches, leaves and flowers.

Stems are either soft (herbaceous) or hard (woody) and in British agriculture it is only the soft and green herbaceous stems which are of any importance. These usually die back every year.

All stems start life as buds and the increase in length takes place at the tip of the shoot called the terminal bud. If a Brussels sprout is cut lengthwise and examined it will be seen that the young leaves arise from the bud axis. This axis is made up of different types of cell tissue, which is continually making new cells and thus growing (Fig. 15).

Stems are usually jointed, each joint forming a node, the part between two nodes being called the internode. At the nodes the stem is usually solid and thicker, and this swelling is caused by the storing up of material at the base of the leaf (Fig. 16).

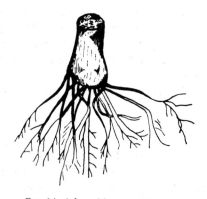

FIG. 14. Adventitious root system.

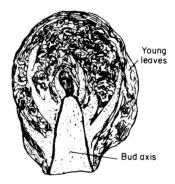

FIG. 15. Longitudinal section of a Brussels sprout.

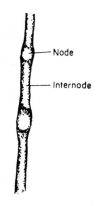

FIG. 16. Jointed stem.

The bud consists of closely packed leaves arising from a number of nodes. It is, in fact, a condensed portion of the stem which develops by a lengthening of the internodes.

Axillary buds are formed in the angle between the stem and leaf stalk. These buds, which are similar to the terminal bud, develop to form lateral branches, leaves and flowers.

Modified stems

(1) A stolon is a stem which grows along the ground surface.

Adventitious roots are produced at the nodes, and buds on the runner can develop into upright shoots, and separate plants can be formed, e.g. strawberry plants (Fig. 17).

(2) A rhizome is similar to a stolon but grows under the surface of the ground, e.g. couch grass (Fig. 17).

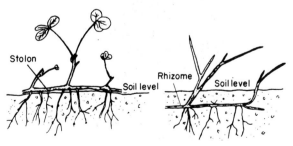

FIG. 17. Modified stems.

(3) A tuber is really a modified rhizome. The end of the rhizomes swell to form tubers. The tuber is therefore a swollen stem. The potato is a well-known example with "eyes" (buds) which develop shoots when the potato tuber is planted.

(4) A tendril is found on certain legumes, such as the pea. The terminal leaflet is modified as in the diagram. This is useful for climbing purposes to support the plant (Fig. 18).

Corms and suckers are other examples of modified stems.

FIG. 18. Modified stems.

The leaf

Leaves in all cases arise from buds. They are extremely important organs, being not only responsible for the manufacture of sugar and starch from the atmosphere for the growing parts of the plant, but they are also the organs through which transpiration of water takes place.

A typical leaf of a dicotyledon consists of three main parts:

(1) The blade.
(2) The stalk or petiole.
(3) The basal sheath connecting the leaf to the stem. This may be modified as with legumes into a pair of wing-like stipules (Fig. 19(a)).

The blade is the most obvious part of the leaf and it is made up of a network of veins.

There are two main types of dicotyledonous leaves:

(1) A prominent central midrib, from which lateral veins branch off on either side. These side veins branch into smaller and smaller ones (Fig. 19(a)).
(2) No single midrib, but several main ribs spread out from the top of the leaf stalk; between these the finer veins spread out as before, e.g. horse-chestnut leaf (Fig. 19(b)).

The veins are the essential supply lines for the process of photosynthesis. They consist of two main parts—the xylem for bringing the required raw material up the leaf and the phloem which carries the finished product away from the leaf.

Leaves can show great variation in shape and type of margin, as in Fig. 19. They can also be divided into two broad classes as follows:

(1) Simple leaves. The blade consists of one continuous piece (Fig. 19(a)).
(2) Compound leaves. Simple leaves may become deeply lobed and when the division between the lobes reaches the midrib it becomes a compound leaf, and the separate parts of the blade are called the leaflets (Fig. 19(b)).

The blade surface may be smooth (glabrous) or hairy, according to variety. This is important in legumes because it can affect their palatability.

Monocotyledonous leaves are dealt with in Chapter 10 (Grassland).

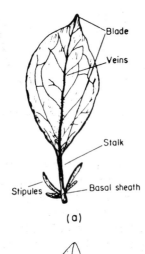

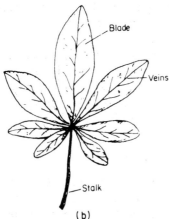

FIG. 19. (a) Simple leaf, (b) compound leaf.

Modified leaves

(1) Cotyledons or seed leaves are usually of a very simple form.
(2) Scales are normally rather thin, yellowish to brown membranous leaf structures, very variable in size and form. On woody stems they are present as bud scales which protect the bud; they are also found on rhizomes such as couch grass.
(3) Leaf tendrils. The terminal leaflet like the stem can be modified into thin threadlike structures, e.g. the pea plant.

Other examples of modified leaves are leaf-spines and bracts.

The flower

In the centre of the flower is the axis which is simply the continuation of the flower stalk. It is known as the receptacle and on it are arranged four kinds of organ:

(1) The lowermost is a ring of green leaves called the calyx, made up of individual sepals.
(2) Immediately above the calyx is a ring of petals known as the corolla.
(3) Above the corolla are the stamens, again arranged in a ring. They are similar in appearance to an ordinary match, the swollen tip being called the anther which, when ripe, contains the pollen grains.
(4) The highest position on the receptacle is occupied by the pistil which is made up of one or a series of small green bottle-shaped bodies—the carpel, which is itself made up of three parts: the stigma, style and the ovary (containing ovules). It is within the ovary that the future seeds are produced (Figs. 20 and 21).

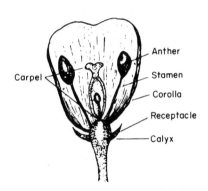

FIG. 20. Longitudinal section of a simple flower.

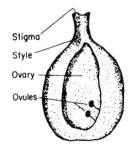

FIG. 21. Carpel.

Most flowers are more complicated in appearance than the above, but basically they consist of these four main parts.

The formation of seeds

Pollination precedes fertilization which is the union of the male and female reproductive cells. When pollination takes place the pollen grain is transferred from the anther to the stigma. This may be self-pollination where the pollen is transferred from the anther to the stigma of the same flower, or cross-pollination when it is carried to a different flower (Fig. 22).

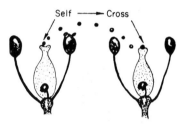

FIG. 22. Self- and cross-pollination.

With fertilization the pollen grain grows down the style of the carpel to fuse with the ovule. After fertilization, changes take place whereby the ovule develops into the embryo, and endosperm may be formed according to the species. This makes up the seed. The ovary also changes after fertilization to form the fruit, as distinct from the seed.

With the grasses and cereals there is only one seed formed in the fruit and, being so closely united with the inside wall of the ovary, it cannot easily be separated from it.

The one-seeded fruit is called a grain.

The inflorescence

Special branches of the plant are modified to bear the flowers, and they form the inflorescence. There are two main types of inflorescence:

(1) Where the branches bearing the flowers continue to grow, so that the youngest flowers are nearest the apex and the oldest farthest away, an indefinite inflorescence (Fig. 23(a)).

A well-known example of this inflorescence is the spike found in many species of grasses.

(2) Where the main stem is terminated by a single flower and ceases to grow in length; any further growth takes place by lateral branches, and they eventually terminate in a single flower and growth is stopped, a definite inflorescence, e.g. stitchworts (Fig. 23(b)).

There are many variations of these two main types of inflorescence.

(a)

FIG. 23(a). Indefinite inflorescence.

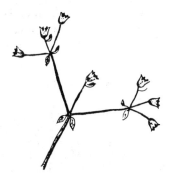

FIG. 23(b). Definite inflorescence.

PLANT REQUIREMENTS

To grow satisfactorily a plant needs warmth, light, water, carbon dioxide and other chemical elements which it can obtain from the soil.

Warmth

Most crop plants in this country start growing when the average daily temperature reaches 6°C. Growth is best between 16°C and 27°C. These temperatures apply to thermometer readings taken in the shade about 1.5 m above ground. Crops grown in hotter countries usually have higher temperature requirements.

Cold frosty conditions may seriously damage plant growth. Crop plants differ in their ability to withstand extreme cold. For example, winter rye and wheat can stand colder conditions better than winter oats. Potato plants and stored tubers are easily damaged by frost. Sugar beet plants may go to seed (bolt) if there are frosts after germination. Frost in December and January may destroy crops left in the ground.

Light

Without light, plants cannot produce carbohydrates and will soon die. The amount of photosynthesis which takes place daily in a plant is partly due to the length of daylight and partly to the intensity of the sunlight. Bright sunlight is very important where there is dense plant growth.

The periods of daylight and darkness will vary according to the distance from the equator and also from season to season. This affects the flowering and seeding of crop plants and is one of the limiting factors in introducing new crops into a country.

Water

Water is an essential part of all plant cells and it is also required in extravagant amounts for the process of transpiration. Water carries nutrients from the soil into and through the plant; it also carries the products of photosynthesis from the leaves to wherever they are needed. Plants take up about 200 tonnes of water for every tonne of dry matter produced.

Carbon dioxide

Plants need carbon dioxide for photosynthesis. This is taken into the leaves through the stomata and so the amount which can go in is affected by the rate of transpiration. Another limiting factor is the small amount (0.03%) of carbon dioxide in the atmosphere. The percentage can increase just above the surface of soils rich in organic matter where soil bacteria are active and releasing carbon dioxide. This is possibly one of the reasons why crops grow better on such soils.

Chemical elements required by plants

A number of chemical elements are needed by the plant in order that it may live and flourish. Most soils supply the majority of these nutrients and in farming practice it is only with regard to nitrogen, phosporus, potassium and magnesium (the major elements) that there is any widespread necessity to supplement the natural supplies from the soil. Other major elements may be required in certain situations. Calcium is an essential plant food but, as lime, it is regarded more as a soil conditioner. Sodium is highly desirable for maritime crops such as sugar beet and fodder beet when it can replace some of the potassium requirements. Sulphur, in areas away from industry, could be needed for the grass crop when it is cut more than once in the year.

The main plant foods are discussed more fully in Chapter 3 (Fertilizers and Manures) and see also Table 1.

Those elements required only in small amounts by the plant are known as the minor or trace elements. They are, nevertheless, essential and a shortage in the soil, especially of boron and

TABLE 1

The major nutrients	Use	Source
Carbon (C) Hydrogen (H) Oxygen (O)	Used in making carbohydrates	The air and water
Nitrogen (N)	Very important for building proteins	Organic matter (including FYM); rainfall Nitrogen-fixing soil micro-organisms Nitrogen fertilizers such as ammonium and nitrate compounds and urea
Phosphorus (P) (phosphate)	Essential for cell division and many chemical reactions	Small amounts from the mineral and organic matter in the soil Mainly from phosphate fertilizers, e.g. superphosphate, ground rock phosphate, basic slag and compounds, and residues of previous fertilizer applications
Potassium (K) (potash)	Helps with formation of carbohydrates and proteins Regulates water in and through the plant	Small amounts from mineral and organic matter in the soil. Potash fertilizers, e.g. muriate and sulphate of potash
Calcium (Ca)	Essential for development of growth tissues, e.g. root tips	Usually enough in the soil. Applied as chalk or limestone to neutralize acidity
Magnesium (Mg)	A necessary part of chlorophyll	If soil is deficient, may be added as magnesium limestone or magnesium sulphate, also FYM
Sulphur (S)	Part of many proteins and some oils	Usually sufficient in the soil. Atmospheric sulphur absorbed by the soil and plant. Added in some fertilizers (e.g. sulphate of ammonia and superphosphate)

manganese (often as a result of liming), will cause deficiency diseases in particular crops.

Other trace elements are chlorine, iron, molybdenum and zinc, but these rarely cause trouble on most farm soils (page 49).

LEGUMES AND NITROGEN FIXATION

Legumes are plants which have a number of interesting features such as:

(1) A special type of fruit called a legume, which splits along both sides to release its seeds, e.g. pea pod.
(2) Nodules on the roots containing special types of bacteria (rhizobia) which can "fix" (convert) nitrogen from the air into nitrogen compounds. These bacteria enter the plant through the root hairs from the surrounding soil.

This "fixation" of nitrogen is of considerable agricultural importance. Many of our farm crops are legumes, for example, peas, beans, vetches, lupins, clovers, lucerne (alfalfa), sainfoin and trefoil. The bacteria obtain carbohydrates from the plant and in return they supply nitrogen compounds as ammonium which is released into the soil. This is changed to nitrate and taken up by neighbouring plants, e.g. by grasses in a grass and clover sward, or by the following crop, e.g. wheat after clover or beans. The amount of nitrogen which can be fixed by legume bacteria varies widely—estimates of 50–450 kg/ha of nitrogen have been made. Some of the reasons for variations are:

The type of plant. Some crop plants fix more nitrogen than others, e.g. lucerne and clovers (especially if grazed) are usually better than peas and beans.

The conditions in the soil. The bacteria usually work best in soils which favour the growth of the plant on which they live. A good supply of calcium and phosphate in the soil is usually beneficial, although lupins grow well on acid soils.

The strains of bacteria present. The majority of soils in this country contain the strains of bacteria required for most of the leguminous crops which are grown. Lucerne is an exception and, unless it has been grown in the field within the previous three years, it is necessary to inoculate the seed, i.e. treat it with an inoculum containing *Rhizobium meliloti* before sowing to encourage effective nodulation.

THE NITROGEN CYCLE

The circulation of nitrogen (in various compounds such as nitrates and proteins), as found on the farm, is illustrated in Fig. 24.

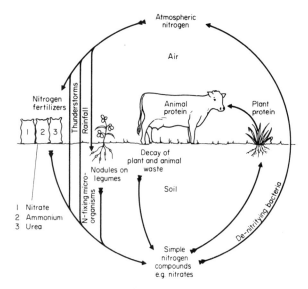

FIG. 24. The nitrogen cycle.

Atmospheric nitrogen is fixed in compounds by rhizobia, thunderstorms and in the manufacture of nitrogen fertilizers.

Simple nitrogen compounds (mainly nitrates) are taken up by plants to form plant proteins which may then be eaten by animals to form animal proteins. Dead plants and animals, and the faeces and urine of animals are broken down by micro-organisms to leave simple nitrogen compounds in the soil.

The de-nitrifying bacteria change nitrogen compounds back to free nitrogen. This is most likely to happen where nitrates are abundant and oxygen is in short supply, e.g. in water-logged soils.

FURTHER READING

Agricultural Botany, Gill and Vear, Duckworth.
Textbook of Botany, Lowson, Oxford Univ. Tutorial Press.

2

SOILS

SOILS are very complex natural formations which make up the surface of the earth. They should provide a suitable environment in which plants may obtain water, nutrients and oxygen for growth and development. Most soils also have enough depth to allow plant roots to provide a firm anchorage. Mineral soils are formed initially by the weathering of parent rock, often followed by deposition of the material by ice, water and/or wind. Organic material is added by the growth and decay of living organisms. If a farmer is to provide the best possible conditions for crop growth, it is necessary for him to understand what soils are, how they were formed and how they should be managed.

The topsoil is a layer about 8–45 cm deep which may be taken as the greatest depth which a farmer would plough or cultivate and in which most of the plant roots are found. Loose, cultivated, topsoil is sometimes called mould.

The subsoil, which lies underneath, is an intermediate stage in the formation of soil from the rock below. Deep rooting plants use it for moisture extraction.

A soil profile is a section taken vertically through the soil. In some cases this may consist

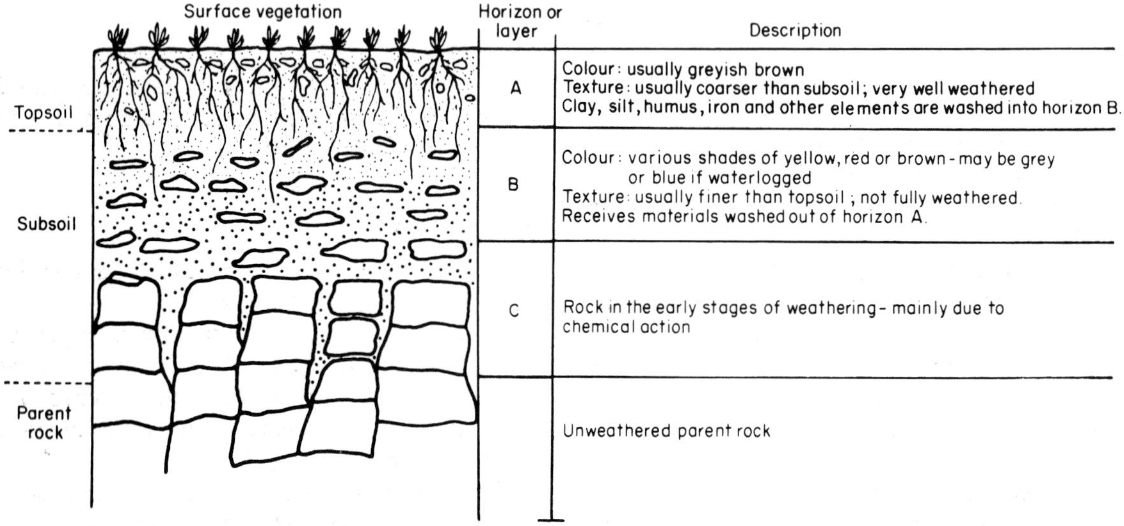

	Horizon or layer	Description
Topsoil	A	Colour: usually greyish brown Texture: usually coarser than subsoil; very well weathered Clay, silt, humus, iron and other elements are washed into horizon B.
Subsoil	B	Colour: various shades of yellow, red or brown - may be grey or blue if waterlogged Texture: usually finer than topsoil; not fully weathered. Receives materials washed out of horizon A.
	C	Rock in the early stages of weathering - mainly due to chemical action
Parent rock		Unweathered parent rock

FIG. 25. Soil profile showing the breakdown of rock to form various soil layers (horizons).

only of a shallow surface soil 10–15 cm on top of a rock such as chalk or limestone. In deeper well-developed soils there are usually three or more definite layers (or horizons) which vary in colour, texture and compaction (Fig. 25).

The soil profile can be examined by digging a trench or by taking out cores of soil from various depths with a soil auger.

A careful examination of the layers can be useful in deciding how the soil was formed, its natural drainage and how it might be farmed. Some detailed soil classifications are based on soil profile (Page 281).

SOIL FORMATION

There are very many different types of top-soils and subsoils. The differences are mainly due to the kind of rocks from which they are formed. However, other factors such as climate, topography, plant and animal life, the age of the developing soil material and farming operations affect the type of soil which develops.

The more important rock formations

Igneous or primary rocks, e.g. granite (coarse crystals) and basalt (fine crystals), were formed from the very hot molten material which made up the earth millions of years ago. The minerals (chemical compounds) in these rocks are mostly in the form of crystals which are the primary source of the minerals found in all our soils.

Igneous rocks are very hard and weather very slowly. Clay and sand are breakdown products.

Sedimentary or transported rocks have been formed from weathered material (e.g. clay, silt and sand) carried and deposited by water and wind. The sediments later became compressed by more material on top and cemented to form new rocks such as sandstones, clays and shales.

The chalks and limestones were formed from the shells and skeletons of sea animals of various sizes. These rocks are mainly calcium carbonate but in some cases are magnesium carbonate. The calcareous soils are formed from them (page 27).

Metamorphic rocks, e.g. marble (from limestone) and slate (from shale), are rocks which have been changed in various ways.

Organic matter

Deep deposits of raw organic matter are found in places where waterlogged soil conditions did not allow the breakdown of dead plant material by micro-organisms and oxidation.

Peats have been formed in waterlogged acidic areas where the vegetation is mainly mosses, rushes, heather and some trees.

Black Fen (muck soil) has been formed in marshy river estuary conditions where the water was hard (lime rich) and often silty; the plants were mainly reeds, sedges, rushes and some trees.

Glacial drift

Many soils in the British Isles are not derived from the rocks underneath but are deposits carried from other rock formations by glaciers, e.g. boulder clays. This makes the study of such soils very complicated.

Alluvium

This is material which has been deposited recently, for example, by river flooding. It has a very variable composition.

Weathering of rocks

The breakdown of rocks is mainly caused by physical and chemical effects.

Physical weathering is caused by changes of temperature which cause the various mineral crystals in rocks to expand and contract by different amounts, resulting in the occurrence of cracking and shattering.

Water can cause pieces of rock surfaces to split off when it freezes and expands in cracks and crevices. Also, the molecules (small particles) of water in the pores and fissures of the rock exert expansion and contraction forces similar to those of freezing and thawing. The pieces of rock broken off are usually sharp-edged, but if they are carried and knocked about by glaciers, rivers or wind, they become more rounded in shape, e.g. sand and stones in a river bed.

Wetting and drying of some rocks such as clays and shales cause expansion and contraction, resulting in cracking and flaking.

Chemical weathering is the breakdown of the mineral matter in a developing soil brought about by the action of water, oxygen, carbon dioxide and nitric acid from the atmosphere, and by carbonic and organic acids from the biological activity in the soil. The soil water, which is a weak acid, dissolves some minerals and allows chemical reactions to take place.

Water can also unite with substances in the soil (hydration) to form new substances which are more bulky and so can cause shattering of rock fragments.

Clay is produced by chemical weathering. In the case of rocks such as granite, when the clay-producing parts are weathered away, the more resistant quartz crystals are left as sand or silt.

In the later stages of chemical weathering the soil minerals are broken down to release plant nutrients. This is a continuing process in most soils.

In poorly drained soils, which become water-logged from time to time, various complex chemical reactions (including a reduction process) occur and are referred to as gleying. This process, which is very important in the formation of some soils (especially upland soils), results in ferrous iron, manganese and some trace elements moving around more freely and producing colour changes in the soil. Gley soils are generally greyish in colour (but may also be green or blue) in the waterlogged regions. However, rusty-coloured deposits of ferric iron also occur in root and other channels, and along the boundaries between the waterlogged and aerated soil, so producing a mottled appearance. Glazing or coating of the soil structure units with fine clay is also associated with gleying.

OTHER FACTORS IN SOIL FORMATION

Climate

The rate of weathering partly depends on the climate. For example, the wide variations of temperature, and the high rainfall of the tropics, makes for much faster soil development than would be possible in the colder and drier climatic regions.

Topography

The depth of soil can be considerably affected by the slope of the ground. Weathered soil tends to erode from steep slopes and build up on the flatter land at the bottom. Level land is more likely to produce uniform weathering.

Biological activity

Plants, animals and micro-organisms, during their life cycles, leave many organic substances in the soil. Some of the substances may dissolve some of the mineral material; dead material may partially decompose to give humus.

The roots of plants may open up cracks in the soil. Vegetation such as mosses and lichens can attack and break down the surface of rocks.

Holes made in the soil by burrowing animals such as earthworms, moles, rabbits, etc., help to break down soft and partly weathered rocks.

Biological activity is usually increased with higher temperatures and decreased by water-logging and acidity.

Farming operations

Deep ploughing and cultivation, artificial drainage, liming, etc., can speed up the soil formation processes very considerably.

THE PHYSICAL MAKE-UP OF SOIL AND ITS EFFECT ON PLANT GROWTH

The farmer must consider the soil from the point of view of its ability to grow crops. To produce good crops the soil must provide suitable conditions in which plant roots can grow. It must also supply nutrients, water and air. The temperature must also be suitable for the growth of the crop.

The soil is composed of:

ing the breakdown process are very beneficial in restoring and stabilizing soil structure.

The amount of humus formed is greatest from plants which have a lot of strengthening (lignified) tissue, i.e. straw. Humus is finally

Solids	Mainly mineral matter (stones, sand, silt, clay, etc.) and organic matter (remains of plants and animals).
Liquids	Mainly soil water (a weak acid).
Gases	Soil air which competes with water to occupy the spaces between the soil particles.
Living organisms	Micro-organisms (bacteria, fungi, small soil animals, earthworms, etc.).

Mineral matter

This weathered material, especially the clay part, is mainly responsible for making a soil difficult or easy to work. It may provide many plant nutrients, but **not** nitrogen. The farmer normally cannot alter the mineral matter in a soil (except possibly by claying (page 40)).

The amounts of clay, silt and sand which a soil contains can be measured in the laboratory by mechanical analysis (Table 2).

Organic matter

The organic matter content of a soil varies with texture. It usually increases with clay content, e.g. ordinary heavy soils have about 3–4% organic matter compared with 1–1.5% in very light soils. Most fertile soils contain about 3–5% on a dry weight basis. Peaty soils contain 20–35% organic matter and peats (including Black Fen) have over 35%, although some entirely consist of organic matter.

Organic matter may remain for a short time in the undecayed state and as such can help to "open up" the soil; this could be harmful on sandy soils. However, the organic matter is soon attacked by the soil organisms, i.e. bacteria, fungi, earthworms, insects, etc. When they have finished eating and digesting it and each other, a complex, dark-coloured, structureless material called humus remains. Materials produced dur-

broken down by an oxidation process which is not fully understood.

The amount of humus which can remain in a soil is fairly constant for any particular type of soil. The addition of more organic matter often does not alter the humus content appreciably because the rate of breakdown increases. Organic matter is broken down most rapidly in warm, moist soils which are well limed and well aerated. Breakdown is slowest in waterlogged, acid conditions.

Like clay, humus is a colloid, i.e. it is a gluey substance which behaves like a sponge; it absorbs water and swells up when wetted and shrinks on drying. The humus colloids are not so gummy and plastic as the clay colloids but they can improve light (sandy) soils by binding groups of particles together. This reduces the size of the pores (spaces between the particles) and increases the water-holding capacity. Humus can also improve clay soils by making them less plastic and by assisting in the formation of a crumb structure (lime must also be present). Earthworms help in this soil improvement by digesting the clay and humus material with lime.

Plant nutrients, particularly nitrogen and phosphorus, are released for uptake by other plants when organic matter breaks down. The humus colloids can hold bases such as potassium and ammonia in an available form. In these ways it has a very beneficial effect in promoting steady crop growth.

Organic matter in the soil may be maintained or increased by growing grass, working-in straw and similar crop residues, farmyard manure, composts, etc. The roots and stubble are usually sufficient to maintain an adequate organic matter content in a soil growing cereals continuously, always provided that it is the right type of soil.

In areas where erosion by wind and water is common, mineral soils are less likely to suffer damage if they are well supplied with humus. Where it is possible to grow good leys and utilize them fully, this is one of the best ways of maintaining a high level of organic matter and a good soil structure.

Increasing the organic matter content of a soil is the best way of increasing its water-holding capacity; 50 tonnes/ha of well-rotted farmyard manure may increase the amount of water which can be held by 25% or more.

Water in the soil

There are various reasons why soils vary in their capacity to hold water.

The soil is a mass of irregular-shaped particles forming a network of spaces or channels called the pore space, which may be filled with air or water or both. If the pore space is completely filled with water, the soil is waterlogged. It is then unsuitable for most plant growth because the roots need oxygen for respiration. Ideally, there should be about equal volumes of air and water.

In a clay soil about half the total volume is pore space, whereas in a sandy soil only about one-third is pore space. These volumes refer to dry soils.

The pore space may be altered by a change in the:

(a) Grouping of the soil particles (i.e. structure).
(b) Amount of organic matter present.
(c) Compaction of the soil.

The fact that a clay soil has a greater pore space than a sandy soil partly explains why it can hold more water.

Another important factor is the surface area of the particles. Water is held as a thin layer or film around the soil particles. The smaller the particles, the stronger are the attractive forces holding the water. Also, the smaller the particles, the greater is the surface area per unit volume. A comparison for pure materials is set out in Table 2.

The surface area of the particles in a cubic metre of clay may be over 1000 hectares.

The organic matter in the soil also holds water. The water in the soil comes from rainfall, irrigation or groundwater.

When water falls on a dry soil it does **not** become evenly distributed through the soil. The topmost layer is saturated first and, as more water is added, the depth of the saturated layer increases. In this layer most of the pore space is filled with water. However, a well-drained soil cannot hold all this water for very long; after a day or two some of it will soak into the lower layers or run away in drains. The soil will eventually reach a stage when the amount of any additional water added by rain or irrigation is matched by the loss of water from the profile through natural or artificial drainage. This is known as "field capacity". The amount of water which can be held in this way varies according to the texture, and structure of the soil. The weight of water held by a clay soil may be half the weight of the soil particles, whereas a sandy soil may hold less than one-tenth of the weight of the particles. The water-holding capacity of a soil is usually expressed in millimetres, e.g. a clay soil may have a field capacity of 4 mm/cm in depth.

The ways in which water is retained in the soil can be summarized as follows:

(a) As a film around the soil particles.
(b) In the organic matter.
(c) Filling some of the smaller spaces.
(d) Chemically combined with the soil minerals.

Most of this water can be easily taken up by plant roots, but as the soil dries out the remaining water is more firmly held and eventually a stage is reached when no more water can be extracted by the plant. This is the permanent wilting point because plants remain wilted and soon die. This permanent wilting should not be confused with temporary wilting which sometimes occurs on very hot days. This is because the rate of transpiration is greater than the rate of water absorption through the roots. In these cases the plants recover as it gets cooler on the same day. The water which can be taken up by the plant roots is called the available water. It is the difference between the amounts at field capacity and wilting point. In clay soils only about 50–60% of "field capacity" water is available; in sandy soils up to 90% or more may be available. Although plants may not die until the wilting point is reached, they will suffer from shortage of water as it becomes more difficult to extract (Fig.26).

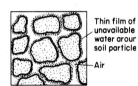

Thick film of water around particles

Air bubbles

Field capacity—pore space filled with water and air which is ideal for plant growth

Thin film of unavailable water arour soil particle

Air

Permanent wilting point – plants wilt and will soon die due to lack of water

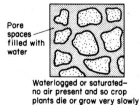

Pore spaces filled with water

Air

Waterlogged or saturated— no air present and so crop plants die or grow very slowly

Dry—no water present and so plants die – this is unlikely to happen in a field

FIG. 26. Highly-magnified particles and pores showing how water and air may be found in the soil.

Water in the soil tends to hold the particles together and thus lumps of soil may stick together. When a loam or heavy soil is at or above half field capacity, it is possible to form it into a ball which will not fall apart when handled. At wilting point, the soil is crumbly and will not hold together.

Some of the water in soils with very small pores and channels can move through the soil by capillary forces, i.e. surface tension between the water and the walls of the fine tubes or capillaries. This is a very slow movement and may not be fast enough to supply plant roots in a soil which is drying out. Heavy rolling of a soil may reduce the size of the pores and so set up some capillary action.

Water is lost from the soil by evaporation from the surface and by transpiration through plants. It moves very slowly from the body of the soil to the surface; after the top 20–50 mm have dried out, the loss of water by evaporation is very small. Cultivations increase evaporation losses. Most of the available water in a soil is taken up by plants during the growing season, and air moves in to take its place. Air moves easily where the soil has large pore spaces but the movement into the very small pore channels in clay soils is slow until the soil shrinks and cracks—vertically and horizontally—as water is removed by plants.

The water which enters the soil soon becomes a dilute solution of the soluble soil chemicals. It dissolves some of the carbon dioxide in the soil and so becomes a weak acid.

Soil aeration

Plant roots and many of the soil animals and micro-organisms require oxygen for respiration and give out carbon dioxide. The air found in the soil is really atmospheric air which has been changed by these activities (also by various chemical reactions), and so it contains less oxygen and more carbon dioxide. After a time this reduction in oxygen and increase in carbon dioxide becomes harmful to the plant and other organisms.

Aeration is the replacement of this stagnant soil air with fresh air. The process is mainly brought about by the movement of water into and out of the soil, e.g. rainwater soaks into the soil filling many of the pore spaces and driving out the air. Then, as the surplus water soaks down to the drains or is taken up by plants, fresh air is drawn into the soil to refill the pore spaces. Additionally, oxygen moves into the soil and carbon dioxide moves out by a diffusion process similar to that which happens through the stomata in plant leaves.

The aeration process is also assisted by:

(1) Changes in temperature.
(2) Changes in barometric pressure.
(3) Good drainage.
(4) Cultivations—especially on clay soils where a soil cap has formed.
(5) Open soil structure.

Sandy soils are usually well aerated because of their open structure. Clay soils are usually poorly aerated, especially when the very small pores in such soils become filled with water.

Good aeration is especially important for germinating seeds and seedling plants.

Soil micro-organisms

There are thousands of millions of organisms in every gramme of fertile soil. Many different types are found, but the main groups are:

(1) *Bacteria* are the most numerous group and are the smallest type of single-celled organisms. Most of them feed on and break down organic matter. They need nitrogen to build cell proteins, but if they cannot get this nitrogen from the organic matter they may use other sources such as the nitrogen applied as fertilizers. When this happens (e.g. where straw is ploughed in) the following crop may suffer from a temporary shortage of nitrogen. Some types of bacteria can convert (fix) the nitrogen from the air into nitrogen compounds which can be used by plants (Legumes and the Nitrogen Cycle, page 15).

(2) *Fungi* are simple types of plants which feed on and break down organic matter. They are mainly responsible for breaking down lignified (woody) tissue. They have **no** chlorophyll or proper flowers. The fungi usually found in mineral soils are very small, but larger types are found in other soils, e.g. peats. Fungi can live in more acid and drier conditions than bacteria. Disease-producing fungi can develop in some fields, e.g. those causing "take-all" in cereals.

(3) *Actinomycetes* are organisms which are intermediate between bacteria and fungi and have a similar effect on the soil. They need oxygen for growth and are more common in the drier, warmer soils. They are not as numerous as bacteria and fungi. Some types can cause plant disease, e.g. common scab in potatoes (worst in light, dry, alkaline soils).

(4) *Algae* in the soil are very small simple organisms which contain chlorophyll and so can build up their bodies by using carbon dioxide from the air and nitrogen from the soil. Algae growing in swampy (waterlogged) soils can use dissolved carbon dioxide from the water and release oxygen. This process is an important source of oxygen for crop plants such as rice. Algae are important in colonizing bare soils in the early stages of weathering.

(5) *Protozoa*. These are very small, single-celled animals. Most of them feed on bacteria and similar small organisms. A few types contain chlorophyll and so can produce carbohydrates like plants.

Earthworms

It is generally believed that earthworms have a beneficial effect on the fertility of soils, particularly those under grass. There are several different kinds found in our soils, but most of their activities are very similar. They live in holes in the soil and feed on organic matter—either living plants or, more often, dead and decaying matter. They carry down into the soil fallen leaves and twigs, straw and similar materials. Earthworms do not thrive in acid soils because they need plenty of calcium (lime) to digest the organic matter they eat. Their casts, which are usually left on the surface, consist of a useful mixture of organic matter, mineral matter and lime. The greatest numbers are found in loam soils (under grass) where there is usually a good supply of air, moisture, organic matter and lime.

The many holes they make allow water to enter and drain from the soil very easily; this in turn draws fresh air in as it soaks downwards. This may not always be a good thing, because the holes often have a smooth and, in places, impervious lining which may allow the water to go through to the drains too easily instead of soaking into the soil.

Earthworms are the main food of the mole which does so much damage by burrowing and throwing up heaps of soil.

Other soil animals

In addition to earthworms, there are many species of small animals present in most soils. They feed on living and decaying plant material and micro-organisms. Some of the common ones are slugs, snails, millipedes, centipedes, ants, spiders, eelworms, beetles, larvae of various insects such as cutworms, leatherjackets and wireworms. The farmer is only directly concerned with those which damage his crops or livestock.

SOIL TEXTURE AND STRUCTURE

Soil texture is that characteristic which is determined by the amounts of clay, silt, sand and organic matter which the soil contains. This property normally cannot be altered by the farmer. Soil texture can be measured by mechanical analysis and classified accordingly (Table 2) and also by a "feel" test (page 25).

The texture of a soil has a major influence on its properties. Soils with high clay contents

are wetter, colder and harder to work with implements than those with more sand in them. They restrict the types of crops that can be grown on them and usually have a shorter growing season (Clay soils, page 25).

Soil structure is the arrangement of the soil particles individually (e.g. grains of sand), in groups (e.g. crumbs or clods) or as a mixture of the two. It can be altered by weather conditions (e.g. lumps changed to crumbs by frost action or alternate wetting and drying), penetration of plant roots and cultivations, etc. The structure of the soil can be easily damaged by harmful operations, e.g. heavy traffic in wet conditions.

Crumb structure is formed by the grouping together (aggregation) of the particles of clay, sand and silt. This aggregation is possible because there are positive and negative electric charges (forces) acting through the surface of the particles. These forces are strongest in clay and very weak in sand. The more clay there is, the stronger will be the forces holding the particles together. This strong adhesive property of clay particles is the reason why clay soils are more difficult to work than sandy soils.

Water has special electric properties and its presence is necessary for the grouping (crumbing) of soil particles. The electric forces in the water and in the soil particles make the water stick as a thin film around these particles. As this film becomes thinner (e.g. when soil is drying out), the particles are drawn closer together to form groups (crumbs). They may come apart again if the soil becomes very wet.

There must be lime present in the water if clay particles are to stick together to form porous crumbs (Clay soils, page 25).

If organic matter or ferric hydroxide is present, then the particles in the crumbs may remain cemented together and have a more lasting effect on soil structure. Too much ferric hydroxide can have a harmful effect because tightly cemented crumbs are very difficult to wet again after they have dried out. This condition occurs in the so-called "drummy" soils found in the fen district.

Some soil structures are more stable than others. Soils containing fine sand and silt easily lose their structure and are difficult to work if they are low in organic matter.

Tilth is a term used to describe the condition of the soil in a seedbed. For example, the soil may be in a finely-divided state or it may be rough and lumpy. Also, the soil may be damp or it may be very dry. Whether a tilth is suitable or not partly depends on the crop to be grown. In general, small seeds require a finer tilth than large seeds.

SOIL FERTILITY AND PRODUCTIVITY

Soil fertility is a rather loose term used to indicate the potential capacity of a soil to grow a crop (or sequence of crops). The productivity of a soil is the combined result of fertility and management.

The fertility of a soil at any one time is partly due to its natural make-up (inherent or natural fertility) and partly due to its condition (variable fertility) at that time. Natural fertility has an important influence on the rental and sale value of land. It is the result of factors which are normally beyond the control of the farmer, such as:

(1) The texture and chemical composition of the mineral matter.
(2) The topography (natural slope of the land) which can affect drainage, temperature and workability of the soil.
(3) Climate and local weather, particularly the effects on temperature, and rainfall (quantity and distribution).

Soil condition is largely dependent on the management of the soil in recent times. It can be built up by good husbandry but, if this standard is not maintained, the soil will soon return to its natural fertility level. The application of fertilizers can raise soil fertility by increasing the quantities of plant food in the growth and decay cycle.

Soil management can control the following production factors:

(1) The amount of organic matter in the soil (page 19).
(2) The amount of water in the soil by drainage and irrigation.
(3) The loss of soil by erosion (removal by wind and water).
(4) The pH of the soil (Liming, page 30).
(5) The amount of plant nutrients in the soil (Fertilizers, page 47).
(6) The soil structure (Cultivations, page 40).

Good management of the above factors should maintain or increase soil fertility and at the same time be commercially profitable.

TYPES OF SOIL

There are wide variations in the types of soil found on farms. They may be classified in various ways, e.g. according to texture. The amount of clay, silt and sand which soils contain can be found by mechanical analysis. This is an elaborate separation of the particles by settling from a water suspension and sieving in the laboratory. This can give accurate measurements of the amount of clay, silt and sand particles present. Gravel and stones are not included in a sample for mechanical analysis. The size of particles for each material is shown in Table 2.

TABLE 2

Material	Diameter of particles (mm)	Surface area
Clay	Less than 0.002	100,000 x a
Silt	0.002–0.06	1,000 x a
Fine sand	0.06–0.02	100 x a
Medium sand	0.2–0.6	10 x a
Coarse sand	0.6–2.0	a
Gravel	More than 2.0	

A given amount of clay has a greater effect on the characteristics of a soil than the same amount of sand or silt. Soils with a high clay content usually have a relatively high organic matter content. The organic matter is also higher in soils in cooler and higher rainfall areas.

Soils are classified into textural groups for scientific research and some advisory work (page 284).

The following is a general farming classification based on the clay, silt, sand and organic matter content of clay soils, sandy soils, calcareous soils, silts, peats and peaty soils, and Black Fen soils.

Clay soils

These soils have a high proportion of clay and silty material—usually over 60%, at least half of which is pure clay, which is mainly responsible for their characteristic qualities. Particles of clay have very important properties:

(1) They can group together into small clusters (flocculate) or become scattered (deflocculated).
(2) They can combine with various chemical substances (base exchange) such as calcium, sodium, potassium and ammonia, and in this way plant nutrients can be held in the soil.

Grouping or flocculation of the particles is very important in making clay soils easy to work. Clay particles combined with calcium (lime) will flocculate easily, whereas those combined with sodium will not, and so salt (sodium chloride) must be used carefully on clay soils. Deflocculation can occur when clays are worked in a wet condition and if they are flooded with seawater. The adhesive properties of clay are very beneficial to the soil structure when the groups of particles are small (like crumbs), but they can be very harmful when large lumps (clods) are formed. Frost action, and alternating periods of wetting and drying, will help to restore them to the flocculated crumb condition.

Characteristics

(1) Clay soils feel very sticky when wet and can be moulded into various shapes.
(2) They can hold more total water than most other soil types and, although only about half of this is available to plants, crops seldom suffer from drought.
(3) They swell when wetted and shrink when dried.
(4) They lie wet in winter and so stock should be taken off the land to avoid poaching (the compaction of soils by animals' hooves).
(5) They are very late in warming up in the spring because water heats up more slowly than mineral matter.
(6) They are normally fairly rich in potash, but are deficient in phosphates.
(7) Lime requirements are very variable; a clay soil which is well limed usually has a better structure and so is easier to work. Overliming will not cause any troubles such as trace element deficiency.

Management

They should not be worked in spring when wet because they become puddled and later dry into hard lumps. This can only be broken down by well-timed cultivations following repeated wetting (swelling) and drying (shrinking). Some air is drawn into cracks caused by shrinkage. These cracks remain when the clod is wetted again and so lines of weakness are formed which eventually break the clod. In dry weather irrigation may be used to wet the clods. In prolonged dry weather, these cracks may become very wide and deep which later may be very beneficial for drainage.

Clays are often called heavy soils because for ploughing and subsequent cultivations, compared with light (sandy) soils, two to four times the amount of tractor power may be required. All cultivations must be very carefully timed (often restricted to a short period) so that the soil structure is not damaged. This means that more tractors and implements are required than on similar sized loam or sandy soil farms. Autumn ploughing, to allow for a frost tilth, is essential if good seedbeds are to be produced in the spring.

Good drainage is very necessary. Some clay fields are still in "ridge and furrow". This was set up by ploughing—making the "openings" and "finishes" in the same respective places—until a distinct ridge and furrow pattern was formed. The direction of the furrows is the same as the fall on the field so that water can easily run off into ditches. If the ridges and furrows are levelled out, a mole drainage system using piped main drains should be substituted (page 34). This change is well worthwhile where arable crops are grown.

In many clay-land areas, especially where rainfall is high, the fields are often small and irregular in shape because the boundaries were originally ditches which followed the fall of the land. The hedges and deciduous trees which were planted later grow very well on these fertile, wet soils.

The close texture and an adequate water supply often restrict root development on clay soils.

Organic matter, such as straw-rich farmyard manure, ploughed-in straw or grassland residues, make these soils easier to work.

Cropping

Because of the many difficulties to be overcome in growing arable crops on these soils they are often left in permanent grass and only grazed during the growing season. Where arable crops are grown, a three- or four-year ley is often included in the rotation. Winter wheat is the most popular arable crop; winter beans are also grown. Both these crops are planted in the autumn (late September/October) when more liberties can be taken with seedbed preparation than would be permissible in spring. For the same reason, August-sown winter oilseed rape is also popular. Sugar beet and potatoes are grown in some districts. However, there can be difficulties in seedbed preparation, weed control and harvesting these crops, especially in a wet season. The best sequence of cropping with these root crops is following a period under grass. Then the soil structure is more stable and the soil easier to work.

Sandy soils

Characteristics

(1) In many ways these are the opposite of clays and are often called light soils because, when working these soils, comparatively little power is required to draw cultivation implements.
(2) They can be worked at any time, even in wet weather, without harmful effects.
(3) They are normally free-draining, but some drains may be required where there is clay or other impervious layer underneath.
(4) They have a high proportion of sand and other coarse particles but very little clay, usually less than 5%. They feel gritty.
(5) They warm up early in spring but crops are very liable to "burn-up" in a dry period because the water-holding capacity is low.
(6) They are unstable and can be easily eroded by water (on slopes) and by wind.
(7) They have little natural structure of their own and often need subsoiling at regular intervals to loosen compacted layers (pans).

Management

Sandy soils are naturally very low in plant nutrients, and fertilizers (especially nitrogen) are easily washed out. Adequate amounts of fertilizer must be applied to every crop. Liming is necessary but must be used carefully—"little and often".

Organic matter, especially as humus, is very beneficial because it helps to hold water and plant nutrients in the soil. On properly limed fields it breaks down very rapidly. This is because the soil micro-organisms are very active in these open-textured soils which have a good air supply.

Irrigation can be very important on the coarser sands, but very fine sandy soils have good available water capacity.

In some sandy areas the surface soil is liable to "blow" in dry, windy weather which could destroy a young crop. Where possible, the remedy is to add about 400 tonnes/ha of clay (Claying, page 40). Shelter belts are helpful where clay is not readily available.

Cropping

A wide range of crops can be grown, but yields are very dependent on a good supply of water and adequate plant foods. Market gardening is often carried on where a good sandy area is situated near a large population.

On the lighter sands in low rainfall areas and where irrigation is not possible, the main crops grown are rye, carrots, sugar beet and lucerne; lupins are grown in a few areas where the soil is very poor and acid.

On the better sandy soils, and particularly where the water supply (from rain or irrigation) is reasonably good, the main arable farm crops grown are barley, peas, rye, sugar beet, potatoes and carrots.

Because of the inherent low fertility of this type of land, the farms and fields are usually larger than on more fertile soils. Hedges are not very common because there is not enough water for good growth. The trees are usually drought-resistant coniferous types.

Livestock can be outwintered on sandy soils with very little risk of damage by poaching, even in wet weather.

Loams

Characteristics

(1) Loams are intermediate in texture between the clays and sandy soils and, in general, have most of the advantages and few of the disadvantages of these two extreme types.

They may feel gritty but also somewhat sticky.

(2) The amount of clay present varies considerably and so this group is sometimes divided into heavy or clay loams (resembling clays in many respects), medium loams, and sandy or light loams (resembling the better sandy soils).

(3) They warm up reasonably early in the spring and are fairly resistant to drought.

Management

Loams are easily worked but should not be worked when wet, especially clay loams. They usually require to be drained, but this is not difficult (Drainage, page 32).

Cropping

Loams are generally regarded as the best all-round soils because they are naturally fertile and can be used for growing any crop provided the depth of soil is sufficient. Crop yields do not vary much from year to year.

These soils can be used for most types of arable or grassland farming but, in general, mixed farming is carried on. Cereals, potatoes and sugar beet are the main cash crops, and leys provide grazing and winter bulk foods for dairy cows, beef cattle or sheep.

Calcareous soils

Characteristics

These are soils derived from chalk and limestone rocks and contain various amounts of calcium carbonate, between 5% and 50%. The depth of soil and subsoil may vary from 8 cm to over a metre. In general, the deep soils are more fertile than the shallow ones. The ease of working and stickiness of these soils depends on the amount of clay and chalk or limestone present; they usually have a loamy texture. Sharp edged flints of various sizes, found in soils overlying some of the chalk formations, are very wearing on cultivation implements and rubber tyres, as well as being destructive, when picked up by harvesting machinery. In some places the flints are found mixed with clay, e.g. clay-with-flint soils.

The soils are free-draining except in a few small areas where there is a deep clayish subsoil. Those overlying chalk are generally more productive than over limestone because plant roots can penetrate the soft chalk and explore for water. The limestones are harder and mainly impenetrable. Limestone rock pieces, loosened by cultivations, are a more severe problem to the farmer than the pieces of chalk that work their way to the surface on the downland arable farms. Some Cotswold fields have more rock in the profile than soil.

Dry valleys are characteristic of the downlands and wolds. The few rivers rise from underground streams, and the deliberate flooding of watermeadows in the river valleys used to be a common practice. Watercress beds flourish along some chalk streams.

There are very few hedges and the trees are mainly beech and conifers.

Walls of local stone form the field boundaries in some limestone areas, e.g. the Cotswolds.

Management

The soils are usually deficient in phosphates and potash, but only the deeper ones are likely to need liming. Organic matter can be beneficial, but it breaks down fairly rapidly and may be expensive to replace.

The farms and fields on this type of land are usually large, especially on the thinner soils.

Some areas are still unfenced and have no water laid on for stock. However, this is changing as mixed farming systems with grazing animals replace the folded sheep flocks.

Cropping

Barley and wheat (on deeper soils) are the best crops for these soils. The combine drill has been very important in producing good crops on some of the poorer, thinner soils, but now the level of fertility has been raised on most farms and the fertilizer is often broadcast. Continuous barley and/or wheat production is a common practice, but this could change because of the increasing costs of controlling grass weeds. Roots such as sugar beet, fodder beet and potatoes (some for seed) are grown on the deeper soils on some farms. Other crops such as oilseed rape, peas, beans, linseed, and leys for grazing, conservation or seed production provide a break from cereals. Kale is grown on some farms for stock and pheasants!

Calcareous soils can be found at fairly high altitudes and the fields are often exposed. Great care should therefore be taken when growing crops which shed their seed easily, e.g. oilseed rape. Harvesting must be carried out carefully to minimize yield loss.

Areas of black puffy soil (18–25% organic matter) are found on some chalkland farms and they require special treatment, e.g. for copper deficiency.

Silts

Characteristics

The pure silts (e.g. Lower Weald in Sussex) contain about 80% silt, have a very silky, buttery feel, and are very difficult to drain and manage for arable cropping. They are best left in permanent grass.

The deep, silty, alluvial types of soil consist mainly of fine sand and silt particles and are naturally very fertile. The lighter ones are some of the most fertile soils in this country. Some have been reclaimed from sea marshes, e.g. around the Wash, Romney Marsh, Solway Firth and parts of Essex, and the warplands of the Lower Trent and Humber. Other soils are the brickearths (wind-blown material) in North Kent, the Sussex coastal plain, south-east Essex, parts of Hertfordshire and the Mendip plateau. There are also the silty shale soils derived from Old Red Sandstone and Silurian rocks. The high fertility is mainly due to their great depth, very good available water capacity and good working properties, provided that organic matter is maintained above 3%.

Management

Although very fertile, silts can also be difficult to manage. The two main problems are capping and compaction (Soil capping, page 45).

Capping occurs when heavy rain falls on a very fine seedbed. Silt and clay particles go into suspension in a surface slurry and, as this dries out, it forms an impenetrable layer on the surface of the soil. It is particularly damaging if seeds have been sown but not yet germinated and emerged. Crops with small seedlings, such as sugar beet or brassicae, will sometimes need to be redrilled if capping occurs.

Leaving a rougher seedbed and increasing surface organic matter can decrease the risk of capping.

The lighter silts, as with the sands, can be damaged by compaction if worked in unsuitable conditions. Correction by deep cultivations will be necessary, adding to the expense of growing the crop.

Cropping

In the west, mainly grass and cereals are grown. However, in the east, the main crops are wheat, potatoes, sugar beet, peas, onions, seed production of root crops, bulbs and market gardening.

Peats and peaty soils

Characteristics

Peaty soils contain about 20–25% of organic matter, whereas there is about 50–90% in true peats.

The acid or peat bog peats and peaty soils have been formed in waterlogged areas where plants such as mosses, cotton grass, heather, molinia and rushes grew. The dead material from these plants was only partly broken down by the types of bacteria which can survive under these acidic

waterlogged conditions. This "humus" material built up slowly, about 30 cm every century.

When reclaimed, these soils break down easily to release nutrients, particularly nitrogen, but they are very low in phosphates and potash. Old tree trunks have to be dug out from time to time as the level of the soil falls due to organic matter breakdown.

Management

Before reclaiming this land for cropping, much of the peat is often cut away for fuel or sold as peat moss for horticultural purposes.

Good drainage must then be carried out by cutting deep ditches through the area. Deep ploughing also helps to drain the soil. Heavy applications—up to 25 tonnes/ha of ground limestone—may be required to neutralize the acidity.

In the first year, about 15 tonnes/ha of farmyard manure improves the yields of pioneer crops (usually potatoes, sometimes oats or rye). The reason for this may be that the farmyard manure introduces beneficial types of bacteria.

Cropping

In exposed areas these soils are often sown down to grass and clover.

Good swards can be established, but these must not be overgrazed or "poached" in wet weather otherwise the field will quickly revert to rushes and weed grasses. Under cultivation, most arable crops can be grown, but potatoes and oats are very suitable, being most tolerant of acid conditions.

Black Fen soils

Characteristics

Black Fen soils are found in part of the Fen district of East Anglia and are amongst the most fertile soils in the British Isles.

These soils were formed in marshy river estuary conditions as a result of water coming from limestone and chalk areas, carrying calcium carbonate and, in flood times, considerable amounts of silty material. The remains of the vegetation (mainly reeds, sedges and other estuary plants) did not break down completely because of the waterlogged conditions and so built up as humus. The soils vary from district to district, but most of them consist almost entirely of organic matter.

Management

After building strong sea walls, the area has been reclaimed by draining with deep ditches and underground drains. Most of the land is below sea level and so the water in the ditches has to be pumped over the sea walls or into the main drainage channels.

The soil breaks down readily and the level is falling about 2 cm per year and eventually will reach the clay or gravel subsoil. Tree trunks have to be dug out occasionally.

"Blowing" in spring is a serious problem on these dry sooty-black, friable soils. Several plantings of crop seedlings, together with the top 5–8 cm of soil and fertilizers, may be blown into the ditches. This can be prevented by applying 400–800 tonnes of clay per hectare (Claying, page 40), or by deep ploughing and/or cultivation to mix the underlying clay and organic topsoil. Husbandry techniques such as straw planting and strip cultivation are also used to prevent wind erosion.

These soils are rich in nitrogen, released by the breakdown of the organic matter, but are very poor in phosphates and potash and also trace elements such as manganese and copper.

Cropping

This is an intensive arable area where the main crops are wheat, potatoes and sugar beet; also smaller areas of celery, peas, onions, carrots and market garden crops.

In some parts of the Fens, leys have been introduced, with limited success, in an attempt to check the rapid rate of breakdown of the soils.

SOIL IMPROVEMENT

Liming

Most farm crops will not grow satisfactorily if the soil is very acid. This can be remedied by applying one of the commonly used liming materials.

Soil reaction. All substances in the presence of water are either acid, alkaline or neutral. The term reaction describes the degree or condition of acidity, alkalinity or neutrality. Acidity and alkalinity are expressed by a pH scale on which pH 7 is neutral, numbers below 7 indicate acidity and those above 7 alkalinity. Most cultivated soils have a pH range between 4.5 and 8.0 and may be grouped as in Table 3.

TABLE 3

	pH	Description
Over	7	Alkaline
	7	Neutral
	6.0–6.9	Slightly acid
	5.2–5.9	Moderately acid
Below	5.2	Very acid

Lime requirement. This is the amount of lime required to raise the pH to approximately 6.5 in the top 15 cm layer of soil. This amount varies considerably with the degree of acidity and the type of soil. Heavy (clay) soils and soils rich in organic matter require more lime to raise the pH than other types of soil. For example, to raise the pH from 5.5 to 6.5 on a sandy loam may require about 5 tonnes/ha of ground limestone, but on a clay soil 10–12.5 tonnes/ha of ground limestone may be necessary. The sandy loam, however, will need to be limed more frequently than the clay soil. The actual lime requirement can be calculated from chemical tests in the laboratory. It is unnecessary to lime soils which have a pH of more than 6.5.

Indications of soil acidity (i.e. the need for liming):

(1) Crops failing in patches, particularly the acid-sensitive ones such as barley and sugar

beet. The plants usually die off or are very unthrifty in the seedling stage.

(2) On grassland, there are poor types of grasses present such as the bents. Often a mat of undecayed vegetation builds up because the acidity reduces the activities of earthworms and bacteria which break down such material.

(3) On arable land, weeds such as sheep's sorrel, corn marigold and spurrey are common.

The main benefits of applying lime are:

(1) It neutralizes the acidity of the soil.
(2) It supplies calcium (and sometimes magnesium) for plant nutrition.
(3) It improves soil structure. In well-limed soils, plants usually produce more roots and grow better, and bacteria are more active in breaking down organic matter. This usually results in an improved soil structure and the soil can be cultivated more easily.
(4) It affects the availability of plant nutrients. Nitrogen, phosphate and potash are freely available on properly limed soils. Too much lime is likely to make some minor nutrients unavailable to plants, e.g. manganese, boron, copper and zinc. This is least likely to happen in clay soils.

pH and crop growth. To give crops the best opportunity to grow well, the soil pH should be near or above the following.

	pH
Barley, sugar beet and lucerne	6.5
Red clover, maize, oilseed rape, beans	6.0
Wheat, oats, peas, turnips and swedes	5.5
Potatoes	5.0
Rye and lupins	4.5

Lime is removed from the soil by:

Drainage. Lime is fairly easily removed in drainage water. 125–2000 kg/ha of calcium carbonate may be lost annually. The rate of loss

is greatest in industrial, smoke-polluted areas, areas of high rainfall, well-drained soils and soils rich in lime.

Fertilizers and manures. Every 1 kg of sulphate of ammonia removes about 1 kg of calcium carbonate from the soil. Poultry manure may also remove some lime.

Crops. The approximate amounts of calcium carbonate removed by crops are:

Cereals	1–3 kg/tonne of grain.
	5–7 kg/tonne of straw.
Potatoes	28 kg/40 tonnes of tubers.
Sugar beet	58 kg/40 tonnes of roots.
	230 kg/35 tonnes of tops.
Swedes	100 kg/60 tonnes of roots.
Kale (carted off)	450 kg/50 tonnes of crop.
Lucerne	58 kg/tonne of hay.

Stock also remove lime, e.g. a 500-kg animal sold off the farm removes about 16 kg of calcium carbonate in its bones, a 40-kg lamb about 1.3 kg of calcium carbonate, and 5000 litres of milk about 18 kg of calcium carbonate.

Soil analysis will be used to determine the pH and lime requirements of a soil. Portable testing equipment, using colour charts, can be used by the grower to test for pH.

Materials commonly used for liming soils

Ground limestone or chalk (also called carbonate of lime and calcium carbonate, $CaCO_3$). This is obtained by quarrying the limestone or chalk rock and grinding it to a fine powder. It is the commonest liming material used at present. *Burnt lime* (also called quicklime, lump lime, shell lime and calcium oxide, CaO). This is produced by burning lumps of limestone or chalk rock with coke or other fuel in a kiln. Carbon dioxide is given off and the lumps of burnt lime which are left are sold as lumps, or are ground up for mechanical spreading. This concentrated form of lime is especially useful for application to remote areas where transport costs are high. Burnt lime may scorch growing crops because it readily takes water from the leaves. When lumps of burnt lime are wetted they break down to a fine powder called hydrated lime.

Hydrated lime (also called slaked lime, $Ca(OH)_2$). This is a good liming material but it is usually too expensive for the farmer to use.

Waste limes. These are liming materials which can sometimes be obtained from industrial processes where lime is used as a purifying material. These limes are cheap but usually contain a lot of water. Some of the sources are: sugar beet factories, waste from manufacture of sulphate of ammonia, soap works, bleaching, tanneries, etc. Care is needed when using these materials because some may contain harmful substances. Sugar beet waste lime is also a valuable source of plant nutrients, but because of rhizomania (Table 72) its use is now very much restricted.

A comparison of the various liming materials shows that one tonne of burnt lime (CaO) is equivalent to:

1.37 tonnes hydrated lime $Ca(OH)_2$
1.83 tonnes ground limestone $CaCO_3$
or at least 2.5 tonnes waste lime (usually $CaCO_3$).

The supplier of lime must give a statement of the neutralizing value (NV) of the liming material which is the same as the calcium oxide equivalent.

Magnesian or dolomite limestone. This limestone consists of magnesium carbonate ($MgCO_3$) and calcium carbonate, ($CaCO_3$). It is commonly used as a liming material in areas where it is found. Magnesium carbonate has an approximate 20% better neutralizing value than calcium carbonate. Additionally, the magnesium may prevent magnesium deficiency diseases in crops (e.g. interveinal yellowing of leaves in potatoes, sugar beet and oats).

Cost. The cost of liming is largely dependent on the transport costs from the lime works to the farm. By dividing the cost per tonne of the liming material by the figure for the neutralizing value, the **unit cost** is obtained. In this way it is possible to compare the costs of various liming materials.

Most farmers now use ground limestone or chalk and arrange for it to be spread mechanically by the suppliers. Where large amounts are required (over 7 tonnes/ha) it may be preferable to apply it in two dressings, e.g. half before and half after ploughing.

Rates of application. 2.5–25 tonnes/ha of calcium carbonate, or its equivalent, may be needed to correct the lime deficiency of a soil. Maintenance dressings of about 2.5–4 tonnes/ha calcium carbonate, every four years, should be adequate to replace average losses.

Drainage (see also Water in the Soil, page 20)

Normally, the soil can only hold some of the rainwater which falls on it. The remainder either runs off or is evaporated from the surface or soaks through the soil to the subsoil. If surplus water is prevented from moving through the soil and subsoil, it soon fills up the pore space; this will kill or stunt the growing crops.

The water table is the level in the soil or subsoil below which the pore space is filled with water. This is not easy to see or measure in clay soils but can be seen in open textured soils (Fig. 27).

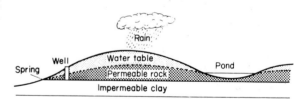

FIG. 27. Position of water-table and the effect on water levels in wells and ponds.

The water table level fluctuates throughout the year and in the British Isles is usually highest in February and lowest in September. This is because of a higher amount of evaporation, more transpiration and usually lower rainfall in the summer.

In chalk and limestone areas, and in most sandy and gravelly soils, water can drain away easily into the porous subsoil. These are free-draining soils.

On most other types of farmland some sort of artificial field drainage is necessary to carry away the surplus water and so keep the water table at a reasonable level. For most arable farm crops the water table should be about 0.6 m or more below the surface; for grassland 0.3–0.45 m is sufficient.

The withdrawal of grants for drainage work has meant a large reduction in the area of land being drained.

Indications of poor drainage are:

(1) Machinery is easily "bogged down" in wet weather.
(2) Stock grazing pastures in wet weather easily poach the sward.
(3) Water remains on the surface for many days following heavy rain.
(4) Weeds such as rushes, sedges, horsetail, tussock grass and meadowsweet are common in grassland. These weeds usually disappear after drainage. Peat forms in places which have been very wet for a long time.
(5) Young plants are pale green or yellow in colour and generally unthrifty.
(6) The subsoil is often coloured in various shades of blue or grey, compared with shades of reddish-brown, yellow and orange in well-drained soil.

The benefits of good drainage are:

(1) Well-drained land is better aerated and the crops grow better and are less likely to be damaged by root-decaying fungi.
(2) The soil dries out better in spring and so warms up more quickly and can be worked earlier.
(3) Plants are encouraged to form a deeper and more extensive root system. In this way they can often obtain more plant food.
(4) Grassland is firmer, especially after wet periods. Good drainage is essential for high density stocking and where cattle are outwintered if serious poaching of the pasture is to be avoided.
(5) Disease risk from parasites is reduced. A good example is the liver fluke. Part of its life cycle is in a water snail found on badly drained land.
(6) Inter-row cultivations and harvesting of root crops can be carried out more efficiently.

The main methods used to remove surplus water and control the water table are:

(1) Ditches or open channels.
(2) Underground pipe drains—tiles and plastic.
(3) Mole drains.
(4) Ridge and furrow (page 26).

Ditches and open drains

Ditches may be adequate to drain an area by themselves, but they usually serve as outlets for underground drains. They are capable of dealing with large volumes of water in very wet periods. The size of a ditch varies according to the area it serves (Fig. 28). Ditches ought to be cleaned out to their original depth once every three or four years. The spoil removed should be spread well clear of the edge of the ditch. Many different types of machines are now available for making new ditches and cleaning neglected ones.

Small open channels 10–60 m apart are used for draining hill grazing areas. These are either

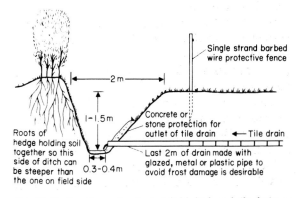

FIG. 28. Section through a typical field ditch and tile drain.

dug by hand using a spade designed for the purpose, or made with a special type of plough drawn by a crawler tractor. Similar open channels are used on lowlying meadow land where underground drainage is not possible.

Underground drains

The distance between drains which is necessary for good drainage depends on the soil texture. In clay soils the small pore spaces restrict the movement of water. Therefore, the drains must be spaced much closer together than on the lighter types of soil where water can flow freely through the large pore spaces (Figs. 29a, 29b).

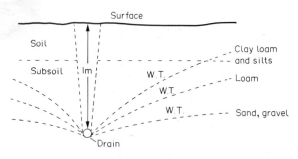

FIG. 29a. The steepness of a water-table (W.T.) varies with different soil types.

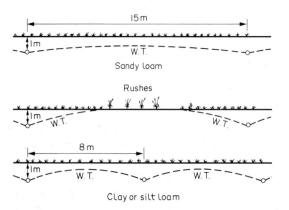

FIG. 29b. Diagram showing the effect of spacing of drains on the water-table (W.T.) (a) sandy loam, (b) and (c) silt or clay loam.

In the past, many different materials have been used in trenches to provide an underground passage for water, e.g. bushes, turf, stones, and flat tiles before cylindrical clay tiles became popular. (Concrete pipes have also been used.) However, increasingly, the flexible plastic pipe is replacing the clay tile for all new drainage work. Underdrainage is very expensive and cost must be related to likely benefits. Consequently, when designing a scheme, the spacing of the drains must be carefully considered. Digging some holes and examining the subsoil can be very helpful in deciding how the field should be drained. If cost is a limiting factor, it would, for example, be advisable to lay the drains about 40 m (two chains) apart in an area where 20 m would be an ideal spacing. This should be quite satisfactory over most of the field and only a few extra drains may be required later in the wetter areas. This should not be difficult if a detailed map of the drainage layout has been made.

The slopes on a field determine the way the drains should run to be efficient in removing water (Fig. 30). Water will not run uphill, unless it is pumped, and so the drains must have a fall. If the slope is more than 2% (1 : 50) the laterals (side drains) should run across the slope. Ideally, the minimum fall on laterals should be 1 : 250 and, in mains, 1 : 400. It is usually preferable that the laterals go into a main drain before entering a ditch; this means fewer outfalls to maintain. In flat areas, and where the soil is silty (causing silting of the drains), it is preferable to let each lateral run straight into the ditch. A high-pressure jetting device can be used to clean them periodically. Iron ochre deposits can also be cleared in this way.

Where the land is lowlying and there is no natural fall to a river, the area can be drained in the usual way and the outlet water from the drainage system can lead to a ditch from which it is pumped into a river.

Underground drainage is best carried out in reasonably dry weather. It is fairly common practice now to lay drains through a growing crop of winter cereals in a dry period in

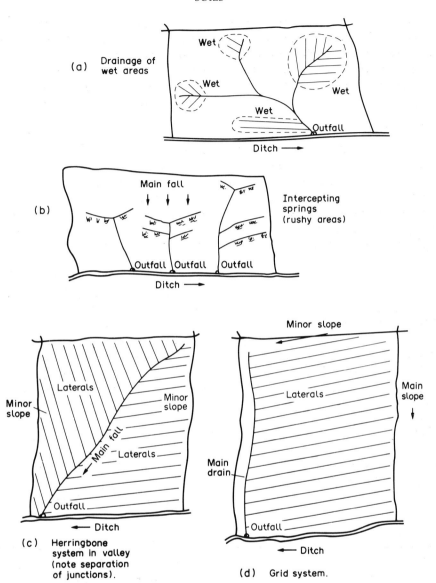

(a) Drainage of wet areas

(b) Intercepting springs (rushy areas)

(c) Herringbone system in valley (note separation of junctions).

(d) Grid system.

FIG. 30. Drain layouts.

the spring. With modern machinery the little damage to the crop is more than offset by the beneficial effects of the drains.

Underground piped drains—tiles and plastic

Tiles are burnt clay pipes about 300 mm long which are laid end to end. They have internal diameters of 75 mm for laterals and 100, 150 and 220 mm for main drains. Most of the water enters the pipes at the butt joint between pipes. They are heavy and costly to transport.

Plastic pipes come in various configurations. Most of them are corrugated with slits cut at intervals to allow water to enter easily. They are usually supplied in 200-metre rolls which are easily transported. Various diameter sizes are available, e.g. 60, 80 mm for laterals, 100, 125, 150 and 250 mm for mains.

The size of pipe required will depend on the rainfall, area to be drained by each pipe, fall and soil structure.

Various types of machine are available for laying tiles and plastic pipes, and the porous backfill (e.g. washed gravel, clinker) required on the heavier soils. The depth of porous fill is more important than the width in the trench and so much less of this expensive material is required when using plastic pipe machines. Modern machines can lay pipes very rapidly, especially when a laser beam system is used automatically to adjust the fall on the drain.

Underground drains—mole drainage

Mole drainage is a cheap method which can be used in some fields. Although mole drains are sometimes used on peat soils, the system is normally applied to fields which have a:

(1) Clay subsoil with no stones, sand or gravel patches.
(2) Suitable fall 5–10 cm/20 m.
(3) Reasonably smooth surface.

A mole plough, which has a torpedo or bullet-shaped "mole" attached to a steel coulter or blade, forms a cylindrical channel in the subsoil (Fig. 31).

The best conditions for mole draining are when the subsoil is damp enough to be plastic

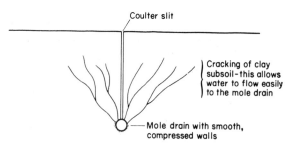

FIG. 32. Section through mole drain and surrounding soil.

and forms a good surface on the mole channel. It should also be sufficiently dry above to form cracks as the mole plough passes (Fig. 32). Furthermore, if the surface is dry the tractor hauling the plough can get a better grip. The plough should be drawn slowly (about 3 km/h) otherwise the vacuum created is likely to spoil the mole. Reasonably dry weather after moling will allow the surface of the mole to harden and so it should last longer.

Mole drains are drawn 3–4 m apart.

For best results the moles should be drawn through the porous backfilling of a tiled or

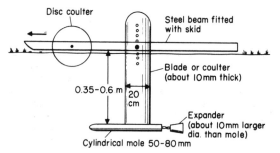

FIG. 31. Mole plough.

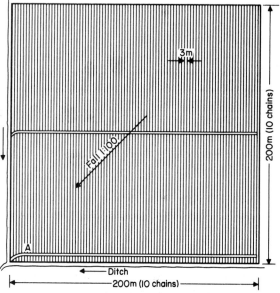

FIG. 33. Layout of mole drainage (with field main) in a 4-hectare area.

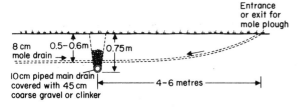

Entrance
or exit for
mole plough

8 cm   0.5–0.6m   0.75m
mole drain

10cm piped main drain
covered with 45 cm |← 4–6 metres →|
coarse gravel or clinker

FIG. 34. Section through tile drain shown in Fig. 33 to show how water from the mole drain can enter the pipe drain through the porous backfilling.

TABLE 4 *Water requirements for green crops—southern UK*

Period	per day		per month	
	mm	m³/ha	mm	m³/ha
April and September	1.5	15	50	500
May and August	2.5	25	80	800
June and July	3.5	35	100	1000

plastic main drain (Figs. 33,34). A new set of mole drains can be drawn every three to ten years as required. They will last much longer in a clay soil than in a sandy soil where the sides of the moles collapse quite quickly.

Installing a new drainage system is the first step in removing surplus water from a field. However, the structure of the soil and subsoil must, if necessary, be improved so that water can move easily to the drains.

IRRIGATION

Irrigation is the term used when water is applied to crops which are suffering from drought.

It is usual to measure irrigation water, like rainfall, in millimetres:

1 mm on 1 hectare = 10 m³ = 10 tonnes.
25 mm on 1 hectare = 250 m³ (a common application at one time).
100 mm on 1 hectare = 1000 m³ = 1 km³ (about a season's requirement).
100 mm on 50 ha = 50 km³ (a guide to the amount of water required).

Green plants take up water from the soil and transpire it through their leaves, although a small amount is retained to build up the plant structure. Most green crops which are covering the ground have the same water requirements which vary from day to day and place to place according to climate and weather conditions. It is called potential transpiration (Table 4).

When the rainfall is well distributed, crops are less likely to suffer than in wet periods followed by long dry periods. During such dry periods the crops have to survive on the available water held in the soil (Water in the Soil, page 20).

The difference between the actual amount of water in the soil and the amount held at field capacity is known as the soil moisture deficit (SMD). It is the main factor in determining the need for irrigation.

An approximate soil moisture deficit can, at any time, be calculated from the rainfall and the figures given in Table 6. Example—if the soil were at field capacity at the end of April, with a green crop covering the ground, and with no rain in May, then the SMD would be 80 mm at the end of May. However, if 50 mm of rain fell during May, then the deficit would be 30 mm. The amount of water available to a crop at any one time depends on the capacity of the soil per unit depth, and the depth of soil from which the roots take up water. The figures in Table 5 are average rooting depths. They would be greater on deep, well-drained soils and less on shallow, badly drained or panned soils. This can be checked in the field by digging a soil profile. Normally, most of the plant roots are

TABLE 5

Crop	Rooting depth (cm)
Cereals, pasture	35–100
Potatoes (varietal differences)	30–70
Beans, peas, conserved grass	45–75
Sugar beet	55–100
Lucerne (alfalfa)	over 120

concentrated in the top layers and only a small proportion penetrate deeply into the soil.

More than three-quarters of all the farm soils in this country have available water capacities between 60 and 100 mm (12–20% by volume) within the root range of most crops (about 500 mm depth). Sandy, gravelly and shallow stony soils have less than 60 mm (12%), whereas deep silty soils, very fine sands, warpland, organic and peaty soils have well over 100 mm (20%) of available water within root range. The available water capacity of the topsoil is often different from that of the subsoil. This should be taken into consideration when calculating the amount present within the root range of a crop.

It is estimated that the roots of a plant cannot take up much more than half of the available water within the depth of its deepest roots. 70% of roots are in the top half of the rooting zone and they do not penetrate all the pore spaces and so, where available, irrigation should be started when about half the available water has been taken up. This is known as the *critical deficit*.

Decisions on when to start irrigating must be influenced by the time it will take to apply 25 mm of water over the area involved, usually five to ten days.

Some fertilizer companies, Irrigation Management Services and ADAS Irriguide, provide helpful computerized predictions on when irrigation should be started. These take into consideration all the relevant factors and are tailored for each field and its crop.

The greatest need for irrigation is in the south-east counties where the lower rainfall and higher potential evapo-transpiration means that it would be beneficial about nine years out of ten. The need is much less in the wetter western and northern areas.

Water for irrigation is strictly controlled by licences issued by the National Rivers Authority (NRA). The charge for the water varies between areas and it usually costs up to ten times more for taking water directly from rivers in summer than for taking it to fill reservoirs in winter.

Water is generally very scarce in summer in the areas which need it most, but annual rainfall is adequate for most irrigation needs provided the winter surplus is stored in reservoirs or underground aquifers. The main sources of irrigation water are rivers and streams, ponds and lakes, reservoirs and boreholes. Some is taken from mains water but is very expensive. Boreholes are also expensive; they may have to be lined, e.g. through sand or gravel, and require special deep well pumps and sometimes filters. However, they may be the most reliable source in a very dry summer. Above-ground reservoir construction is specialized work and must be designed and supervised by a qualified engineer, if over 25,000 m^3 above ground level. (A hole in the ground is simpler and may be partly replenished by groundwater seepage.) The lining of a reservoir may be necessary and this could double the cost.

All water used for irrigation should be tested for impurities which may be harmful to crops. (ADAS and other services are available for this testing.)

Irrigation can be expensive, so it is important to use it as efficiently as possible, for example:

Early **potatoes** will respond with earlier, and so more profitable, production if irrigation is used when the SMD reaches 25 mm. Second earlies and maincrops are not normally irrigated until the tubers are fingernail size, otherwise some varieties, e.g. King Edward, may produce too many small tubers and a low saleable yield. On the better soils (with a high available water capacity) irrigation can be delayed until a 40-mm deficit occurs. On light soils where common scab causes problems, irrigation is recommended each time a 15-mm deficit occurs when the tubers are forming. If the soil is very dry and hard at harvest time, some irrigation can make harvesting easier and there is much less tuber damage (page 127).

With **sugar beet** it may be necessary to irrigate a very dry seedbed to obtain a fine tilth, even germination and more effective use of soil-acting herbicides. After this it is not usually necessary to irrigate until the leaves start meeting across the rows (sugar beet is deep-rooted). It can start with applications of about 25 mm, and later up to 55 mm, at one application on the high available water capacity soils when the SMD is over 100 mm.

For **peas**, unless the soil is very dry, it is preferable to wait and apply the water at the very responsive stages, namely 25 mm at the start of flowering (increases numbers of peas) and 25 mm when the pods are swelling (increases pea size), always provided that the SMD is about 25 mm at these times.

With **cereals**, irrigation can only be justified on very light soils in low rainfall areas, unless it is a very dry season everywhere. The most critical time is about GS 37 when the flag leaf is showing and the ear is developing rapidly. 25 mm application should be adequate.

Grass and other **leafy crops** respond throughout the season and about 25 mm of water can be applied every time the deficit reaches 30 mm. Although intensive dairying is the most profitable way to utilize the extra production, it is unusual to irrigate the grass crop.

In the above suggestions, for times and amounts of water to apply, it is assumed that it is a dry period with little or no sign of rain. To irrigate up to field capacity shortly before rain is to risk nitrogen losses by leaching.

Different textured soils absorb water at different rates (Table 6). When irrigation water or rain falls on a dry soil it saturates the top layers of soil to full field capacity before moving further down. However, if the soil is deeply cracked (e.g. clay), then some of the water will run down the cracks. Very dry soils can absorb water at a faster rate than shown in Table 6 but only for a short period.

TABLE 6

Soil type	Time (hr) required to absorb 25 mm of water
Sandy	1–2
Loams	3
Clay	4–5

Irrigation water is applied by:

(1) *Rotary sprinklers*. Each rotating sprinkler covers an area about 21–36 m in diameter. The sprinklers are usually spaced 10.5 m apart along the supply line which is moved about 18 m for each setting. Special nozzles can be used to apply a fine spray at 2.5mm per hour for frost protection of fruit crops and potatoes. Icicles form on the plants, but the latent heat of freezing protects the plant tissue from damage.

(2) *Rain guns*. These are now the most widely used machines. They are used on all types of crops. The droplets are large and the diameter of area covered may be between 60 and 120 m. Some types are used for organic irrigation (slurry).

(3) *Travelling rain gun with hose reel*. This is similar in many respects to a mobile machine, but the hose reel remains stationary—usually on a headland while the rain gun and its hose (nylon/PVC) are towed out to the start of the run (up to 500 m). It irrigates as it is wound back, using water pressure to turn the hose reel. At the end of the run it switches itself off automatically, and it can then be moved to the next position. About 80% of all irrigation water is now put on using this system.

(4) *Centre pivot and linear irrigators*. These are large machines which cover up to 160 hectares. They are mainly centre pivots which move round a pivot (the water source) in a complete circle or part of a circle.

(5) *Spraylines*. These apply the water gently and are used mainly for horticultural crops.

(6) *Trickle*. Water is applied through small bore pipes and special emitters on or under the soil surface.

(7) *Surface channels*. This method requires almost level or contoured land. It is wasteful of land, water and labour, but capital costs are very low.

(8) *Underground pipes*. On level land, water can be dammed in the ditches and allowed to

flow up the drainage pipes into the subsoil and lower soil layers. This is drainage in reverse!

On sloping fields, problems can arise with run-off if water is applied too quickly. It may run down between potato ridges, for example, without really wetting the ridges where the crop roots are growing. There is less run-off if a bed system is used (page 132).

Reservoirs with a capacity of over 25,000 m³ above the natural level of adjoining land must now be registered with the County Council. Inspection and repairs can be very expensive to ensure that they are safe.

The need for irrigation can be reduced by improving soil structure, breaking pans, correcting acidity (soil and subsoil), increasing organic matter in the soil and selecting crops and varieties with deep-rooting habits.

WARPING

This is a process of soil formation where land, lying between high and low water levels, alongside a tidal river, is deliberately flooded with muddy water. The area to be treated is surrounded by earth banks fitted with sluice gates. At high tide, water is allowed to flood quickly on to the enclosed area and is run off slowly through sluice gates at low tide. The fineness of the material deposited will depend on the length of time allowed for settling. The coarse particles will settle very quickly, but the finer particles may take one or more days to settle. The depth of the deposit may be 0.4 m in one winter. When enough alluvium is deposited, it is then drained and prepared for cropping.

These soils are very fertile and are usually intensively cropped with arable crops such as potatoes, sugar beet, peas, wheat and barley.

Part of the land around the Humber estuary is warpland and the best is probably that in the lower Trent valley. Most of this work was done last century. It is too expensive to consider now.

CLAYING

The texture of "blowaway" sandy soils and Black Fen soils can be improved by applying 400–750 tonnes/hectare of clay or marl (a lime-rich clay). If the subsoil of the area is clay, it can be dug out of trenches and roughly scattered by a dragline excavator. In other cases, the clay is dug in pits and transported in special lorry spreaders. Rotary cultivators help to spread the clay. If the work is done in late summer or autumn, the winter frosts help to break down the lumps of clay. As a practice, it is no longer considered economical.

TILLAGE AND CULTIVATIONS

Cultivations are field operations which attempt to alter the soil structure. The main object is to provide a suitable seedbed in which a crop can be planted and will grow satisfactorily. Cultivations are also used to kill weeds and/or bury the remains of previous crops. The cost of the work can be reduced considerably by good timing and the use of the right implements. Ideally, a good seedbed should be prepared with the minimum amount of working and the least loss of moisture. On heavy soil and in a wet season, some loss of moisture can be desirable. On the medium and heavy soils full advantage should be taken of weathering effects. For example, ploughing in the autumn will allow frost to break the soil into a crumb structure; wetting and drying alternately will have a similar effect. The workability of a soil is dependent on its consistency, which is a reflection of its texture and moisture content.

Seedbed requirements

(1) *Cereals.*
    (a) *Autumn planted.* The object here is to provide a tilth (seedbed condition) which consists of fine material and egg-sized lumps. It should allow for the seed to be drilled and easily covered with the surface remaining rough after planting. The lumps on the surface prevent the siltier soils from

"capping" in a mild wet winter and they also protect the base of the cereals from the harmful effects of very cold winds. Harrowing and/or rolling may be carried out in the spring to break up a soil cap which may have formed. It will also firm the soil around plants which have been heaved by frost action.

(b) *Spring planted.* A fairly fine seedbed is required in the spring (very fine if grasses and clovers are to be undersown). If the seedbed is dry or very loose after drilling it should be consolidated by rolling. This is especially important if the crop is undersown and where the soil is stony.

(2) *Root crops,* e.g. sugar beet, swedes and carrots: also *kale.* These crops have small seeds and so the seedbed must be as fine as possible, level, moist and firm. This is very important when precision drills and very low seed rates are used. Good, early, ploughing with uniform, well-packed and broken furrow slices will reduce considerably the amount of work required in the spring when, if possible, deep cultivations should be avoided to keep frost mould on top, leaving unweathered soil well below the surface.

(3) *Direct seeding and/or reseeding of grasses and clovers.* The requirements will be the same as for the root crops.

(4) *Beans and peas.* The requirements will generally be similar to the cereal crop, although the tilth need not be so fine. Peas grown on light soils may be drilled into the ploughed surface if the ploughing has been well done. Winter beans are often broadcast on to stubble and ploughed in.

(5) *Potatoes.* This crop is usually planted in ridges 60–90 cm wide and so deep cultivations are necessary. The fineness of tilth required depends on how the crop will be managed after planting. A fairly rough, damp seedbed is usually preferable to a fine, dry tilth which has possibly been worked too much. Some crops are worked many times after planting—such as harrowing down the ridges, ridging-up again, deep cultivations between the ridges and final earthing-up. The main object of these cultivations is to control weeds, but the implements often damage the roots of the potato plants. Valuable soil moisture can also be lost.

Most annual weeds can be controlled by spraying the ridges when the potato sprouts start to appear. This can replace most of the inter-row cultivations and reduce the number of clods produced by the rubber-tyred tractor wheels.

(6) *Oil seeds.* Winter oilseed rape is sown in August and moisture conservation at this dry, hot time is of paramount importance. It is usual to carry out minimum cultivations to achieve a fine seedbed for the very small seed.

Spring oilseed rape and linseed also need fine seedbed conditions to give intimate contact between soil and seed. It is usual to plough in late autumn and then work the resultant frost tilth carefully with straight-tined shallow implements.

Non-ploughing techniques

Direct drilling (slit-seeding, zero tillage) and *minimal cultivation* techniques have now become alternatives to conventional cultivations on many farms and on a wide range of soil types, including difficult clays.

Several types of special drills are available for direct drilling, using such developments as heavily weighted discs for cutting slits, strong cultivator tines or modified rotary cultivators.

For it to be successful, direct drilling has relied on achieving a good straw and stubble burn to remove surface trash. Now that burning is banned, the practice has declined. With cereals it should only be used on very clean stubbles where the previous crop has been cut and the straw removed from the field. Even in these circumstances, there will be a risk of disease spread from the unburnt stubble to the new crop.

It can still be seriously considered for seeding grass and for crops such as kale.

Minimal cultivation systems are those using various types of cultivators instead of the plough, and in such a way that only the minimum depth of soil is moved to allow drilling to take place.

Tillage implements

The main implements used for tillage are:

(1) *Ploughs.* Ploughing is the first operation in seedbed preparation on most farms and is likely to remain so for some time yet. However, many farmers are now using rotary cultivators, heavy cultivators with fixed or spring tines, and mechanically-driven digging or pulverizing machines, as alternatives to the plough. Good ploughing is still the best method of burying weeds and the remains of previous crops. It can also set up the soil so that good frost penetration is possible. Fast ploughing produces a more broken furrow slice than slow steady work. The mounted or semi-mounted plough has replaced the trailed type on most farms because of ease of handling. General-purpose mouldboards are commonly used. The shorter digger types (concave mouldboards) break the furrow slices better and are often used on the lighter soils. Deep digger ploughs are used where deep ploughing is required, e.g. for roots or potatoes. The one way (reversible) type of plough has become popular for crops such as roots and peas. It has right-hand and left-hand mouldboards and so no openings or finishes have to be made when ploughing; the seedbed should therefore be that much more level. Round-and-round ploughing with the ordinary plough has almost the same effect, although this is not a suitable method on all fields.

The proper use of skim and disc coulters and careful setting of the plough for depth, width and pitch can greatly improve the quality of the ploughing. The furrow slice can only be turned over satisfactorily if the depth is less than about two-thirds the width. The usual widths of ordinary plough bodies vary from 20 to 35 cm. If possible, it is desirable to vary the depth of ploughing from year to year to avoid the formation of a plough pan. Very deep ploughing, which brings up several centimetres of poorly weathered subsoil, should only be undertaken with care: the long-term effects will probably be worthwhile but, for a few years afterwards, the soil may be rather sticky and difficult to work. Buried weed seeds, such as wild oats which have fallen down cracks, may be brought to the surface and may spoil the following crops. "Chisel ploughing" is a modern term used to describe the work done by a heavy duty cultivator with special spring or fixed tines; unlike the ordinary plough, it does not move or invert all the soil. Disc ploughs have large saucer-shaped discs instead of shares and mouldboards. Compared with the ordinary mouldboard plough, they do not cut all the ground or invert the soil so well, but they can work in harder and stickier soil conditions. They are more popular in dry countries. Double mouldboard ridging ploughs are sometimes used for potatoes and some root crops in the wetter areas.

(2) *Cultivators.* These are tined implements which are used to break up the soil clods (to ploughing depth). Some have tines which are rigid or are held by very strong springs which only give when an obstruction, such as a strong tree root, is struck. Others have spring tines which are constantly moving according to the resistance of the soil. They have a very good pulverizing effect and can often be pulled at a high speed. The shares on the tines are of various widths. The pitch of the tines draws the implement into the soil. Depth can be controlled by tractor linkage or wheels. The timing of cultivations is very important if the operation is to be effective.

(3) *Harrows.* There are many types of harrow. The zigzag type, which has staggered tines, is the commonest. Harrow tines are usually straight, but they may vary in length and strength, depending on the type of harrow. Drag harrows have curved ends on the tines.

These implements are often used to complete the work of the cultivator. Besides breaking the soil down to a fine tilth, they can have a useful consolidating effect due to shaking the soil about and rearranging particle distribution.

"Dutch" harrows have spikes fitted in a heavy wooden frame and are useful for levelling a seedbed as well as breaking clods.

Some harrows, e.g. the chain type, consist of flexible links joined together to form a rectangle. These follow an uneven surface better and do not jump about as much on grassland as the zigzag type. Most chain harrows have spikes fitted on one side.

Special types of harrows fitted with knife-like tines are used for improving matted grassland by tearing out surface trash.

Power-driven reciprocating harrows, on which rows of tines are made to move at right angles to the direction of travel, and rotary harrows, can result in much better movement of the soil in one pass. They are most valuable on the heavier soils when preparing fine seedbeds for potatoes, sugar beet and other crops.

(4) *Hoes*. These are implements used for controlling weeds between the rows in root crops. Various shaped blades and discs may be fitted to them. Most types are either front-, mid- or rear-mounted on a tractor. The front- and mid-mounted types are controlled by the steering of the tractor drive. The rear-mounted types usually require a second person for steering the hoe.

(5) *Disc harrows* consist of "gangs" of saucer-shaped discs between 30–60 cm in diameter. They have a cutting and consolidating effect on the soil. This is particularly useful when working a seedbed on ploughed-out grassland; some discs have scalloped edges to improve the cutting effect.

The more the discs are angled, the greater will be the depth of penetration, the cutting and breaking effect on the clods and the draught.

Disc harrows are widely used for preparing all kinds of seedbeds, but it should be remembered that they are expensive implements to use. They have a heavy draught and many wearing parts, such as discs, bearings and linkages, and so should only be used when harrows would not be suitable. They tend to cut the rhizomes of weeds such as couch and creeping thistle into short pieces which are easily spread. Discing of old grassland before ploughing will usually allow the plough to get a better bury of the turf. Heavy discs, and especially those with scalloped edges, are very useful for working in chopped straw after combining.

(6) *Rotary cultivators* (e.g. rotavator). This type of implement consists of curved blades which rotate round a horizontal shaft set at right angles to the direction of travel. The shaft is driven from the power take-off of the tractor; depth is controlled by a land wheel or skid. This implement can produce a good tilth in difficult conditions and, in many cases, it can replace all other implements in seedbed preparation. A light fluffy tilth is sometimes produced which may "cap" easily if wet weather follows. The fineness of tilth can be controlled by the forward speed of the tractor, for example, a fast speed can produce a coarse tilth. It is a very useful implement for mixing crop remains into the soil. The rotating action of the blades helps to drive the implement forward and so extra care must be taken when going down steep slopes.

In wet heavy soils the rotating action of the blades may have a smearing effect on the soil. This can usually be avoided by having the blades properly angled. Rotavating of ploughed or cultivated land when the surface is frozen in winter can produce a good seedbed for cereals in the spring without any further working or loss of moisture. Narrow rotary cultivator units are available for working between rows of root crops.

(7) *Rolls*. These are used to consolidate the top few centimetres of the soil so that plant roots can keep in contact with the soil particles, and the soil can hold more moisture. They are also used for crushing clods and breaking surface crusts. Rolls should not be used when the soil is wet; this is especially important on the heavier soils. The two main types of rolls are the flat roll, which has a smooth surface, and

the Cambridge or ring roll, which has a ribbed surface and consists of a number of heavy iron wheels or rings (about 7 cm wide) each of which has a ridge about 4 cm high. The rings are free to move independently and this helps to keep the surface clean. The ribbed or corrugated surface left by the Cambridge roll provides an excellent seedbed on which to sow grass and clover seeds or roots. Also, it is less likely to "cap" than a flat rolled surface.

A furrow press is a special type of very heavy ring roller (usually with three or four wheels) used for compressing the furrow slices after ploughing. It is usually attached to and pulled alongside the plough. It used to be commonly used when ploughing an old grass field.

Pans

A pan is a hard, cement-like layer in the soil or subsoil which can be very harmful because it prevents surplus water draining away freely and restricts root growth.

Such a layer may be caused by ploughing at the same depth every year. This is a plough pan and is partly caused by the base of the plough sliding along the furrow. It is more likely to occur if rubber-tyred tractors are used when the soil is wet, and there is some wheelslip which has a smearing effect on the bottom of the furrow. Plough pans are more likely to form on the heavier types of soil. They can be broken up by using a subsoiler or deeper ploughing.

Pans may also be formed by the deposition of iron compounds, and sometimes humus, in layers in the soil or subsoil. These are often called chemical or iron pans and may be destroyed in the same way as plough pans. Clay pans form in some soil formation processes.

Soil loosening

Soil loosening (which is justified only when necessary) is a term now used to cover many different types of operation (including subsoiling). This aims to improve the structure of the soil and subsoil by breaking pans and having a general loosening effect.

An ordinary subsoiler can be used. It is very effective when worked at the correct depth. A subsoiler is a very strong tine (usually two or more are fitted on a toolbar) which can be drawn through the soil and subsoil (about 0–5 m deep and 1 m apart) to break pans and produce a heaving and cracking effect. This will only produce satisfactory results when the subsoil is reasonably dry and drainage is good. The modern types with "wings" fitted near the base of the tine produce a very good shattering effect.

Other implements which may be used include the Paraplow. The angled mainframe carries three or four sloping tines which have adjustable rear flaps; there are also adjustable sloping discs. The effect is to give a fairly uniform lifting and cracking to about 35 cm. The Shakaerator has five strong tines which are made to vibrate in the ground by a mechanism driven from the tractor PTO. Modified rotary cultivators with fixed tines attached can also be described as soil looseners.

The soil loosening process is very necessary in situations where shallow cultivations—up to 150 mm—have been used for many years and the soil below has become consolidated and impermeable to water and plant roots. Rain water collects on this layer and causes plant death by waterlogging and/or soil erosion.

When attempting to loosen a compacted soil/subsoil, it is advisable to start with shallow cultivations, then to work deeper with stronger tines and then, if necessary, with a winged subsoiler. This can be achieved by several passes over the field. However, the one-pass method can also be used. The unit has a toolbar fitted with shallow tines in front, strong spring tines in the middle and winged subsoilers attached to the rear. It obviously requires greater tractor power, but it is an energy-efficient method. This work should only be carried out when the soil crumbles and does not smear.

Soil erosion

Soil erosion by water is an increasing problem on many soils, especially sandy, silty and chalk soils which are in continuous arable cultivation, and where the organic matter is below 2%. Rain splash causes capping of such soils and heavy rain readily runs off instead of soaking into the soil. There may also be panning problems. Erosion can be serious on sloping fields (especially large fields) where the cultivation lines, crop rows, tramlines, etc., run in the direction of the slope, and the wheelings are compacted. In these situations, in a wet time or a thunderstorm, deep rills and gullies, up to a metre deep, can be cut in the fields and up to 150 tonnes/ha of soil washed away. Sometimes crops can be covered at the lower end of a field with soil washed from the upper slopes. Fitting tines behind tractor wheels to rectify compaction, e.g. when spraying, can reduce erosion damage.

Erosion by wind can cause very serious damage on Black Fen soils and on some sandy soils. Various controls are being used, e.g. straw planting by special machines, sowing nursecrops such as mustard or cereals which are selectively killed by herbicides when the crop (e.g. sugar beet) is established, claying or marling and applying sewage sludge.

Soil capping

A soil cap is a hard crust, often only about 2–3 cm thick, which sometimes forms on the surface of a soil.

It is most likely to form on soils which are low in organic matter. Heavy rain or large droplets of water from rain guns, following secondary cultivations, may cause soil capping. Tractor wheels (especially if slipping), trailers and heavy machinery can also cause capping in wet soils.

Although a soil cap is easily destroyed by weathering (e.g. frost, or wetting and drying) or by cultivations, it may do harm while it lasts because it prevents water moving into the soil, as well as preventing air moving into and out of the soil in wet weather. It also hinders the development of seedlings from small seeds such as grasses and clovers, roots and vegetables.

Chemicals such as cellulose xanthate can be sprayed on seedbeds to prevent soil capping. They enhance seedling establishment and do not affect herbicidal activity in the soil, but are very expensive.

Control of weeds by cultivation

The introduction of chemicals which kill weeds has reduced the importance of cultivation as a means of controlling weeds. The cereal crops are now regarded by many farmers as the cleaning crops instead of the roots and potato crops, mainly because chemical spraying of weeds in cereals is very effective, if expensive.

However, weeds should be dealt with in every possible way and there are still occasions when it is worthwhile to use cultivation methods.

Annual weeds can be tackled by:

(1) Working the stubble after harvest (e.g. discing, cultivating or rotavating) to encourage seeds to germinate. These young weeds can later be destroyed by harrowing or ploughing. Unfortunately, this allows wild oats to increase (page 232).
(2) Preparing a "false" seedbed in spring to allow the weed seeds to germinate. These can be killed by cultivation before sowing the root crop.
(3) Inter-row hoeing of root crops which can destroy many annual weeds and some perennials.

It should be noted that in organic farming (page 75) mechanical weed control is the only option and various fine-tined implements have been developed for use in the growing crop.

Perennial weeds such as couch grass, creeping thistle, docks, field bindweed and coltsfoot, can usually be controlled satisfactorily by the fallow (i.e. cultivating the soil periodically through the growing season instead of cropping). However, this is very expensive. A fair amount of control can be obtained by short-term working in dry weather.

The traditional method of killing couch has been to drag the rhizomes to the surface in late summer. This is then followed by rotary cultivation—three or four times at two to three week intervals—to chop up the rhizomes. To get a good kill of the plant, adequate growing conditions are necessary for the couch rhizome to respond to being chopped up by sending out more green shoots and thus to hasten its exhaustion. A much more reliable, if expensive, method is to use glyphosate.

The deeper rooted bindweed, docks, thistles and coltsfoot cannot be controlled satisfactorily by cultivations, although periodic hoeing and cultivating between the rows of root crops can generally reduce the problem.

Thorough cultivations which provide the most suitable conditions for rapid healthy growth of the crop can often result in the crop outgrowing and smothering the weeds.

FURTHER READING

Direct Drilling and Reduced Cultivations, Allen, Farming Press.

Field Drainage—Principles and Practices, Castle, McCunnall and Tring, Batsford.

Land Drainage, Farr and Henderson, Longmans.

Modern Farming and the Soil, MAFF Report, HMSO.

Soil Assessment, SAWMA (Soil and Water Management Association).

Soil Management, Davis, Eagle and Finney, Farming Press.

Soils of the British Isles, Curtis, Courtney and Trudgill, Longmans.

Soils, Simpson, Longmans.

3

FERTILIZERS AND MANURES

If good crops are to be grown in a field, there should be at least as many nutrients returned to the soil as have been removed. The only satisfactory way of achieving this is by the sensible use of chemical fertilizers (inorganic manures), where possible in conjunction with organic manures such as farmyard manure and slurry. Table 7 indicates average figures for nutrients removed by various crops.

Nitrogen is supplied by fertilizers, organic

TABLE 7 *Nutrients removed by crops (kg/ha)*

Crop (good average yield)		N	P_2O_5	K_2O	Mg	S	Notes
Wheat							
grain 7 tonnes/ha		130	60	42	9	35	
straw 5 tonnes/ha		17	7	40			
	Total	147	67	82	9	35	
Barley							
grain 6 tonnes/ha		100	50	33	7	22	
straw 4 tonnes/ha		25	6	46			
	Total	125	56	79	7	22	
Oats							
grain 6 tonnes/ha		100	50	33			
straw 5 tonnes/ha		15	9	74			
	Total	115	59	107			
Beans							
grain 4 tonnes/ha		176	24	80	8	40	
Potatoes							The response to phosphatic fertilizers is greater than these figures suggest.
tubers 50 tonnes/ha		150	75	300	15	20	
dry haulm 3 tonnes/ha		65	8	143			
	Total	215	83	443	15	20	
Sugar beet							
roots 45 tonnes/ha		80	43	86	27	33	
fresh tops 35 tonnes/ha		120	39	202			If sugar beet tops or kale are eaten by stock on the field where grown, some of the nutrients will be returned to the soil.
	Total	200	82	288	27	33	
Kale							
fresh crop 50 tonnes/ha		224	67	202			

manures, nodule bacteria on legumes (e.g. clovers, peas, beans, lucerne), and nitrogen-fixing micro-organisms in the soil. (It is difficult to estimate how much nitrogen is produced by legumes and micro-organisms; clovers in grassland may supply up to 250 kg/ha and micro-organisms about 60 kg/ha per annum.)

Phosphates and potash are supplied by the soil minerals, fertilizers and organic manures.

A decision has to be made as to what fertilizers should be used for each crop. The amount applied should be based on soil analysis and previous cropping.

Soil analysis will show:

(1) Soil texture.
(2) pH (usual range 4–8). This is a useful guide to the lime requirement needed to bring acid soils up to an optimum level for the particular crop (page 31).
(3) Available nutrients. This is the level of **phosphate, potash** and **magnesium** and it is indicated by index ratings 0–9. 0 indicates a deficiency level and 9 an excessively high level (never reached under field conditions). An index of 2 and 3 is satisfactory for farm crops and standard fertilizer rates can be applied. For other indices, higher or lower rates are recommended.

The field sampling for analysis must be done in a methodical manner if a reasonably accurate result is to be obtained. At least 25 soil cores, 15 cm deep (7.5 cm on old pasture), should be taken to make up a sample weighing at least 0.5 kg. Unless the area is very uniform, samples

TABLE 8. *ADAS N index system*

Nitrogen index based on last crop grown		
Nitrogen index 0	Nitrogen index 1	Nitrogen index 2
Cereals	Beans	Any crop in field receiving large frequent dressings of farmyard manure or slurry
Forage crops removed	Forage crops grazed	
Leys (1–2 year) cut	Leys (1–2 year) grazed, high N[b]	
Leys (1–2 year) grazed, low N[a]	Long leys, low N[a]	Long leys, high N[b]
Maize	Oilseed rape	Lucerne
Permanent pasture – poor quality, matted	Peas	Permanent pasture, average
Sugar beet, tops removed	Potatoes	Permanent pasture, high N[b]
Vegetables receiving less than 200 kg/ha N	Sugar beet, tops ploughed in	
	Vegetables receiving more than 200 kg/ha N	

Nitrogen index following lucerne, long leys and permanent pasture					
Crop	1st crop	2nd crop	3rd crop	4th crop	5th crop
Lucerne	2	2	1	0	0
Long leys, low N[a]	1	1	0	0	0
Long leys, high N[b]	2	2	1	0	0
Permanent pasture – poor quality, matted	0	0	0	0	0
Permanent pasture – average	2	2	1	1	0
Permanent pasture – high N[b]	2	2	2	1	1

[a] Low N = less than 250 kg/ha N per year and low clover content.
[b] High N = more than 250 kg/ha N per year or high clover content.

should be taken for every four hectares. The analysis is carried out on the fine part of the soil and this must be borne in mind when interpreting the results for very stony soils. In this case the indices are usually too high.

An index of 0–2 for **nitrogen** is based on previous cropping and manuring of the field (Table 8). However, its value can be limited because it is impossible to predict accurately the rate at which the nitrogen is likely to be released in the soil. It depends so much on the weather and its effect on soil conditions.

A number of organizations will undertake an analysis of available nitrogen. This test is commonly used on the Continent but has not yet been proven in the United Kingdom.

Table 9 shows the major plant foods for crop growth.

TRACE ELEMENTS

The need for trace elements is likely to be greatest:

(1) On very poor soils.
(2) Where soil conditions, such as a high pH, make them unavailable.
(3) Where intensive farming (with high yields) is practised.
(4) On soils where organic manures such as farmyard manure and slurry are not used.

The importance of trace elements in plant nutrition is most appreciated when plants are grown in culture solutions circulated past their root systems (hydroponics). This is a form of horticultural crop production which is rapidly increasing, especially where the solutions can be replenished automatically with nutrients.

Supplying trace elements to plants is not always easy and care is required to prevent overdosing which may damage or kill the crop. To facilitate their application and availability to the plants, trace elements such as copper, iron, manganese and zinc can now be used in a chelated form. These *chelates* are "protected" water-soluble complexes of the trace elements with organic substances such as EDDHA (eth-

ylene diamine dihydroxyfenic acetic acid). They can be safely applied as foliar sprays for quick and efficient action, or to the soil, sometimes as a supplement with a compound fertilizer, for root uptake without wasteful "fixation" because they do not ionize. They are compatible when mixed with many spray chemicals. Trace elements may also be applied as *frits*, which are produced by fusing the elements with silica to form glass which is then broken into small particles for distribution on the soil.

It is very important, whenever possible, to obtain expert advice before using trace elements. This is because crops which are not growing satisfactorily may be suffering for reasons other than a trace element deficiency, e.g. major nutrient deficiency, poor drainage, drought, frost, mechanical damage, viral or fungus diseases.

Trace element deficiencies can be diagnosed by either leaf or soil analysis, depending on the nutrient.

UNITS OF PLANT FOOD

Since metrication, the kilogram is now the unit of plant food and recommendations are given in terms of kilograms per hectare (kg/ha). For example, the recommendation for a spring cereal crop may be to apply

100 kg/ha N, 50 kg/ha P_2O_5 and 50 kg/ha K_2O.

To convert this into numbers of bags of fertilizer per hectare, it is necessary to use the percentage analysis figures for the fertilizer in question. This is clearly stated on each bag, e.g. 20 : 10 : 10 means that the particular fertilizer contains 20% N, 10% P_2O_5 and 10% K_2O, always given in that order.

100 kg of 20 : 10 : 10 fertilizer contains 20 kg N, 10 kg P_2O_5 and 10 kg K_2O; therefore a 50-kg bag contains 10 kg N, 5 kg P_2O_5 and 5 kg K_2O, i.e. the number of kilograms of plant food in a 50-kg bag of fertilizer is half the percentage figures.

In the example of the plant food requirements for the spring cereal crop, i.e. 100 kg N, 50 kg

TABLE 9 *The need for and effects of nitrogen, phosphorus, potassium, magnesium and sulphur*

Plant nutrient	Crops which are most likely to suffer from deficiency	Field conditions where deficiency is likely to occur	Deficiency symptoms	Effects on crop growth	Effects of excess	Time and method of application
Nitrogen (N)	All farm crops except legumes (e.g. beans, peas, clover) Especially important for leafy crops, e.g. grass, cereals, kale and cabbages	On all soils except peats, and especially where organic matter is low, and after continuous cereals	Thin, weak, spindly growth; lack of tillers and side shoots; small yellow/pale green leaves, sometimes bright colours	Increases leaf size, rate of growth and yield; leaves dark green	Lodging in cereals; delayed ripening; soft growth susceptible to frost and disease; may lower sugar and starch content	N fertilizers in seedbed or top-dressed
Phosphorus (P)	Root crops (e.g. sugar beet, swedes, carrots, potatoes), clovers, lucerne and kale	Clay soils; acid soils especially in high rainfall areas, chalk and limestone soils and peats Poor grassland	Similar to nitrogen except that leaves are a dull, bluish-green colour with purple or bronze tints	Speeds up growth of seedlings and roots; hastens leaf growth and maturity Encourages clover in grassland Improves quality	Might cause crops to ripen too early and so reduce yield if not balanced with nitrogen and potash fertilizers	Phosphorus fertilizers applied in seedbed for arable crops; "placement" in bands near or with the seed is more efficient Broadcast on grassland
Potassium (K)	Potatoes, carrots, beans, barley, clovers, lucerne, sugar beet and fodder beet	Light sandy soils, chalk soils, peat, badly drained soils, grassland repeatedly cut for hay, silage or "zero" grazing	Growth is squat and growing points "die-back", e.g. edges and tip of leaves die and appear scorched	Crops are healthy and resist disease and frost better Prolongs growth Improves quality Balances N and P fertilizers	May delay ripening too much May cause magnesium deficiency in fruit and glasshouse crops and "grass-staggers" in grazing animals	K fertilizers broadcast or "placed" in seedbed for arable crops; broadcast on grassland in autumn or mid-summer

TABLE 9 (cont.) *The need for and effects of nitrogen, phosphorus, potassium, magnesium and sulphur*

Plant nutrient	Crops which are most likely to suffer from deficiency	Field conditions where deficiency is likely to occur	Deficiency symptoms	Effects on crop growth	Effects of excess	Time and method of application
Magnesium (Mg)	Cereals, potatoes, sugar beet, peas, beans, kale	Light sandy soils, chalk soils, often of a temporary nature due to poor soil structure, excessive potash, etc.	Chlorotic patterns on leaves (short of chlorophyll)	Associated with chlorophyll, and potassium metabolism	Unlikely, requirements about same as phosphorus, and one-tenth that of nitrogen and potash	Use FYM or slurry; lime with magnesium limestone; Epsom salts; fertilizer containing kieserite
Sulphur (S)	Most crops but especially brassicae, e.g. kale, oilseed rape; grass cut for conservation; bread wheat	Light sandy soils – especially if low in organic matter and intensively cropped; modern fertilizers used; no smoke pollution	Yellowing leaves and less vigorous growth, like nitrogen deficiency	Nitrogen made more efficient	Unlikely to occur	Use FYM or sulphur foliar spray or to soil as gypsum

P_2O_5, 50 kg K_2O, this would be supplied by 10 bags of 20 : 10 : 10 fertilizer. In some cases, the figures may not work out exactly and a compromise has to be accepted. For example:

	N	P_2O_5	K_2O
	kg/ha	kg/ha	kg/ha
Spring cereals recommendation	100	50	50
Four bags 10 : 25 : 25 combine-drilled, supply	20	50	50
Difference	80		

The extra 80 kg of nitrogen would be top-dressed and may be supplied by approximately 4½ bags per hectare of ammonium nitrate (34.5%): $80 \div 17.25 = 4.6$.

If bulk fertilizer is used, each tonne contains 10 times the percentage of each plant food:

1 tonne (1000 kg) $= 100 \times 10$ kg; therefore, 1 tonne of 20 : 10 : 10 contains 200 kg N, 100 kg P_2O_5 and 100 kg K_2O.

In the case of **liquid fertilizers**, 1000 litres weigh approximately 1 tonne, and so the amount required can be worked out as above and litres substituted for kilograms.

Kilogram cost

The cost of plant food per kilogram can be calculated from the cost of fertilizers which contain only one plant food such as nitrogen in ammonium nitrate, phosphate in triple super-phosphate, and potash in muriate of potash.

If 1 tonne of ammonium nitrate (34.5% N) costs £107 and contains 345 kg N, then

1 kg of N costs $\dfrac{10,700}{345}$ pence, i.e. 31p.

If 1 tonne of triple super (45%) costs £135 and contains 450 kg P_2O_5, then

1 kg of P_2O_5 costs $\dfrac{13,500}{450}$ pence, i.e. 30p.

If 1 tonne of muriate of potash (60%) costs £105 and contains 600 kg K_2O, then

1 kg of K_2O costs $\dfrac{10,500}{600}$ pence, i.e. 17.5p.

1 tonne of (9 : 24 : 24) compound contains:

N	P_2O_5	K_2O
90 kg	240 kg	240 kg

The value of this, based on the costs of a kilogram of nitrogen, phosphate and potash, is:

N	90×31	$=$	2790
P	240×30	$=$	7200
K	240×17.5	$=$	4200

Total		14,190 = £141.90

Normally, the well-mixed granulated compounds cost more than the equivalent in "straights". Bulk blends of "straights" are usually cheaper (page 56). Fertilizer prices have varied considerably in recent years; actual costs at any time should be substituted in the calculations shown.

It is also possible to compare the values of "straight" fertilizers of different composition on the basis of cost per kilogram of plant food, e.g. the cost of a kilogram of nitrogen in ammonium nitrate (34.5% N) is:

345 kg N/tonne $\quad \dfrac{£107 \text{ per tonne}}{345} = 31p.$

This compares with the cost of a kilogram of nitrogen in urea (46% N) which is

460 kg N/tonne $\quad \dfrac{£110 \text{ per tonne}}{460} = 24p.$

Therefore, apart from less handling costs, the urea is the lower-priced fertilizer although, compared with ammonium nitrate, it does have some limitations.

STRAIGHT FERTILIZERS

Straight fertilizers supply only one of the major plant foods.

Nitrogen fertilizers

The nitrogen in many straight and compound fertilizers is in the ammonium (NH_4 ion) form but, depending on the soil temperature, it is quickly changed by bacteria in the soil to the nitrate (NO_3 ion) form. Many crop plants, e.g. cereals, take up and respond to the NO_3 ions quicker than the NH_4 ions, but other crops, e.g. grass and potatoes, are equally responsive to NH_4 and NO_3 ions.

The ammonium, as a base, is held in the soil complex at the expense of calcium and other loosely-held bases which are lost in the drainage water. This will have an acidifying effect on the soil.

Nitrogen fertilizers in common use are:

(1) *Ammonium nitrate* (33.5–34.5% N). This is a very widely used fertilizer for top-dressing. Half the nitrogen (as nitrate—NO_3) is very readily available. It is marketed in a special prilled or granular form to resist moisture absorption. It is a fire hazard but is safe if stored in sealed bags and well away from combustible organic matter. Because of the ammonium present, it has an acidifying effect.

(2) *Ammonium nitrate lime* (21–26% N). This granular fertilizer is a mixture of ammonium nitrate and lime. It is sold under various trade names. Because of the calcium carbonate present it does not cause acidity when added to the soil.

(3) *Urea* (46% N). This is the most concentrated solid nitrogen fertilizer and it is marketed in the prilled form. It is sometimes used for aerial top-dressing. In the soil, urea changes to ammonium carbonate which may temporarily cause a harmful local high pH. Nitrogen, as ammonia, may be lost from the surface of chalk or limestone soils, or light sandy soils when urea is applied as a top-dressing. When it is washed or worked into the soil, it is as effective as any other nitrogen fertilizer. Chemical and bacterial action change it to the ammonium and nitrate forms. If applied close to seeds, urea may reduce germination.

(4) *Sulphate of ammonia* (21% N). At one time, as a fertilizer, this was the main source of nitrogen. However, sulphate of ammonia is seldom used now. It consists of whitish, needle-like crystals and it is produced synthetically from atmospheric nitrogen. Bacteria change the ammonium nitrogen to nitrate. It has a greater acidifying action on the soil than other nitrogen fertilizers. Some nitrogen may be lost as ammonia when it is top-dressed on chalk soils.

(5) *Nitrate of soda* (16% N, 26% Na). This fertilizer is obtained from natural deposits in Chile and is usually marketed as moisture-resistant granules. The nitrogen is readily available and the sodium is of value to some market garden crops. It is expensive and is not widely used.

(6) *Calcium nitrate* (15.5% N). This is a double salt of calcium nitrate and ammonium nitrate in prilled form. It is mainly used on the Continent.

(7) *Anhydrous ammonia* (82% N). This is ammonia gas liquefied under high pressure, stored in special tanks and injected 12–20 cm into the soil from pressurized tanks through tubes fitted at the back of strong tines. Strict safety precautions must be observed; it is a contractor rather than a farmer operation.

The ammonia, as ammonium hydroxide, is rapidly absorbed by the clay and organic matter in the soil and there is very little loss if the soil is in a friable condition and the slit made by the injection tine closes quickly. It is not advisable to use anhydrous ammonia on very wet or very cloddy or stony soils. It can be injected when crops are growing, for example into winter wheat crops in spring, between rows of Brussels sprouts and into grassland.

The cost of application is much higher than for other fertilizers, but the material is cheap, so the applied cost per kilogram compares very favourably with other forms of nitrogen. On grassland it is usually applied twice—in spring and again in midsummer—at up to 200 kg/ha each time. In cold countries it can be applied in late autumn for the following season, but the mild periods in winters in this country usually cause heavy losses by nitrification and leaching.

At one time it was fairly popular in the United Kingdom. However, because the main marketing source ceased, this is no longer the case, although there is no reason why it should not be used again.

(8) *Aqueous ammonia* (21–29% N). This is ammonia dissolved in water under slight pressure. It must be injected into the soil (10–12 cm), but the risk of losses is very much less than with anhydrous ammonia. Compared with the latter, cheaper equipment can be used, but it is usually a contractor operation.

(9) *Aqueous nitrogen solutions* (26–32% N). These are usually solutions of mixtures of ammonium nitrate and urea, and are commonly used on farm crops (page 57, liquid fertilizers).

Various attempts have been made to produce slow-acting nitrogen fertilizers. Reasonable results have been obtained with such products as resin-coated granules of ammonium nitrate (26% N), sulphur-coated urea prills (36% N) (soil bacteria slowly break down the yellow sulphur in the soil), and urea formaldehydes (30–40% N). At present this type of fertilizer is considered too expensive for farm cropping.

Organic fertilizers such as *Hoof and Horn* (13% N), ground-up hooves and horns of cattle, *Shoddy* (up to 15% N), waste from wool mills, and *Dried Blood* (10–13% N), a soluble quick-acting fertilizer, are usually too expensive for farm crops and are mainly used by horticulturists.

It should be noted that much of the nitrogen now supplied to farm crops comes from compound fertilizers in which it is usually present mainly as ammonium phosphate.

Phosphate fertilizers

Phosphate fertilizers can be classed as:

(1) Those containing water-soluble phosphorus.
(2) Those with no water-soluble phosphorus. The insoluble phosphorus is soluble in the weak soil acids.

By custom and by law, the quality or grade of phosphate fertilizers is expressed as a percentage of phosphorus pentoxide (P_2O_5) equivalent.

The main phosphate fertilizers used in agriculture are:

(1) *Superphosphate*. This contains 18–21% water-soluble P_2O_5 produced by treating ground rock phosphate with sulphuric acid. It also contains a small amount of unchanged rock phosphate in addition to gypsum ($CaSO_4$), which may remain as a white residue in the soil. It is suitable for all crops and all soil conditions, but is not widely used now.

(2) *Triple superphosphate*. This contains approximately 47% water-soluble P_2O_5 in a granulated form. It is produced by treating rock phosphate with phosphoric acid. One bag triple superphosphate is approximately equivalent to $2\frac{1}{2}$ bags ordinary superphosphate.

(3) *Ground mineral phosphate* (ground rock phosphate). This contains 25–40% insoluble P_2O_5. It is the natural rock ground to a fine powder, i.e. 90% should pass through a "100-mesh" very fine sieve (16 holes/ mm^2). It should only be used on acid soils in high rainfall areas and then preferably for grassland.

A softer rock phosphate is obtained from North Africa. It is sometimes known as Gafsa. It can be ground to a very high degree of fineness, 90% passing through a "300-mesh" sieve (48 holes/mm^2). This means that it will dissolve more quickly in the soil, although it should still only be used under the same conditions and for the same crops as ordinary ground mineral phosphate. To be classified as a soft rock phosphate under EC Regulations, at least 55% of its total P_2O_5 must dissolve in a 2% solution of formic acid.

Ground mineral phosphate fertilizers are now granulated.

(4) *Basic slags*. These contain 5–22% insoluble P_2O_5. They are by-products from the manufacture of steel. However, because of improved methods of steel manufacture, very little of this valuable fertilizer is now available to farmers in the United Kingdom. Small quantities of low grade slag are sometimes imported.

Potash fertilizers

The quality or grade of potash fertilizers is expressed as a percentage of potassium oxide (K_2O) equivalent. The main potash fertilizers used in agriculture are:

(1) *Muriate of potash* (potassium chloride). As now sold, it usually contains 60% K_2O. It is the most common source of potash for farm use and is also the main potash ingredient for compound fertilizers containing potassium. As a straight fertilizer it is normally granulated, but some muriate of potash is marketed in a powdered form.

(2) *Sulphate of potash* (potassium sulphate). This is made from the muriate and so is more expensive per kilogram K_2O. It contains 48–50% K_2O and, ideally, should be used for quality production of crops such as tomatoes and other market garden crops. However, the cost limits its use.

(3) *Kainit and potash salts*. These are usually a mixture of potassium and sodium salts and, depending on the source, magnesium salts. They contain between 12–30% K_2O and 8–20% sodium (Na). They are most valuable for sugar beet and similar crops for which the sodium is an essential plant food.

Magnesium fertilizers

(1) *Kieserite* (16–17% Mg). This fertilizer is quick acting and is particularly useful on severely magnesium-deficient soils where a magnesium-responsive crop such as sugar beet is to be grown.

(2) *Calcined magnesite* (48% Mg). This is the most concentrated magnesium fertilizer, but it is only slowly available in the soil. A 300 kg/ha applied, say, every four years in a predominantly arable cropping situation will help to maintain the magnesium status of a light sandy soil which is naturally low in magnesium.

(3) *Epsom salts* (10% Mg). This is a soluble form of magnesium sulphate and is used as a foliar spray where deficiency symptoms may have appeared on a high value crop.

Sodium fertilizers

Sodium is not an essential plant food for the majority of crops. However, for some, notably sugar beet and similar crops, it is highly beneficial and should replace up to half the potash requirements. The adverse effects it has on weak-structured soils such as the Lincolnshire silts should be noted but, generally, on other soils this has been exaggerated.

Agricultural salt (sodium chloride—37% Na) is the main sodium fertilizer used. It is now available in a granular form.

Sulphur fertilizers

Sulphur is an important plant nutrient (involved in the build-up of amino acids and proteins in the plant), a deficiency of which can limit the response of the plant to nitrogen.

Sulphur deficiency has become more pronounced in some, but not all, crops in the last decade. The natural build-up of sulphur in the plant is now much less because purer fertilizer is being used and the plant itself is growing in a less polluted atmosphere. It is, however, generally only necessary in areas away from industry and then perhaps for certain crops such as second and third cut grass for silage.

The main sulphur fertilizers (which should be applied in the spring) are gypsum (calcium sulphate), potassium sulphate and sulphur contained in compound fertilizers.

Phosphorus and Potassium Plant Nutrients

Increasingly, plant nutrients are now expressed in terms of the elements P (phosphorus) and K (potassium) instead of the commonly used oxide terms P_2O_5 and K_2O respectively. Throughout this book the oxide terms are used, but these can be converted to the element terms by using the following factors:

$P_2O_5 \times 0.43 = P$, e.g. 100 kg P_2O_5 = 43 kg P,
$K_2O \times 0.83 = K$, e.g. 100 kg K_2O = 83 kg K.

COMPOUND AND BLENDED FERTILIZERS

The main constituents for compound fertilizers used in the United Kingdom are urea, mono and di-ammonium phosphate and potassium chloride. These compounds supply two or three of the major plant foods (nitrogen, phosphorus and potassium). Other plant foods, e.g. trace elements, as well as pesticides, can also be added, although this is not commonly done now.

Because of the use of more concentrated basic ingredients, compound fertilizers have become much more concentrated in the last 30 years. For example:

In 1948 the total N, P, K content averaged 24% and in 1990 the total N, P, K content averaged 50% with some concentrations at 60%, e.g. 10% N, 25% P_2O_5, 25% K_2O.

Approximately 75% of all fertilizer now used is as a compound or blend.

Compound fertilizers are normally made by drying a wet slurry (containing the appropriate raw materials) in a heated drum to produce granules (each containing the appropriate nutrients in the correct ratio) size 2–5 mm diameter. Ammonium nitrate and urea fertilizers are produced by spraying a solution of the fertilizer into a "prilling tower". As the droplets fall down the tower they become round and solid, producing prills 1–3 mm diameter.

Blending of the straight fertilizers to make a mixture is becoming more popular again. It should not be compared with the farm mixture which was quite common 40 years ago.

In modern blends it is important that the individual single ingredients are, as far as possible, matched in physical characteristics (granular size and density). This is to avoid segregation out (separation) of the ingredients. It is a dry mix.

High quality blended fertilizers can be made to specific plant food ratios. They are cheaper than the compounds, but generally the handling and, most important, spreading qualities are not as good as the compounds. However, the future could well see the distribution of straight fertilizers (two or more at the same time) in the field. Triple hoppers could be developed with individual metering devices so that the "blend" could be applied to any required ratio. This would overcome the problem of segregation out of the different sized granules.

The fertilizer manufacturers have invested very heavily in compound fertilizer production and it is unlikely that they will be prepared to change to bulk blending on a large scale, certainly not in the foreseeable future.

Plant Food Ratios

Fertilizers containing different amounts of plant food may have the same plant food ratios. For example:

	Fertilizer	Ratio	Equivalent rates of application
(a)	12 : 12 : 18	1 : 1 : 1½	5 parts of (a)
(b)	15 : 15 : 23	1 : 1 : 1½	= 4 parts of (b)
(c)	15 : 10 : 10	1½ : 1 : 1	7 parts of (c)
(d)	21 : 14 : 14	1½ : 1 : 1	= 5 parts of (d)
(e)	12 : 18 : 12	1 : 1½ : 1	5 parts of (e)
(f)	10 : 15 : 10	1 : 1½ : 1	= 6 parts of (f)

Some examples of compounds and possible uses are shown in Table 10.

TABLE 10

Compound (N : P : K)	Crop	N kg/ha	P$_2$O$_5$ kg/ha	K$_2$O kg/ha
12 : 12 : 18	Potatoes	150	150	225
20 : 10 : 10	Spring cereal	100	50	50
0 : 20 : 20	Autumn cereal	0	50	50
9 : 25 : 25	Autumn cereal	22	60	60

Fertilizers are supplied in various ways:

Solids

(1) *50-kg bags*. Although a small amount of fertilizer is supplied as individual bags, most of it is delivered on 30-bag (1.5-tonne) pallets to facilitate handling. Special fork-lift equipment is required, as manhandling into spreaders is slow and can be dangerous.

(2) *Big bags*, e.g. 500- and 1000-kg bags with a top-lift (hook) facility are easy to load into spreaders, but they can be difficult to stack. 750-kg "dumpy" bags on pallets stack well but they may be not quite so easy for loading spreaders.

Care must be taken to avoid damage to the bags which can result in the loss of fertilizer, and lumpy or sticky material reducing the efficiency of the spreader.

(3) *Loose bulk* (1 tonne occupies about 1 m^3). This system is not widely used now. The fertilizer can be stored in dry, concrete bays and covered with polythene sheets. It can be moved by tractor hydraulic loaders or augers into spreaders.

Liquids

In the United Kingdom as a whole the liquid fertilizer share of the fertilizer market is about 9%. However, in the typical arable areas where the system has, until recently, been concentrated, liquid nitrogen has a share in the range of 10–15%, and compounds account for between 5 and 10% of the total tonnage of compound fertilizer used.

Liquid fertilizers are non-pressurized solutions of the raw material which can be used for solid fertilizers. They ought to be called solution and not liquid fertilizers (the term "fluid" is also being used now), and they should be distinguished from pressurized solutions, such as aqueous ammonia and anhydrous ammonia. The basic constituents for liquids are either water-soluble, held in suspension or as fluids containing free ammonia. This no longer means that higher quality and more expensive sources of phosphate have to be used. This in turn means that, unlike in the past, there is very little price differential in the kilogram cost of plant food in the solid or liquid form. Liquid compounds are based on ammonium polyphosphate or ammonium phosphate, urea and potassium chloride, whilst ammonium nitrate and urea are the main constituents for liquid nitrogen fertilizers.

The concentration of liquid nitrogen is, for all practical purposes, the same as solid nitrogen fertilizer, although liquid compounds are only about two-thirds the concentration of solids. Potash is the main constraint on the concentration of the compounds as it is the least soluble of the important plant foods. However, liquids have bulk densities in the range 1.2–1.3 kg/litre compared with 0.9–1.0 kg/litre for solid fertilizers (i.e. 100 litres = approximately 125 kg). The higher bulk density plus quicker handling and application of liquids will compensate for the lower concentration. It is an obvious consideration when fertilizer work rates are assessed on the weight of nutrients applied in a given time and not the weight of the product.

Liquids are stored in steel tanks (up to 60 tonnes capacity) on the farm. A cheaper form of storage—glass-reinforced plastic tanks—is available as an alternative to mild steel. Storage capacity is generally based on up to 33% of the annual farm requirements.

There is a range of equipment available for applying liquid fertilizer. The broadcasters, which range from a 600-litre tractor-mounted to 2000-litre capacity trailed applicators, have spray booms of 12–24 m wide. Although the broadcaster is specially designed for fertilizer, by changing the jets it can be used for other agricultural chemicals, and so the liquid system fits in well with tramlining because only two pieces of equipment have to be matched. Equipment is also available for the placement of fertilizer for the sugar beet, potato and brassica crops, and a combine drill attachment can be used for sowing the liquid and cereal together. For top-dressing, and to minimize scorch, a special jet can easily be fitted by hand; this produces larger droplets for a better foliar run-off. Alternatively, a dribble-bar which dribbles the fertilizer on to the soil can be used.

Volume of product to apply

Example, if 100 kg nitrogen/ha is recommended, liquid nitrogen, N 37 kg per 100 litres, is used, i.e. $100/37 \times 100$ litres/ha = 270 litres/ha.

Units of sale

Whilst solid fertilizers are applied by weight (kg/ha) and sold by weight as £ per 1000 kg, liquid fertilizers are applied by volume (litres/ha) and sold by volume as £ per 1000/litres (1 m³).

Suspensions

These are liquids containing solids, and they can be produced with a concentration almost up to that of the granular solid. This is achieved either by crushing the solid raw material and adding to it up to 3% of a semi-solid clay to help minimize the settling of the soluble salts, or by reacting phosphoric acid and ammonia followed by the addition of potassium chloride and extra nitrogen.

For long-term storage, crystallization may be a problem and occasional agitation is necessary to prevent the crystals from growing whilst in store. Special applicators are also needed and

this is why suspensions are, at present, usually applied by a contractor.

Suspensions are developing in North America (where, in fact, liquids are more widely used) and to some extent on the Continent.

"Distressed" fertilizer

This is fertilizer which has been damaged, usually in transit from overseas. It is bought by distributors and sold at a heavy discount to farmers who normally make it into a liquid fertilizer, simply by dissolving it in water.

There is no difference in plant growth following the application of fertilizer in liquid, suspended or solid form.

APPLICATION OF FERTILIZERS

The main methods used are:

(1) *Broadcast distributors* using various mechanisms such as:

Pneumatic types. The principle involved is that the fertilizer is metered into an airflow which conveys it through flexible tubes to individual outlets placed over deflector plates to distribute the fertilizer evenly.

Hopper capacity is from 1 to 6 tonnes with a spreading width of 12–24 m (selected to match tramline widths). Application rates vary from 5 kg/ha (for broadcasting seeds) to 2500 kg/ha.

Roller feed, conveyor and brush feed. These are used (although not so widely now) with full-width hoppers. Basically, the *roller feed* consists of a roller rotating on the floor of the hopper to push the fertilizer out through slots. The *conveyor and brush feed* consists of a belt conveyor which moves the fertilizer through an opening to a revolving brush which distributes the fertilizer.

Spinning disc types. These consist of a hopper placed above a rotating disc. The fertilizer is fed on to the disc from where it is distributed. Wide areas can be covered, but the accuracy of distribution can vary quite considerably.

(2) *Combine drills*. Fertilizer and seed (e.g. cereals) from separate hoppers are fed down the same or an adjoining spout. A star-wheel feed mechanism is normally used for the fertilizer and this usually produces a "dollop" effect along the rows. In soils low in phosphate and potash, this method of placement of the fertilizer is much more efficient than broadcasting. It is known as combine drilling and is sometimes referred to as "contact placement". Because of possible scorch, combine drilling should only be used for cereal crops.

(3) *Placement drills*. These machines can place the fertilizer in bands 5–7 cm to the side and 3–5 cm below the row of seeds. It is more efficient than broadcasting for crops such as peas and sugar beet. Other types of placement drills attached to the planter are used for applying fertilizers to the potato crop and some brassicas.

(4) *Broadcast from aircraft*. This is useful for top-dressing cereals, especially in a wet spring. It can also be used for applying fertilizers in inaccessible areas such as hill grazings. Highly concentrated fertilizers should be used, e.g. urea.

(5) *Liquids injected under pressure* into the soil, e.g. anhydrous and aqueous ammonia.

(6) *Liquids (non-pressurized)* broadcast or placed.

ORGANIC MANURES

Farmyard manure (FYM)

This consists of dung and urine, and the litter used for bedding stock. It is not a standardized product, and its value depends on:

(1) *The kind of animal that makes it*. If animals are fed strictly according to maintenance and production requirements, the quality of dung produced by various classes of stock will be similar. But in practice it is generally found that, as cows and young stock utilize much of the nitrogen and phosphate in their food, their dung is poorer than that produced by fattening stock.

(2) *The kind of food fed to the animal that makes the dung*. The more proteins and minerals in the diet, the richer will be the dung. But it is uneconomical to feed a rich diet just to produce a richer dung. This used to be the case some 50 years ago on arable farms when yarded in-wintered cattle were fattened on expensive protein. This was then used for a high value cash crop.

(3) *The amount of straw used*. The less straw used, the more concentrated will be the manure and the more rapidly will it break down to a "short" friable condition.

Straw is the best type of litter available, although bracken, peat moss, sawdust and wood shavings can be used. About 1.5 tonnes of straw per animal is needed in a covered yard for six months, and 2–3 tonnes in a semi-covered or open yard.

(4) *The manner of storage*. There can be considerable losses from FYM because of bad storage, although it is appreciated that expensive, elaborate storage is no longer viable these days.

Dung from cowsheds, cubicles and milking parlours should, if possible, be put into a heap which is protected from the elements to prevent the washing out and dilution of a large percentage of the plant food which it contains. Dung made in yards should preferably remain there until it is spread on the land, and then, to prevent further loss, it is advisable to plough it in immediately.

Farmyard manure is important chiefly because of the valuable physical effects on the soil of the organic matter it contains. It is also a valuable source of plant foods, particularly nitrogen, phosphate and potash, as well as other elements in smaller amounts. An average dressing of 25 tonnes/ha of well-made FYM will provide about 40 kg N, 50 kg P_2O_5 and 100 kg K_2O in the first year. At least one-third of the nitrogen could be lost before it is ploughed in, although this depends on the time of application. All the plant food in FYM is less readily available than that in chemical fertilizers.

Application. The application of FYM will be dealt with under the various crops.

Liquid manure and slurry

Following the introduction of more intensive livestock enterprises, the cost of straw for bedding, and the need for cheap and effective mechanical methods of dealing with animal excreta, many livestock farmers now have to deal with manure in a liquid or semi-solid (slurry) form instead of the traditional solid form as produced in straw-bedded yards.

Slurry must not be allowed to pollute watercourses. At 10,000–20,000 mg/l it has a high Biological Oxygen Demand (BOD). This is the measure (in milligrams per litre) of the amount of oxygen needed by the micro-organisms to break down organic material (pollution of water is caused when the micro-organisms multiply, and so extract oxygen, to deal with the organic material). Problems can also arise from the nuisance of smells and possible health hazards over a wide area when the slurry is applied. If slurry is stored in anaerobic conditions, dangerous, obnoxious gases (mainly hydrogen sulphide) are produced and are released when it is being spread. The Water Authorities have a duty to prevent the pollution of streams and rivers with farmyard effluents. Legislation and the Code of Good Agricultural Practice should prevent any problems in this respect, but it does mean that the farmer has to find the best possible way to utilize the slurry produced on the farm.

When applying slurry, relevant recommendations from the Code of Good Agricultural Practice should be followed:

(1) A single slurry application should not exceed 50,000 litres/ha. At least a three-week interval should be allowed between applications. When applying slurry through irrigation lines, the precipitation should not exceed 5 mm/hour. Rain guns should be avoided.

Too much slurry can:
(i)   restrict aeration leading to partial oxidation products such as methane and ethylene which are toxic to plants, and
(ii) weaken soil structure by the sealing of the soil surface.

(2) An untreated strip of at least 10 m width should be left next to all watercourses, and slurry should not be applied within at least 50 m of a spring, well or borehole that supplies water for human consumption or is to be used for farm dairies.

(3) Slurry should not be applied:
(i)   to fields which are likely to flood in the month after application;
(ii) to fields which are frozen hard; surface run-off must be avoided;
(iii) when the soil is at field capacity;
(iv) when the soil is badly cracked down to the field drains, or to fields which have been piped, moled or subsoiled over a drainage system in the preceding 12 months.

Table 11 indicates the average amounts and composition of slurry produced by livestock.

TABLE 11

Livestock	Undiluted litres/day	Available nutrients kg/10 m^3 in year of application		
		N	P_2O_5	K_2O
		kg/ha	kg/ha	kg/ha
Dairy cows	45	25	10	45
Fattening pigs	5	40	20	27
Poultry (1000 hens)	114	90	55	55

(1) For slurry diluted 1 : 1 with water, the figures in Table 11 should be divided by 2.

(2) To estimate the dilutions of slurry, a comparison should be made of the volume of slurry in the store with the expected volume of undiluted slurry.

The figures in Table 11 can only be used as a basis and they apply to slurry collected in an undiluted form, e.g. under slatted floors or passageways where washing and rainwater are excluded. The slurry on most farms is obviously diluted and it is not easy to be certain of its composition, even when samples are analysed.

Rota-spreaders and similar machines can handle fairly solid slurry, and modern vacuum tankers and pumps are very efficient in dealing with slurry with less than 10% dry matter.

The injection of slurry into the soil using strong tines fitted behind the slurry tanker can reduce, but not eliminate, wastage and offensive odours. It can also reduce the volatization of nitrogen into the air as ammonia gas. However, it is expensive and it can spoil the surface of a grass field where there are stones present.

It is highly desirable that the valuable nutrients in organic manures should be utilized as fully as possible. This is best achieved if the slurry can be applied at a time when growing crops can utilize it. This means that storage is usually necessary. This is expensive, whether the slurry is stored in a compound (lagoon) or storage tank. The annual cost of storage and handling of slurry is usually more than the plant food value of the slurry.

Nitrogen losses resulting from applying slurry in the autumn/early winter to next year's maize field, for example (ground conditions permitting), can be reduced by the use of nitrogen inhibitors such as Dicyandiamide (e.g. Didin : Enrich). These inhibit the activity of the nitrifying bacteria which delays the change of ammonia to nitrate. Denitrification and leaching is therefore reduced. The cost of inhibitors is approximately £45/ha with slurry applied at 50,000 litres/ha. It is claimed that the increased yield with the maize crop following the use of an inhibitor can be valued at £70/ha.

In addition to forage maize, slurry is best applied to:

(1) *Grass*. At least a six-week, preferably longer, interval should be allowed between a slurry application and taking the crop for silage or hay. This will avoid any disease problems and it should minimize any possible contamination of the conserved crop. It should not be considered for the grazing sward. If there is no other option, it must be well washed off the plants before stock graze the sward.

(2) *Kale*. As this crop (like maize) is not normally sown until late April/May, there is usually a good opportunity to apply slurry from the winter accumulation. There should be no need for any phosphate and potash fertilizer following a slurry application, although extra nitrogen will normally be necessary. This fertilizer recommendation also applies to forage maize.

Although it is not so usual, slurry can also be used on cereals and other forage crops.

Slurry can be separated into solid and liquid fractions by screening or centrifugal action. The solid part (12–30% DM) can be handled like FYM, and the liquid portion is much less likely to cause tainting of pastures; the disease risk is also reduced. Slurry separators are, however, very expensive and they are not easy to justify. Table 14 shows an example of separated cow slurry.

Pig slurry

The composition of pig slurry depends on the feeding system, i.e. dry meal feed has a 10% DM; liquid feed has a 6–10% DM and whey 2–4% DM. Table 12 shows that pig slurries are higher in nitrogen content and lower in potassium than the average cattle slurry. However, in addition, supplementation to the diet means that pig slurry can contain large amounts of

TABLE 12

1000 kg slurry dm, unseparated			Mechanically separated slurry					
			720 kg liquid at 14% dm			280 kg fibre at 26% dm		
N	P_2O_5	K_2O	N	P_2O_5	K_2O	N	P_2O_5	K_2O
4.4	0.6	2.5	3.2	0.5	2.1	1.1	0.1	0.4

See also *MAFF Bulletin* No. 210.

copper and zinc. These elements will build up slowly in the soil, but eventually there could be crop toxicity problems. It should be noted that sheep are particularly susceptible to copper poisoning.

Poultry manure

Poultry manure refers to:
(1) Fresh poultry manure.
(2) Broiler manure.
(3) Deep litter manure.
(4) Dried manure.

Table 13 shows the annual production of poultry manure and its nutrient value.

Fresh poultry manure from battery cages or from slatted or wire floors is free from litter. It is semi-solid, but rather sticky and, to reduce any public nuisance from smell, it should be spread as soon as possible. To reduce nitrogen losses, the preferred time is in spring and summer. Application rates are similar to farmyard manure. Copper and zinc toxicity could be a problem if very high rates of fresh poultry manure are made.

Broiler and deep litter manure are the droppings mixed with litter. This produces a bulky manure, relatively dry and friable and easily handled. It should be treated like farmyard manure.

Dried poultry manure. Fresh poultry manure when dried down to 10–15% moisture content is easy to handle, and although its bulk density is half that of a granular fertilizer, it should be

treated in this way. However, because of the cost, there is little drying now.

Sewage sludge

As a means of disposal, most Water Authorities offer sewage sludge to farmers and growers at a fairly nominal cost.

Apart from the objectionable smell, raw sludge contains pathogenic organisms which, apart from being a possible health hazard, can contaminate both the soil and the growing crop. It should not be used.

However, most sewage sludge is now processed and, as such, it is quite valuable, e.g. 20 tonnes dried digested sludge can contain 75 kg N, 90 kg P_2O_5 and 30 kg K_2O available in the year of application.

Additionally, sewage sludge contains heavy metals such as zinc, copper and nickel. These can build up in the soil to a level which can affect crops indefinitely. All sewage sludge should be analysed.

Seaweed

Seaweed is sometimes used instead of farmyard manure for crops such as early potatoes in coastal areas, e.g. Ayrshire, Cornwall and the Channel Islands. However, it is expensive to handle which is why it is not used to any extent now. 10 tonnes contain about 50 kg of N, 10 kg P_2O_5 and 140 kg K_2O; it also contains about 150 kg salt. The organic matter in seaweed

TABLE 13 Comparison of poultry manures

	Average annual production per 1000 birds (tonnes)	% dry matter	Available nutrients (year of application) (kg/tonne)		
			N	P_2O_5	K_2O
Battery layers	55	30	11	7	5
Broilers	15	71	16	11	10
Deep litter	50	68	11	9	10
Dried manure	10	85	25	21	10

breaks down rapidly because it is mainly cellulose. It should therefore be collected, spread and ploughed in immediately. In this way the loss of potassium, particularly, will be reduced.

Cereal straws

Cereal straw is lignified material and can be a useful source of soil organic matter for maintaining or improving soil fertility. It is particularly beneficial for light, sandy and silty soils in which organic matter breaks down rapidly. However, the amount of organic matter in a soil is closely related to the texture of the soil, and repeated incorporations of straw on various soil types over many years have had little or no effect on the percentage of organic matter in the soils.

The average yield of straw (4 tonnes/ha) supplies about 15 kg N, 5 kg P_2O_5 and 35 kg K_2O, and so the soil should benefit when the straw is ploughed in or otherwise worked into the soil. However, with the sudden influx of straw, the soil population will start to multiply to cope with the problem of breaking down the straw. In so doing, much of the nitrogen in the soil gets used up for the bodily needs (the protein) of the bacteria. The carbon : nitrogen ratio widens to 50 or 60 : 1 from the normal 10 : 1. The protein is released as the straw decomposes and eventually the ratio returns to 10 : 1. But in the meantime, unless extra nitrogen is added (20–25 kg/ha), the following crop can suffer from shortage of nitrogen. This extra nitrogen is only necessary under naturally low nitrogen conditions.

Incorporation of straw is easier to achieve if it is chopped and spread from the combine. Ploughing is the best method, preferably using fairly wide furrows (30 cm) and 15–20 cm deep. There is seldom any great advantage in pre-mixing; increased costs can result. Tines and/or discs can only do the job well with chopped straw, and the result is not as satisfactory as good ploughing. Trash and clods left near the surface encourage slugs which can cause damage to following crops by eating the seed and young plants. Repeated shallow incorporations can build up high organic matter levels in the surface layers of some soils. These can be beneficial to the development of young crop plants, but can also cause problems with the adsorption of soil-acting herbicides. If the straw is incorporated at a wet time, anaerobic conditions may develop and the decomposing straw will release toxic substances, such as acetic acid, which can check or kill seedling crop plants.

Green manuring

This is the practice of growing and ploughing in green crops to increase the organic matter content of the soil. It is normally only carried out on light sandy soils.

White mustard is a very commonly-grown crop for this purpose. Sown at 9–17 kg/ha it can produce a crop ready for ploughing within 6–8 weeks. Fodder radish is becoming more popular for green manuring and, like mustard, it can also provide useful cover for pheasants.

However, it has been shown that a short ley has very little benefit in the way of building up organic matter in the soil. There must, therefore, be even less from quick-growing crops, which break down equally quickly in the soil.

Waste organic materials

Various waste products are used for market garden crops, partly as a source of organic matter and partly as a means of releasing nitrogen slowly to the crop. They are usually too expensive for ordinary farm crops.

Shoddy (waste wool and cotton) contains 50–150 kg of nitrogen per tonne. Waste wool is preferable, and the recommended application rate is 2.5–5 tonnes/ha.

Dried blood, ground *hoof and horn* and *meat and bone meal* are also used; the plant food content is variable.

RESIDUAL VALUES OF FERTILIZERS AND MANURES

The nutrients in most manures and fertilizers are not used up completely in the year of application. The amount likely to remain for use in the following years is taken into account when compensating outgoing farm tenants.

All the nitrogen in soluble nitrogen fertilizers (e.g. ammonium nitrate and some compounds, and in dried blood) is used in the first year.

For nitrogen in bones, hoof and horn, meat and bone meal, allow 1/2 after one crop and 1/4 after two crops.

For phosphate in soluble form, e.g. triple super-phosphate and compounds containing phosphate, allow 2/3 after one crop, 1/3 after two and 1/6 after three crops.

For phosphate in insoluble form, e.g. bone meal and ground mineral phosphate, allow 1/3 after one crop, 1/6 after two and 1/12 after three.

For potash, e.g. muriate or sulphate of potash and compounds containing potash, allow 1/2 after one crop and 1/4 after two crops.

For lime one-eighth of the cost is subtracted each year after application.

FERTILIZERS AND THE ENVIRONMENT

When applying fertilizers, it is economically and environmentally important to use the optimum amounts and not just that required for a maximum yield. Accuracy of application, timing and handling is also important in order to reduce the losses, either by leaching or run-off. Only small amounts of phosphorus can be lost by run-off, but with potassium, losses can be greater both by leaching and run-off. Nitrogen as nitrate is the nutrient which is most likely to be lost by leaching and this is affected by soil type, cropping, cultivation timing and rainfall.

There are now directives from the EC concerning nitrates in drinking water; a maximum of 50 parts per million has been set. In 1990 the government set up two pilot schemes — "Basic" and "Premium" — to run for a period of five years, by establishing Nitrate Sensitive Areas (NSAs) to study the effects of husbandry on nitrate leaching. Initially ten areas were chosen — from Somerset across to Lincolnshire — where the water exceeded or was at risk of containing more than 50 parts per million nitrates. Although the scheme is voluntary, most farmers in the designated areas joined, receiving payments for complying with the regulations.

The Basic scheme involves small changes in husbandry; only the economic optimum, or less, of nitrogen for some crops is allowed, but the timing and amount applied are restricted. Application of organic manures is also affected and a careful record has to be kept. To hold the nitrogen in the soil, winter fallows are avoided by planting cover crops. There are restrictions on timing of cultivations on grassland. Hedgerows and/or woodland must not be removed unless replaced by an equivalent area.

The Premium scheme involves a change in cropping and is only applicable to arable land which is converted to grassland. There are four options in this scheme, depending on whether the grass is fertilized, grazed or planted with trees; also clover must not be included in the grass sward.

GREEN COVER CROPPING

This describes the practice of growing a green crop with the primary objective of absorbing soil nitrogen over the winter months to prevent its leaching. This can benefit the environment as well as saving on expensive fertilizer nitrogen.

The cover crop is sown in late summer to hold any nitrogen in the autumn and winter, to be released in the spring for the following spring-sown crop.

A number of crops are at present under trial and it would appear that those taking up most nitrogen in November (and this month can be significant) do not necessarily hold it until March. For example, because of their rapid growth, white mustard and forage rape are very effective in trapping autumn nitrogen, but natural senescence and frost kill mean that the nitrogen can be lost before any following crop can take it up.

Rye and ryegrass hold nitrogen well in the autumn and winter and retain it until the spring. They can also be used for early cropping in the spring.

FURTHER READING

Crop Nutrition and Fertilizer Use, 2nd ed., J. Archer, Farming Press.

Fertilizers and Manures, K. Simpson, Longmans.

Fertilizer Recommendations for Agricultural and Horticultural Crops, 5th ed., No. 209, HMSO.

Lime and Fertilizer Recommendations No. 1, Arable Crops, MAFF Bulletin No. 2191, HMSO.

Organic Manures, MAFF Bulletin No. 210, HMSO.

4

CROPPING IN THE UNITED KINGDOM

CLIMATE AND WEATHER AND THEIR EFFECTS ON CROPPING

Climate has an important influence on the types of crops which can be grown in the United Kingdom. It may be defined as a seasonal average of weather conditions.

Weather is the state of the atmosphere at any time. It is the combined effect of such conditions as heat or cold, wetness or dryness, wind or calm, clearness or cloudiness, pressure and the electric state of the air.

The daily, monthly and yearly changes of temperature and rainfall give a good indication of the conditions likely to be found.

The climate of the United Kingdom is mainly influenced by:

(1) Its distance from the equator (50–60°N latitude).
(2) The warm Gulf Stream which flows along the western coasts.
(3) The prevailing south-west winds.
(4) The numerous "lows" or "depressions" which cross from west to east and bring most of the rainfall.
(5) The distribution of highland and lowland; most of the hilly and mountainous areas are on the west side.
(6) Its nearness to the continent of Europe, from where hot winds in summer and very cold winds in winter can affect the weather in the southern and eastern areas.

Local variations are caused by altitude, aspect and slope.

Altitude (height above sea level) can affect climate in many ways. The temperature drops about 0.5°C for every 90 m rise above sea level. Every 15 m rise in height can shorten the growing season by two days (one in spring and one in autumn) and it may check the rate of growth during the year. High land is more likely to be buffeted by strong winds and will receive more rain from the moisture-laden prevailing winds which are cooled as they rise upwards.

Aspect (the direction in which land faces) can affect the amount of sunshine (heat) absorbed by the soil. In this country the temperature of north-facing slopes may be 1°C lower than on similar slopes facing south.

Slope. When air cools it becomes heavier and will move down a slope and force warmer air upwards. This is why frost often occurs on the lowest ground on clear still nights whereas the

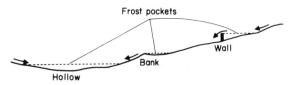

FIG. 35. Frost pockets are formed as cold air flows down a slope.

upper slopes may remain free of frost. "Frost pockets" occur where cold air collects in hollows or alongside obstructing banks, walls, hedges, etc. (Fig. 35). Frost-susceptible crops such as early potatoes, maize and fruit should not be grown in such places.

Rainfall

Rain comes mainly from the moist south-westerly winds and from the many "depressions" which cross from west to east. Western areas receive much more rain than eastern areas; this is partly due to the west to east movement of the rain-bearing air and also because most of the high land is along the western side of the country.

The average annual rainfall on lowland areas in the west is about 900 mm (35 in) and in the east is about 600 mm (24 in). It is much greater on higher land.

Temperature

The temperature changes are mainly due to:

(1) The seasonal changes in length of day and intensity of sunlight.
(2) The source of the wind, e.g. whether it is a mild south-westerly, or whether it is cold polar air from the north or from the Continent in winter.
(3) Local variations in altitude and aspect.
(4) Night temperatures which are usually higher when there is cloud cover which prevents too much heat escaping into the upper atmosphere.

The soil temperature may also be affected by colour—dark soils absorb more heat than light coloured soils. Also, damp soils can absorb more heat than dry soils.

The average January temperature in lowland areas along the west side of the country is about 6°C and about 4°C along the east side. The average July temperature in lowland areas in the southern counties is 17°C and 13°C in the north of Scotland.

CROPPING

Grass grows well in the wetter, western areas where dairying, stock rearing and fattening can be carried on successfully.

The drier areas in the east are best suited to arable crops.

Most of the chalk and oolitic limestone areas (in the south and east) are now used for large-scale cereal production. The leys grown in these areas are mainly used by dairy cattle or sheep or for herbage seed production. Cereals are grown on all types of soil. Maincrop potatoes and root crops, such as sugar beet, are grown on the deeper loamy and silty soils of the Midlands and eastern counties.

The mild, frost-free areas in the south-west of England and Wales (i.e. parts of Devon, Cornwall, and Pembroke) are suitable for early crops of potatoes, broccoli, etc. The Isle of Thanet (Kent) and the Ayrshire coast are also early areas, free from late frosts.

Mixed farming (i.e. both crop and stock enterprises on the same farm) is found on most lowland farms. The proportion of grass (and so stock) to arable crops usually varies according to soil type and rainfall. The heavier soils and high rainfall areas usually have more grass than arable crops.

The more exposed hill and mountain areas are unsuitable for intensive production because of the lower temperatures, very high rainfall, inaccessibility and steep slopes. These are mainly rough grazings used for extensive cattle and sheep rearing. Large areas are now forestry plantations.

Rotations

A rotation is a cropping system in which two or more crops are grown in a fixed sequence. If the rotation includes a period in grass (a ley), which is used for grazing and conservation, the system is sometimes known as alternate husbandry or mixed farming. The term ley farming describes a system where a farm or group of fields is cropped with leys which are reseeded at regular intervals.

Farm crops may be grouped as follows:

Cereals (wheat, barley, oats, rye, triticale, durum wheat, maize for grain). These are exhaustive crops because they are removed from the field and usually sold off the farm, i.e. they are cash crops. Cereals can be grown continuously (mainly barley and wheat, not oats) on the same field, although this can increase problems with weeds, pests and diseases. Cereals have peak labour demands when preparing seedbeds and sowing, and again at harvest, but these are reduced by having large tractors and combines and simplified planting, e.g. minimal cultivations. However, now that straw burning is banned, there will be an increase in cultivation requirements.

Potatoes and root crops (sugar beet, fodder beet, carrots). These are mainly cash crops, requiring deep soils, and they can justify irrigation in dry areas. They have high demands for plant nutrients and can be exhaustive, but they allow the farmer to use large amounts of fertilizers and manures which can restore fertility. Weeds used to be controlled by cultivations before planting or by inter-row hoeing. This is dependent on good weather and it can result in poor seedbeds and considerable loss of moisture. Herbicides are now used pre- and/or post-emergence of the crop and they can be very effective (page 223). It is unwise to grow these crops continuously, mainly because of cyst nematode build-up. In a close rotation there is also an increased volunteer crop problem. The risk of rhizomania in sugar beet is increased by frequent cropping. There are peak labour demands at planting and harvesting which have been overcome by improved machinery, e.g. precision "drilling-to-a-stand" for sugar beet and de-stoning fields for easier potato harvesting. Labour demands for cereals and root crops need not clash; they can, in fact, be complementary. These are high cost, high risk crops, but they can be very profitable.

Pulse crops (peas and beans). In many ways these crops resemble cereals, but they can build up nitrogen in the nodules in their roots (page 14). They should not be grown continuously because of a build-up of fungus diseases, e.g. foot rot, stem rot and ascochyta, and pests such as pea cyst nematode. They are good alternative crops to cereals.

Oilseeds (oilseed rape, linseed). These are good alternative crops to cereals. They should not be grown more often than one year in four or five. Because of sclerotinia, oilseed rape, peas, spring beans and sugar beet should be regarded as similar crops with no more than one of them grown in the same field every five years.

Restorative crops. These are crops which are usually fed-off on the fields and, in this way, nutrients are returned to the soil. Good examples are the ley and kale.

A good rotation should include several crops because this would:

(1) Reduce the financial risk if one crop yielded or sold badly.
(2) Spread the labour requirements more evenly over the year.
(3) Reduce the risk of diseases and pests associated with single cropping (mono-culture).
(4) Give better control of many weeds.
(5) Provide more interest for the farmer.

However, most of these objectives could be obtained without having a rigid system of cropping, and recently there has been a tendency to simplify the cropping programmes as much as possible. This approach has been encouraged by:

(1) The need to economize on labour and capital expenditure.
(2) Better machinery for growing and harvesting crops.
(3) Much better control of pests and diseases, mainly by chemicals and resistant varieties.
(4) Chemical weed control.
(5) Guaranteed price systems for most crops.

Many well-tried rotations have been practised in various parts of the country. One of the earliest and best known was the **Norfolk Four Course Rotation** which was well suited to arable areas in eastern England.

It started as:

Roots (turnips or swedes)—folded off with sheep in winter.
Cereal (spring barley undersown) as a cash crop.
Ley (red clover)—grazed in spring and summer.
Cereal (winter wheat) as a cash crop.

This was a well-balanced rotation for building up and maintaining soil fertility; controlling weeds and pests; employing labour throughout the year and providing a reasonable profit. However, considerable changes have occurred over the years mainly due to the introduction of fertilizers, other crops, better machinery, greater freedom of cropping for tenant farmers and the need for increased profits.

Some of the changes which have occurred are that:

(1) Sugar beet, fodder beet, potatoes and carrots have replaced all or part of the folded roots.
(2) Beans and peas have replaced red clover in some areas or a two- or three-year ley has been introduced.
(3) Two or three successive winter cereal crops have replaced the barley and wheat.

Where there is a big difference in soil type on a farm, it may be advisable to have one rotation for the heavy soil and another for the light soil.

The most suitable rotation or cropping programme must be based on the management plan for the farm. It should provide grazing and other foods for the livestock and also the maximum possible area of cash crops. The cash crops grown will partly depend on the amount of labour available throughout the year.

Continuous cereal production and "break crops"

Continuous cereal production (mainly wheat and barley, not oats) has been practised successfully on many farms on various soil types for many years (some for up to 50 years). In some cases, the whole farm is in cereals—in others, a limited number of fields are continuously growing corn. The main reasons for this practice are:

(1) It is simple and has low labour and capital requirements compared with cash root crops and livestock enterprises.
(2) The availability of herbicides to control the majority of grass weeds.
(3) Improved fungicides to control diseases.
(4) Reasonable profitability.

Intensive cereal growing started with spring barley, mainly on chalk and limestone soils, then winter barley and, more recently, winter wheat on many soil types including heavy clays.

Take-all and eyespot are the main disease problems (Table 72). Cyst nematode (Table 71) has not been as serious a problem as might have been expected because barley and wheat are much more resistant than oats which are not grown continuously.

Continuous winter cereal cropping does encourage weeds—especially annual grasses such as blackgrass and barren brome, and broad-leaved weeds such as cleavers.

The current serious problem of surplus production of cereals and falling prices for wheat and barley have recently resulted in many farmers introducing "combinable" break crops, i.e. crops which can be produced with the same machinery as cereals. The most popular break crops are oilseed rape, linseed, dry-harvested peas, beans (vining peas and beans require high investment in special machinery) and herbage seed crops. Lupins, sunflowers, borage and evening primrose are being tried by some enthusiasts (pages 102–121).

Advantages of these break crops include:

(1) They can be as profitable as some cereals, depending on European Community subsidies.
(2) There can be an average one tonne/ha yield increase in the following cereal crop.
(3) They use the same machinery as the cereal crop.
(4) The workload can be spread because of different times of drilling and harvesting.
(5) They offer a better opportunity to deal with some difficult weed problems, e.g. it is easier to control barren brome in oilseed rape than in cereal crops.
(6) The legumes (peas, beans, lupins) provide their own nitrogen requirements, and so production costs are lower.

There are some disadvantages when including a break crop in a hitherto cereal-only rotation. These include arresting the take-all decline; they can be more difficult to harvest and some break crops are more variable in yield and returns than the cereal crop they have replaced.

SET-ASIDE

The aim of Set-Aside is to keep land in a sound agricultural condition without changing the environment. No crop for human or livestock consumption may be grown on Set-Aside land between December and July or January and August. Crops for industrial use may be grown if a contract is available directly from the processor. Currently, Set-Aside is rotational. Parts of fields or headlands can be set aside if large enough, but these areas must be rotated in Set-Aside one year in six. There are also restrictions on what can be grown on the land. Unless Set-Aside follows a very late harvested crop, the ground should have a green cover crop growing over winter to reduce nitrogen leaching. The cheapest option is to let natural regeneration of cereals take place. Another option is to sow at least two varieties of grass. Additionally, no more than 5% by weight of legumes may be

included in a seeds mixture, unless converting to organic production.

If Set-Aside follows a late harvested crop (after mid-October) it may not be possible to establish a green crop. In this situation, farmers should allow natural regeneration to take place. The land may be lightly cultivated in the spring, though it must be cultivated during May to control weeds. Other than this exception, there are several options for management of Set-Aside. Weed seed production must be minimized. Only spot treatment with herbicides is allowed unless special dispensation is obtained from MAFF. The methods of weed control are either topping or summer fallowing. All cover crops should be topped at least once by early July. From past experience with Set-Aside, on some land up to four or five toppings may be required to minimize production of weed seed. The other option is to convert the cover crop into a bare fallow from May. No fertilizer may be applied to the cover crop except a restricted amount of farmyard manure or slurry. Fungicides and insecticides are also banned from use on Set-Aside land. Research work has shown that crops grown after Set-Aside can yield as well as those following a normal break crop. Even growing volunteer cereals as the cover crop has not appeared to affect disease risk. Pest problems may be increased such as wheat bulb fly following a summer fallow.

Further details of the schemes and future modifications can be obtained from MAFF Regional Service Centres.

FARMING AND WILDLIFE CONSERVATION

Although modern farming and conservation can be compatible to a very great extent, there is still a degree of contention between the farmer and/or landowner and the conservationist. The majority of farmers are conscious of the responsibility they have as custodians of the countryside, and they appreciate the aspirations and concerns of the public who wish to see an enhancement of the amenity and nature conservation aspects of rural areas. Responsible conservationists, on their part, realize the needs of farmers to run efficient agricultural businesses which can make a positive contribution to the rural economy at a time of significant economic pressures.

In 1993, agriculture's contribution to gross domestic product is in excess of £6 billion which represents 1.3% of total United Kingdom GDP. However, our self-sufficiency in terms of all food and feed is only 58.2%, and our self-sufficiency in indigenous food and feed products is 74%. In order to maintain and enhance standards of living in both the United Kingdom and other areas of the world, it is important to continue to preserve the capability to produce adequate food supplies. There is a delicate balance between the supply and demand for food in the world. Shortages, even if they are only temporary, can have devastating effects on both people and national economies. We are at present in a situation of surplus for most of our home-grown food requirements, and it is not practical to transport all of this surplus production to those parts of the world which require long-term food aid.

The industry is now faced with curbs on production of most food produced in the European Community, and in order to maintain income farmers must continually strive for greater efficiency of use of all their resources. This continuing quest for the maintenance and improvement of profit is where there can be a clash between the financial economics of food production and public desire for an enhancement of countryside amenity and the conservation of nature. The farming industry will have to come to terms with the fact that land has multiple uses and that food production cannot always have priority over all other legitimate claims on the land. Nevertheless, if there are to be curbs on inputs, particularly nitrogen, in an effort to control costs and yields, in order to avoid over-production penalties, the farmer will continue to wish to use his best land for agriculture. It is not possible to combine efficient farming and a beautiful well-maintained countryside without public financial support. This could be paid out of national taxation, local taxation and higher food prices. Household expenditure on food and drink as a percentage of total consumers, expenditure is still only around 24%, of which household food is 12% and the remainder is eating out and alcoholic drinks.

Undoubtedly too many hedges, often of great ecological value, have been removed, particularly in the eastern counties, since the Second World War. However, recent surveys show that the annual rate of hedgerow loss has slowed during the last decade, and the amount of new planting has increased significantly. MAFF's Capital Grant Schemes used to pay farmers to remove hedges and woodlands and to underdrain wetland. However, the same Ministry and other public sector institutions are now paying record amounts in percentage terms towards the cost of planting new hedges and woodlands and, indeed, in some cases there are payments to raise the water levels of previously drained land.

Large fields are not inevitable in modern arable farming. The size should obviously blend with the particular area of the countryside involved, but generally it is difficult to justify more than 20 hectares on the one hand, and less than 4 hectares at the other end of the scale. In grassland areas, due to restrictions upon milk production through the introduction of a quota system, it is not expected that dairy herds will get much larger in the foreseeable future. Indeed, the medium term results of the 1992 reform of the Common Agricultural Policy may encourage the reverse to occur. There is, therefore, unlikely to be any significant increase in field size and, in any case, no stock farmer can ignore the value of hedges for shelter. It is also difficult to see very large fields dominating the rural landscape in the United Kingdom due to the unacceptability of prairie farming in the eyes of the non-agricultural public who represent 98% of the British electorate.

The Countryside Commission has developed a vision for the countryside of the twenty-first century which is a multipurpose countryside, managed to sustain its environmental qualities. It wishes to see an environmentally healthy countryside which is both accessible and thriving, and it clearly supports the goal of sustainable use and development of rural resources.

English Nature, and other UK nature conservation authorities, make out a very good case for conservation being important not only for farming but for the nation as a whole. All living organisms are inter-related, and either directly or indirectly, all plants and animals are derived from wild species. This is the only way that natural evolutionary processes can continue. The plant breeder, for example, constantly turns to naturally occurring species in his endeavours to improve crop potential. The advent of genetically modified organisms is, however, likely to have very substantial effects on our ability to redesign existing plants and animals to suit our requirements. The release of such genetically engineered materials into the natural environment and its ecosystems will require the highest degree of responsibility to be exercised by those people in control of this new technology.

For many crop and livestock enterprises, high performance per hectare is no longer necessary, nor indeed desirable, and most arable farmers would agree that even under lowland conditions there are areas of land which are no longer profitable to farm. This new situation is helpful to the conservationist because there is no longer the incentive to cultivate every field on the farm. Nevertheless, farmers still require positive encouragement to manage these areas of land in the interests of wildlife conservation and the maintenance of the rural landscape.

The Wildlife and Countryside Act 1981 (as amended) is a serious attempt to help bridge the gap between the farming industry and conservationists. Voluntary co-operation is still the key element of public policy with regard to this type of legislation, and provided farmers are always seen to be making an effort to achieve compatibility between conservation of both wildlife and landscape and food production

on their farms they should have little to fear. However, if this approach is not seen to be working and is not achieving reasonably desired results, there will be pressure upon the government of the day to bring in tougher statutory controls with, or indeed without, compensation.

Management Agreements have been available for use in certain situations since 1949, but they have become very much more important since the 1981 Act, particularly in respect of Sites of Special Scientific Interest (SSSIs). In England, where English Nature has designated an SSSI, the site is statutorily protected. The owners and occupiers will be given a list of Potentially Damaging Operations (PDOs) and these operations are not permitted without the written consent of English Nature or unless the operation is in accordance with the terms of a Management Agreement with English Nature, or a period of four months has elapsed since the formal request was received by English Nature. Where English Nature refuses permission for the operation to be carried out, it will normally offer to commence negotiations for a Management Agreement. English Nature has powers to make payments under Management Agreements, make discretionary grants or offer to lease or purchase land to secure the conservation of SSSIs. Similar arrangements apply in other parts of the United Kingdom.

Undoubtedly, recent legislation and innovative countryside management schemes are having a very significant effect upon the management of the countryside and large areas of agricultural land. Examples of other types of Management Agreements also available in the United Kingdom include Environmentally Sensitive Areas (ESAs), Nitrate Sensitive Areas (NSAs), the Countryside Stewardship Scheme, Tir Cymen, and the Set-Aside Schemes. These agreements range from one year in the case of rotational Set-Aside through to 20 years (normally) in the case of SSSIs, with the majority of schemes being five or ten years in duration. The annual rates of payments will depend upon the scheme involved and the degree of participation. Many of the current schemes involve both annual payments for different types of management and also capital payments for approved investments.

The Farming and Wildlife Advisory Group (FWAG) seeks to stimulate the management of an attractive, living countryside by encouraging the integration of sustainable farming with the conservation or enhancement of wildlife habitats and landscape. This organization provides guidance to land users on the importance of the environmental aspects of their production systems and promotes the view that responsible conservation is an essential part of the farming tradition. With the backing of the Countryside Commission, FWAG has established a network of "link farms" around the country, and these farms, along with the Countryside Commission's "demonstration farms", provide practical examples of how agriculture and conservation can be integrated at the farm level. It is important that these activities are encouraged and supported by the whole farming industry.

MAFF and other Government Departments have introduced a number of schemes and measures to help conserve the countryside and the rural environment. Examples of these include the Farm and Conservation Grant Scheme, legislation against the introduction of non-native species of flora and fauna, the regulation of pesticide use, and strict new regulations designed to prevent the pollution of water, air and soil due to agricultural activities. A series of Codes of Good Agricultural Practice has recently been published and the level of fines that can be imposed by Magistrates' Courts for water pollution and odour nuisance offences has been increased tenfold, from £2000 to £20,000 for each offence, by the Environmental Protection Act 1990.

It is certain that there will be particular interest in water protection, and "vulnerable zones" are likely to be designated later in the 1990s in the United Kingdom. In addition, there could be ever more stringent tests on the occurrence of nitrates, phosphates and pesticides in the water supply and groundwater.

There is an increasing trend towards whole farm management plans for environmental land management purposes to include agriculture, forestry, farm waste and other pollution control, and public access. It is inevitable that this trend will continue and that there will be a steady shift of support in financial terms from the quantity of production towards these integrated management plans. Indeed, the Government is adopting the principle of "cross compliance" in all its new schemes and it is encouraging the principle to be adopted across the whole European Community. This implies that in future public financial support for farming and other forms of countryside management will be conditional upon the protection of the environment.

FURTHER READING

Action for the Countryside, Department of the Environment.
Agriculture and the Environment: Towards Integration, Royal Society for the Protection of Birds.
Balance in the Countryside, MAFF for the Central Office of Information.
Caring for the Countryside, Countryside Commission.
Conservation and Diversification Grants for Farmers, MAFF.
FWAG Handbook for Environmentally Responsible Farming, The Environment Research Fund. *FWAG Booklets. Our Farming Future*, MAFF.
What you should know about Sites of Special Scientific Interest, English Nature.

ORGANIC FARMING

During the 1980s there was an increased interest in organic farming (also called biological or ecological farming), both in the United Kingdom and the rest of Europe. It is a farming system which avoids the routine use of readily soluble fertilizers and/or agrochemicals, whether naturally occurring or not. It is a system which aims to use renewable resources where possible and thus it is considered more sustainable. Organic farming is not a return to pre-1940s farming. It uses current technology and relies heavily on good husbandry. It encourages biological cycles as well as consideration for animal welfare, pollution and the wider social and ecological issues.

There is a small market at present for organic

food; several supermarkets stock these products. However, for organic farming to be viable, premiums must be high.

Some organic crops are cheaper to grow than those grown conventionally; field vegetables, on the other hand, are more expensive, mainly because of the labour requirements for weed control. Yields from organic crops are often only 60–70% of those obtained by conventional growing methods.

In the future, limited grants or subsidies will be available for organic farming. If this farming system were better supported, premiums could be reduced and this would probably increase demand. For organic farming to work, the grower has to be committed to the system. There has to be a change in the approach to the husbandry of growing the crops as agrochemicals and fertilizers cannot be used if a serious problem arises.

The organization

The first organic movement in the United Kingdom was the Soil Association founded in 1946. In the 1980s the marketing co-operative, Organic Farmers & Growers Limited, was founded. Now, both organizations are registered under one scheme in the United Kingdom—UKROFS (United Kingdom Register of Organic Food Standards). Advice on conversion and husbandry can be obtained from the Organic Advisory Service at Elm Farm Research Station in Berkshire. The Soil Association, British Organic Farmers and the Organic Growers' Association produce journals and organize conferences and other meetings. All farms growing and selling organic produce have to be inspected annually in order to confirm that the Approved Guidelines are being followed. The Inspectorate is organized by the UK ROFs.

Soil and nutrition

One of the fundamental aims in organic farming is to maintain and improve the soil—both structure, flora and fauna. Soil structure can be affected by cultivations, crops grown and application of organic manures. Shallow cultivations tend to be favoured (with or without subsoiling) to try and maintain the level of nutrients in the topsoil. Only a limited range of fertilizers can be applied. Organic manures can be brought on to an organic holding as long as they come from an approved livestock system. A number of organic units which were originally stockless have now introduced livestock, so that it is a more complete system. Manure management is very important in order to minimize nutrient losses. The majority of the nitrogen in the rotation comes from nitrogen fixation by the rhizobium bacteria associated with the legumes. Green manures, as cover crops, are also used to reduce nutrient loss over the winter. There is limited data to suggest that organic farming increases soil bacteria and fungal activity, and these are very important in mobilizing some of the soil's nutrient reserves.

Rotations

Rotation is the basis of a good organic system. Initially, when changing to this farming system, there is a two-year conversion period. Grass clover leys or Set-Aside can be used to cover this period.

A good rotation will aim to balance nutrient requirements and help eliminate or reduce weed, disease and/or pest problems. Organic rotations use more legumes than conventional cropping such as red and white clover, lucerne and vetches; green manures (e.g. mustard) are also more commonly used. Normally, no more than two years of cereals will be grown in succession. The main cereal crops grown are milling wheat and oats for human consumption. There is a small demand for wheat for biscuit making, malting barley and rye. As organic livestock production increases, there will obviously be more need for feed grain. Other arable crops grown include grain legumes and potatoes. There is no requirement at present for organic oil seed rape or sugar beet.

Some examples of rotations used on mixed organic units are as follows:

(1) Two- to three-year grass/clover ley
Wheat
Potatoes
Wheat
Rye or oats (undersown)
(2) Two- to three-year grass/clover ley
Wheat
Rye or oats
Grain legume or red clover
Wheat
Barley or oats (undersown)

The majority of the organic manures available would be applied to the leys or potatoes.

In a stockless rotation, usually two years of cereals will follow a legume. The leguminous crop can either be field beans, a forage legume for seed (or cut and mulched), or put into Set-Aside. All straw is incorporated and any spring crops grown could follow a green manure.

Crop protection

It is often assumed that, as virtually no crop protection products can be applied to organic crops, there will be very serious weed, pest and disease problems. In practice, this is not the case. Organic farmers rely on good husbandry, including rotations, delayed drilling (when necessary), variety choice and biological control methods to reduce these problems. Weeds can be the most important cause of yield loss. The rotations employed, together with cultivations (both pre-sowing and within crop) and the use of high seed rates, help to keep down annual weeds. It is perennial weeds, such as thistles, which are the bigger problem. With cereals, most growers find that the use of finger weeders two or three times in the spring can be quite effective in keeping down annual weeds, particularly in spring-sown crops. In wide row crops, inter-row hoes, brush weeders and flame weeders (using propane gas)

have helped reduce the requirements for hand labour.

The most serious disease problems encountered in organic crops are blight in potatoes and seedborne disease in cereals. Plant breeders are improving blight resistance in potatoes. As organically grown cereal seed cannot be treated, it is very important that it is tested for seedborne diseases such as fusarium and smuts. Foliar diseases in cereals have rarely been a serious problem compared with conventionally grown crops.

Good husbandry can help limit pest problems, e.g. frit fly damage can be reduced by allowing at least a six-week interval before drilling cereal seeds following a ley. Delaying drilling of winter cereals will also reduce the likelihood of aphid-transmitted Barley Yellow Dwarf Virus. In horticultural crops, good use can be made of pest monitoring services and pheremone traps to reduce pest damage. Biological control methods are also being used such as the control of caterpillars using the bacteria, *Bacillus thuringiensis*, in cabbages. Many farmers are also encouraging pest predators by habitat management.

Other systems

Another ecological farming system which is practised by a few growers in the United Kingdom, but is more popular in other parts of Europe, is **Bio-dynamic Farming**. This is very similar in many ways to the normal organic systems. It was started in the 1920s following the lectures of Rudolf Steiner. It is a whole system involving organic ideals as well as the spiritual approach. Special preparations are used when composting manure. Bio-dynamic produce is sold under the Demeter label.

Conservation Grade Farming is another area where there is increasing interest. This is a system midway between organic and conventional farming. The Guild of Conservation Grade Producers was set up in the mid-1980s. Anything sold under the Guild's label has to be certified as being grown to its standards. The aims of conservation growers are very similar to organic producers; it is to produce quality foods with the use of renewable resources and at the same time protecting wildlife habitats and the health and welfare of animals. A limited amount of agrochemicals and fertilizers are approved for use, but only those which are considered the least threat to health and the environment. The amount of fertilizer and the timing of applications are restricted to reduce the risk of pollution. Unlike the normal time scale for organic growers, Guild standards can be achieved in one and not two years. However, emphasis is still on a farming system that relies on good husbandry and crop rotations to reduce the need for crop protection products and fertilizers.

FURTHER READING

Organic Farming, Nicolas Lampkin, Farming Press.
The Handbook of Organic Husbandry, F. Blake, Crowood Press.
The Guild of Conservation Grade Producers' Regulations.
The Soil Association Symbol Scheme.
UKROFS Regulations.
What is Bio-dynamic Agriculture?, H. H. Koepf, Bio-dynamic Agriculture Association.

5

CEREALS

Cereals (corn or grain) are the most widely grown arable crops in the United Kingdom, occupying approximately 3.5 million hectares. Over the last 40 years there have been big changes in the relative proportion of the main cereals—wheat, barley and oats. In the early 1950s an equal area of each of these cereals was grown. Since then, with the elimination of the working horse, oats have declined and are now a minor cereal crop. The area of barley increased until the 1970s, since when wheat has become more popular, occupying about two million hectares. Yields have increased more rapidly than barley, averaging about 7 tonnes/ha compared to 6 tonnes/ha winter barley and 5 tonnes/ha spring barley. The majority of wheat crops is autumn-sown; approximately 60% of the barley crop is autumn-sown.

Since the early 1980s, with the improvements in cereal yields, the United Kingdom has become a net exporter of cereals. Total production is about 22 million tonnes. Yields have increased, largely due to improvements in varieties and husbandry techniques. Fertilizer use—particularly nitrogen—and application of crop protection chemicals for the control of pests, diseases and weeds, have also been important reasons for an increase in cereal yields (Table 14).

As well as changes in production of cereals, there have been changes in their usage. Over 80% of the wheat used by millers is now homegrown due to improvements in wheat quality and milling technology. Very little Canadian hard wheat is now imported for bread making. Wheat, compared with barley,

TABLE 14 *UK home-grown cereals—main uses 1972–1990*

	million tonnes		
	1972/73	1979/80	1991/92
Total crop size	15.5	17.4	22.6
Used for human/industrial consumption	3.7	5.0	6.7
Used for animal feed	10.3	10.0	8.4
Exports	0.2	1.7	6.5
Exports as % of total production	1	10	29

Source: Homegrown Cereals Authority.

is also used in greater quantities by the feed compounders. Over six million tonnes of wheat and barley are now exported. Due to the success of several European Community countries, including the United Kingdom, in improving cereal output in recent years, it is now very expensive either to store (Intervention) or support export subsidies. The recent Common Agricultural Policy reform package, using Set-Aside and reduction in Intervention prices, is aimed at reducing the surplus of cereals in the EC.

Other cereals grown to a limited extent in the United Kingdom include durum wheat, rye, triticale, dredge and/or mixed corn and maize. The majority of the maize grown is for forage, although a small number of farmers in south-east England are growing maize for grain.

(a)

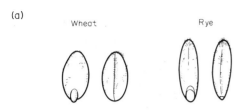

FIG. 36a. "Naked" kernels.

(b)

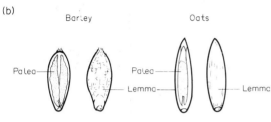

FIG. 36b. "Covered" kernels.

Cereal grains and ears

Wheat, rye and maize grains consist of the seed enclosed in a fruit coat—the pericarp—and are referred to as "naked" caryopses (kernels). In barley and oats, the kernels are enclosed in husks formed by the fusing of the glumes (palea and lemma) and are referred to as "covered" caryopses.

The cereals are easily recognized by their well-known grains (Figs. 36a,b), ears or flowering heads (Figs. 37a–e) and in the early leafy stages (Fig. 38).

Harvesting

Threshing is the separation of the grains from the ears and straw. In wheat and rye the chaff is easily removed from the grain. In barley, only the awns are removed from the grain—the husk remains firmly attached to the kernel. In oats, each grain kernel is surrounded by a husk which is fairly easily removed by a rolling process—as in the production of oatmeal; the chaff enclosing the grains in each spikelet threshes off. Varieties of naked or huskless oats have now been bred; they thresh free from their husks. All cereals are now harvested by combine, except for the few crops where straw is required for thatching.

Cereal harvesting can be carried out more efficiently if it is spread out by growing early and late varieties of barley and wheat, starting with winter barley in July, followed by winter wheat and spring barley in August/September.

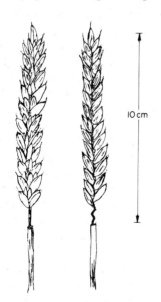

FIG. 37a. Wheat spikelets alternate on opposite sides of the rachis. 1–5 grains develop in each spikelet. A few varieties have long awns.

FIG. 37b. 6-row barley, all three flowers on each spikelet are fertile. Awns are attached to the grains.

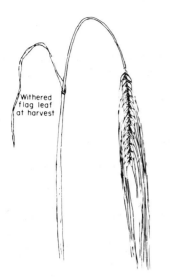

FIG. 37c. 2-row barley heads hang down when ripe; each grain has a long awn; the small infertile flowers are found on each side of the grains.

FIG. 37e. Oats.

FIG. 37d. Grain easily seen in the spikelets of rye.

Spring wheat is the last cereal crop to harvest. Good weed control, to minimize green material at harvest and avoidance of lodging, is also very important. Combining is easier and less or no drying of grain is required. The combine capacity should be adequate to harvest the crops as they ripen. Delay can result in poor-quality grain (worse if it sprouts in wet weather) and shedding losses. It is preferable to have to dry early-harvested crops than to be salvaging damaged crops later. Grain losses of 40–80 kg/ha are reasonable (the latter in a difficult season). Slow combining to reduce losses to 10–20 kg can be false economy and much more can be lost by shedding due to delay. Timing of harvesting will be affected by area and crops grown as well as quality requirements, combine and drying capacity, labour and weather conditions.

Grain monitors, if properly used, can be helpful in checking losses. The average rate of working (tonnes/hour) of a combine is about half the maker's rating, which is usually based on harvesting a heavy wheat crop in ideal conditions. Standard combines vary in drum

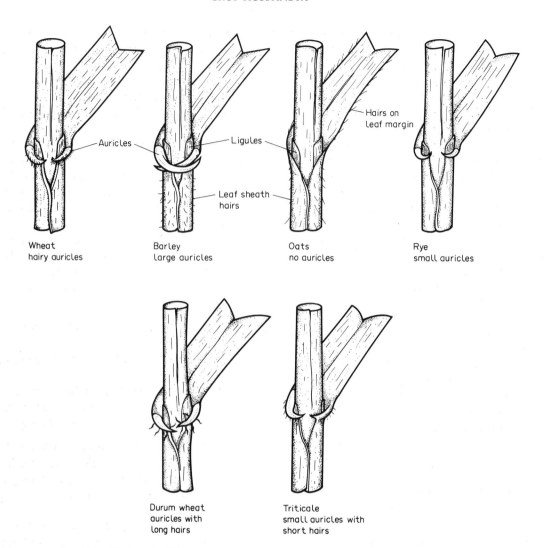

Wheat
hairy auricles

Barley
large auricles

Oats
no auricles

Rye
small auricles

Durum wheat
auricles with
long hairs

Triticale
small auricles with
short hairs

FIG. 38. A method of recognizing cereals in the leafy (vegetative) stage.

width (the main limiting factor) and header width. Wide headers are desirable for farms where the crops are normally light and the fields reasonably level. Pick-up reels are very useful when harvesting laid crops. The standard combine can work up and down slopes reasonably well, provided the speed is adjusted as necessary. However, they do not work satisfactorily going across slopes—the grain moves to the low side of the sieves and little separation takes place because most of the wind escapes on the top side. Some combines are designed to keep the threshing and sieving mechanisms level on sloping ground. The sidehill type has a side-levelling mechanism which can adjust for going across slopes. The hillside type can level in all directions automatically.

The axial-flow combine differs from the conventional machines by having the threshing and separation of grain from straw carried out simultaneously as the material spirals round a large rotor from front to back of the machine. It is less likely to damage grain or become blocked, but sometimes problems can arise in threshing tough barley.

The straw walkers in conventional combines are not the most satisfactory way of separating grain and straw; on some machines these are replaced by a series of drums.

One type of harvester is fitted with a stripper-header which removes the ears only, and the straw is left standing. This can be an advantage for chopping or ploughing-in the straw. The header will be more expensive, but the work-rate can be very high and laid crops do not present serious problems.

Whatever type of machine is used, the manu-facturer's recommendations should be followed for the various settings, e.g. drum speed and clearance and fan speeds. On modern machines adjustments can be made easily.

Combines can be fitted with straw spreaders or choppers which are helpful in dealing with the problem of straw disposal, although the mounted chopper can reduce combining speed.

Cleaning combines to remove weeds, e.g. wild oats, blackgrass, barren brome, is very important before moving to a clean field. It is essential that the combine is clean before moving into a seed crop.

These days most grain is bulk-handled from the combine into and out of store. In some years, when all the cereals are combined at below 15% m.c., no drying will be necessary.

Grain-drying methods

(1) Continuous-flow drier. The principle is that hot air removes the excess moisture and ambient air and then cools the grain to 10–15°C. However, in very warm weather this may not be possible and night air may have to be used to cool the heap after drying. These driers are rated as "x" tonnes/hour taking out 5% moisture at 66°C.

Wet grain (over 22% m.c.) may require two passes through the drier, taking out about 5% moisture each time.

(2) Batch driers. The drying principle is similar to the continuous flow drier, but the grain is held in batches in special containers during the drying process (extraction rate 6% per hour in small types; 6% per day in silos).

(3) Ventilated silos or bins. Cold or slightly heated air is blown through the grain in the silo. This can be a slow process, especially in damp weather (extraction rate $1/3$–1% per day).

(4) Floor drying. A large volume of cold or slightly heated air is blown through the grain to remove excess moisture.

The air may enter the grain in several ways, e.g.

(i) from ducts about a metre apart on, or in, the floor,
(ii) from a single duct in the centre of a large heap,
(iii) through a perforated floor which may also be used to blow the grain to an outlet conveyor when emptying.

Floor drying is very popular because it can be carried out in a general-purpose building. It is a cheap method and requires very little labour when filling. The rate of drying is $1/3$–1% per day.

A common mistake with floor drying is to use heat when drying very wet grain. This over-dries the lower layers around the ducts, and the moisture settles out in the cooler upper layers where it forms a crust which impedes the air flow. The heat should be saved for later stages of drying down to about 14% m.c. Problems can arise where rubbish is allowed to form in "cones" as it is loaded into the store. This could prevent air flowing freely past the grain. The wet and dirty grain should be spread as evenly as possible. Wet grain is more spherical than drier grain; the air spaces are usually larger in the wet grain heap, and so air flows more freely through it than through dry grain. This is a useful self-adjusting phenomenon.

The relative humidity (RH) of the air being blown through grain determines the final moisture content of the grain (Table 15). A 1°C temperature rise from heaters reduces the RH by 4.5%.

TABLE 15

Air RH	50	57	65	72	77	82	86	87	88
Equilibrium grain m.c. (%)	12	13	14	15	16	17	18	19	20

(5) Mobile grain driers. These have become popular in recent years. Most are of the recirculating batch type, but portable versions of well-established static batch or continuous flow machines are also in use.
(6) Membership of co-operatives. The other options which farmers have is to dry and store grain with a co-operative group. This store will provide drying and often grain-wetting facilities (in a dry year) as well as gravity separators for improving grain quality.

Grain intended for specific markets should not be heated above the following **maximum** temperatures (Table 16).

TABLE 16   *Maximum drying temperatures*

Purpose	Moisture content	Temperature
Seed and malting barley	up to 24 (%)	49°C
Seed and malting barley	above 24 (%)	43°C
Milling for human consumption		60°C
Stock feed		100°C

Grain can be stored in bulk for up to one month at 16–17% m.c., but for a long period of storage, beyond the following April, it should be dried down to 14% m.c.

Only fully ripe grain in a dry period is likely to be harvested in this country at 14% moisture. In a wet season, the moisture content may be over 30% and the grain may have to be dried in two or three stages if a continuous-flow drier is used.

Damp grain will heat and may become useless. This heating is mainly due to the growth of moulds, mites and respiration of the grain. Moulds, beetles and weevils may damage grain which is stored at a high temperature, e.g. grain not cooled properly after drying or grain from the combine on a very hot day. Ideally, the grain should be cooled to 18°C. This is difficult, if not impossible, in hot weather.

Before storing grain, it is essential to clean and dry the store thoroughly. When clean and waterproof, the building should be fumigated with a suitable insecticide to kill any remaining pests. Special formulations of the insecticide pirimiphos-methyl are commonly used for this purpose.

The temperature of stored grain should be checked regularly. Any rise in temperature can indicate an insect infestation. Rodents and birds must also be kept out of the store. When grain is sold, any chemical treatments must be declared. The Pesticide Notification Scheme is now mandatory.

Any kind of grain stored for human consumption must be kept under conditions which satisfy the Food Safety Act 1990. This means that the farmer must exercise "due diligence" in keeping stores rodent- and bird-proof. Other commodities, such as fertilizer, must not be stored in the same building.

Moist grain storage

Most grain is stored dry. There are, however, alternative methods of preserving grain which can be simpler and cheaper than more conventional systems. An established method is the storage of damp grain, straight from the combine, in sealed silos. Fungi, grain respiration and insects use up the oxygen in the air spaces and give out carbon dioxide, and the activity ceases when the oxygen is used up. The grain dies but the feeding value does not deteriorate whilst it remains in the silo. This method is best for damp grain of 18–24% m.c., but grain up to 30% or more may also be stored in this way, although it is more likely to cause trouble when removing it from the silo, e.g. "bridging" above an auger. The damp grain is taken out of the silo as required for feeding. This method cannot be used for seed corn, malting barley, or wheat for flour milling.

Another possible method is the storage of damp grain by cooling it. Chilled air is blown through the grain and the higher the moisture content of the grain, the lower the temperature, e.g.

Moisture content of grain %	16	18	20	22
Temperature of grain (approx. °C)	13	7	4	2

The method is cheaper than drying and the grain stores well, i.e. the germination is not affected and mould growth does not develop and so it can be used for seed, malting or milling. Because the moisture content is higher than in dried grain, it is very suitable for rolling for cattle feed.

A further method of storing damp grain safely and economically is by sterilizing it with a slightly volatile acid such as **propionic acid**. The acid is sprayed on to the grain from a special applicator as it passes into the auger conveying it to the storage heap; 5–9 litres per tonne of acid is required. Grain stored in this way is not suitable for milling for human consumption or for seed, but it is very satisfactory for animal feeding and, after rolling or crushing, it remains in a fresh condition for a long time because the acid continues to have a preservative effect.

Other developments include the use of alkali which not only preserves damp grain but also increases its digestibility.

Grain quality in cereals

There are several markets for cereals, including milling, animal feed, malting, seed, export and intervention. Grain quality requirements (pages 96, 100) will be affected by the proposed market.

Standard tests when selling grain include the following:

(1) *Moisture content.* This is very important when storing grain. For long-term storage it should be about 14%. Too high a moisture content will be penalized or not accepted, but no compensation is given for very dry grain. Grain which is overheated when being dried, or in storage, can be spoiled for seed, malting or milling. Overheating can damage the germination ability as well as the protein quality.

(2) *Sample appearance and purity.* Good quality grain is clean and attractive in appearance, free from mould growth, pests and bad odours. Sample purity, freedom from contamination with other cereals or weeds will affect the potential value of grain, as will the amount of shrivelled or broken grain. Careful setting of the combine will help to minimize broken grains and produce a cleaner sample. Depending on quality and quantity of grain being sold, some farmers have a drying system with grain-cleaning facilities.

(3) *Specific weight (bushel weight).* Specific weight is a measure of grain density. A high specific weight is preferred for all markets, particularly export. Wheat has the highest specific weight and oats the lowest. Specific weight can be affected by husbandry (e.g. time of sowing, disease and pest control) as well as weather conditions during grain fill and harvest.

Other standard tests mainly for milling and/or malting include:

(1) *Hagberg Falling Number.* This test measures the amount of breakdown of the grain's starchy endosperm by the enzyme alpha amylase. A high value indicates low enzyme activity which is required for bread

making. If grain has started germinating or sprouting, the Hagberg Falling Number will be low.

(2) *Protein content*. A high protein content is required for bread making compared with a low percentage of grain nitrogen in malting samples. Nitrogen content is analysed using the Kjeldahl method or a near infra-red analyser.

(3) *Protein quality*. The quality of the protein in the grain affects the baking characteristics (when the protein in wheat flour is mixed with water it produces gluten). Protein quality can be assessed by several tests, including baking, gluten washing, SDS sedimentation, Zeleny index or the Chopin alveograph test. Tests undertaken will depend on the target market.

(4) *Dough machinability*. This is a test which measures the stickiness of dough. It is required for wheat being sold into Intervention.

(5) *Germination*. Grain for seed or malting is tested for speed and percentage of germination. Germination should be over 95% for malting.

Husbandry of the crop, variety and time of harvesting can affect several of the above quality requirements. Once harvested and stored, little can be done to change or improve grain quality. Gravity separators are available which can help to raise specific weights and the Hagberg Falling Number.

Table 17 shows the quality standard for selling cereals into Intervention.

Cereal straw

Over 13 million tonnes of recoverable straw are produced on farms in the United Kingdom each year. Approximately 7 million tonnes are baled and removed from the fields for livestock bedding and feed. Other uses include straw for farm and household fuel, mushroom compost, covering over wintered carrots, potato and sugar beet storage. It is also used in horticulture as well as for thatching and insulation board.

Up to the end of harvest 1992, farmers were allowed to burn the straw if not used for other purposes. Now, in England and Wales, straw burning has been banned—mainly because of public pressure. Just under half the straw

TABLE 17   *Minimum quality standards for grain sold into Intervention 1993/94*

	Barley	Wheat	
	Feed	Common (bread making)	Premium
Maximum moisture content (%)	14.5	14.5	14.5
Minimum specific weight (kg/hl)	62	72	72
Maximum total impurities	12	7	7
of which			
– broken grains %	5	5	5
– sprouted grains %	6	6	6
– grain impurities (including – shrivelled etc.) %	12	7	7
– miscellaneous impurities %	3	3	3
– overheated grains %	3	0.05	0.05
– other cereals (including – pest damage) %	5	5	5
– ergot	–	0.05	0.05
Minimum Hagberg F.N.	–	220	240
Zeleny index	–	20*	35
Dough machinability test	–	pass	pass
Minimum protein content % DM	–	11.5	14

Source: HGCA.
*Zeleny >30 machinability pass required in 1993/94.

(mainly wheat) produced has to be incorporated into the soil. It should be chopped, preferably using a combine-mounted chopper. Straw can be incorporated by ploughing should be to at least 15 cm or, by a non-ploughing method, e.g. heavy discs, to a 10-cm depth. Power requirements are greater than for the old method of burning followed by minimal cultivations, particularly on heavy soil types. The straw burning ban has increased the amount of land ploughed on many farms. Non-ploughing techniques will tend to encourage annual grass weeds, especially the bromes and blackgrass as well as volunteer cereals. Light cultivations

TABLE 18 *The Decimal Code for Growth Stages of small grain cereals*

Code 0 Germination. Subdivided—00 dry seed to 09 when first leaf reaches tip of coleoptile.
Code 1 Seedling growth
    10 1st leaf through coleoptile
    11 1st leaf unfolded
    12 2 leaves unfolded
    13 3 leaves unfolded
    14 4 leaves unfolded
    15 5 leaves unfolded
    16 6 leaves unfolded
    17 7 leaves unfolded
    18 8 leaves unfolded
    19 9 or more leaves unfolded
Code 2 Tillering
    20 main shoot only
    21 main shoot and 1 tiller
    22 main shoot and 2 tillers
    23 main shoot and 3 tillers
    24 main shoot and 4 tillers
    25 main shoot and 5 tillers
    26 main shoot and 6 tillers
    27 main shoot and 7 tillers
    28 main shoot and 8 tillers
    29 main shoot and 9 or more tillers
Code 3 Stem elongation
    30 ear at 1 cm (pseudostem erect)
    31 1st node detectable (seen or felt after removing outer leaf sheaths)
    32 2nd node detectable (seen or felt after removing outer leaf sheaths)
    33 3rd node detectable (seen or felt after removing outer leaf sheaths)
    34 4th node detectable (seen or felt after removing outer leaf sheaths)
    35 5th node detectable (seen or felt after removing outer leaf sheaths)
    36 6th node detectable (seen or felt after removing outer leaf sheaths)
    37 flag leaf just visible
    38 —
    39 flag leaf ligule/collar just visible

    For example, a plant having five leaves unfolded (15), a main shoot and three tillers (23), and two nodes detectable (32), would be coded as: 15, 23, 32.

Code 4. 40–49 *Booting* stages in development of ear in leaf sheath to when awns are visible (49).
Code 5. 50–59 stages in development of the inflorescence.
Code 6. 60–69 stages in development of anthesis (pollination).
Code 7. 70–79 milk development stages in grain.
Code 8. 80–89 dough development stages in grain.
Code 9. 90–99 ripening stages in grain (including dormancy stages).

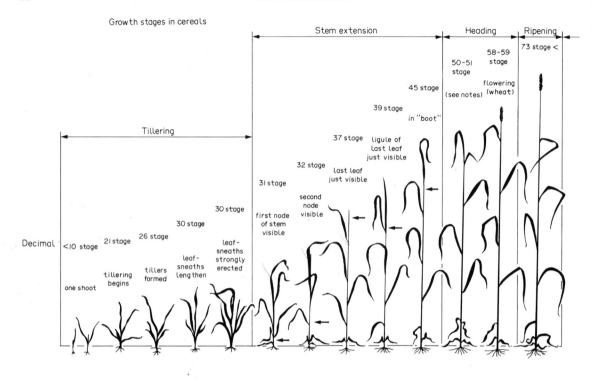

Growth stages in cereals

pre-ploughing can encourage weed seeds to germinate, but they are not essential for successful straw incorporation.

With the introduction of the stripper header there could be more problems incorporating the long unchopped straw. Rolling the straw prior to ploughing can aid incorporation.

Applying extra fertilizer or additives has little effect on straw breakdown. Where straw had been incorporated for several years, there appears to be no problem with its decomposition. In practice, there should be a small increase in soil organic matter.

Cereal growth and yield

A decimal growth stage key is now commonly used to describe the growth and development of the cereal plant (Table 18). There are ten main areas of growth (subdivided into second-

ary stages) from sowing through to the vegetative leaf and tillering stages, stem elongation, ear emergence, flowering to grain filling and ripening. By dissecting out the growing point or apical meristem, using a microscope, the change from vegetative to ear formation can be seen (Figs. 39a, 39b). These changes in internal development do not always coincide with the same growth stage of the cereal plant. Speed of development will depend on type of cereal, variety, temperature, day length and husbandry factors such as time of sowing.

It is very important to be able to identify growth stages correctly as there are only certain stages when pesticides, herbicides and growth regulators should be applied if they are needed. The main stem should always be looked at when checking the growth stage. A major difficulty is deciding when the plants have changed from the tillering to the stem extension stage. The best method is to use a knife and slice the main

(a)

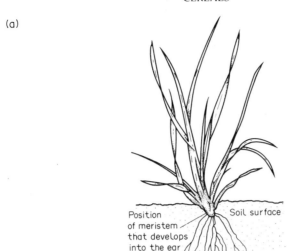

Position of meristem that develops into the ear

Soil surface

FIG. 39a. Leafy winter wheat plant with four tillers.

(b)

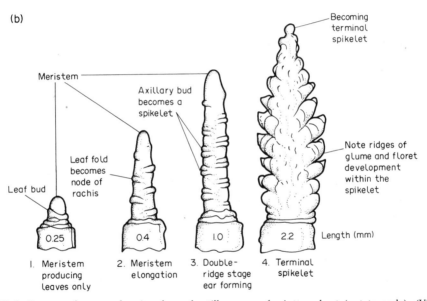

Becoming terminal spikelet

Meristem

Axillary bud becomes a spikelet

Note ridges of glume and floret development within the spikelet

Leaf fold becomes node of rachis

Leaf bud

0.25   0.4   1.0   2.2   Length (mm)

1. Meristem producing leaves only

2. Meristem elongation

3. Double-ridge stage ear forming

4. Terminal spikelet

FIG. 39b. Simplified diagram of an ear forming from the tiller apex of winter wheat (not to scale). (Visible under a microscope or possibly with a good pocket lens.) The apical meristem changes from leaf production to ear formation following cold treatment (vernalization) and increasing day length (photoperiod). Normally these stages occur in February to April and their exact occurrence can be used for efficient timing of pesticides and nitrogen application.

stem in half. Growth stage 30 is when the ear is at 1 cm (Fig. 40). Distinguishing the nodes is another problem. They should only be counted when there are more than 2 cm between each. They, too, are best studied by splitting the stem with a knife; it can be very misleading just to feel the outside of the stem.

The actual yield of a cereal crop is determined by the contributions made by the three components of yield:

number of ears per hectare,
number of grains per ear,
weight (size) of the grains.

These components are inter-related. By increasing the number of ears (e.g. by denser plant populations or more tillering), the number of grains per ear may be reduced and also the size of the grains. Opinions differ on the ideal number or size of each component and obviously it will vary according to the type and variety of cereal, as well as soil and climatic conditions, time of sowing and seed rate, and the occurrence of weeds, diseases and pests.

Winter wheat and 6-row winter barley tend to have the least number of ears/m² (500–700/m²), whereas 2-row winter barley has the highest number of ears (900–1200 ears/m²). The weight of grain is normally heaviest in wheat and lowest

in oats and rye.

Seed rates. Accurate spacing and uniform depth of planting of seed in well-prepared seedbeds are very important if optimum yields are to be obtained. This usually involves doing seed counts/kg and carefully setting the drill. Narrow (10–12 cm) rows are preferable. Broadcasting seed can be successful provided the seed is distributed uniformly and properly covered. Stony soils, wet and cloddy conditions give crops a poor start.

Seed rates should be chosen with the object of establishing the desirable plant population. They can vary between 125 kg/ha and 250 kg/ha. The factors to be taken into account are:

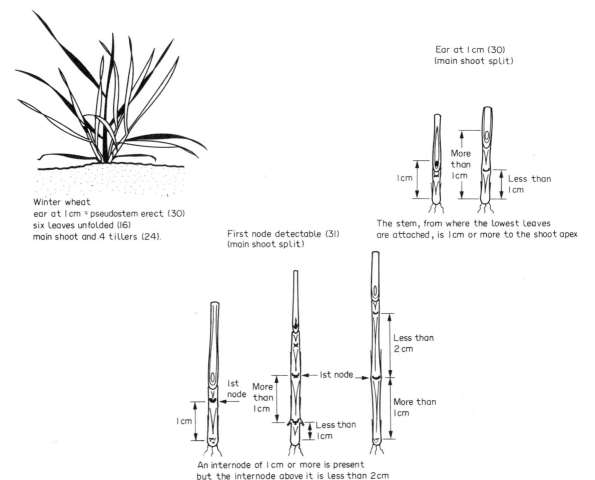

Winter wheat
ear at 1 cm = pseudostem erect (30)
six leaves unfolded (16)
main shoot and 4 tillers (24).

First node detectable (31)
(main shoot split)

Ear at 1 cm (30)
(main shoot split)

The stem, from where the lowest leaves are attached, is 1 cm or more to the shoot apex

An internode of 1 cm or more is present but the internode above it is less than 2 cm

FIG. 40 (a & b). Stem elongation stages in wheat.

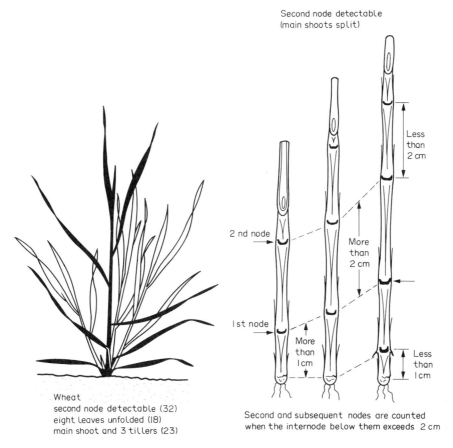

Second node detectable
(main shoots split)

Less than 2 cm

2 nd node

More than 2 cm

I st node

More than I cm

Less than I cm

Wheat
second node detectable (32)
eight leaves unfolded (18)
main shoot and 3 tillers (23)

Second and subsequent nodes are counted
when the internode below them exceeds 2 cm

Fig. 40c.

(1) Crop—there are different optimum plant populations for the various cereal crops.
(2) Seed size, e.g. increase the rate for large seed (low seed counts/kg) and decrease the rate for small seed (high seed counts/kg).
(3) Tillering capacity—some varieties tiller more freely than others and so their seed rate may be reduced. Winter cereals have more time to tiller than spring cereals and so fewer plants need to be established.
(4) Seedbed conditions—the seed rate should be increased in cloddy and stony conditions; this could result in inferior seedling growth; vigorous seed establishes better in poor conditions.
(5) Time of sowing—for late autumn sowings and very early spring sowings the seed rate should be increased.
(6) The possibility of seedling losses by pests such as wheat bulb fly, frit fly or slugs; the more the possibility, the higher the seed rate.
(7) High seed rates can help reduce weed competition.
(8) Price and quality of seed; very expensive seed will often mean a reduced seed rate!

The most desirable plant population to aim for is somewhat debatable; it must be related to the potential yielding capacity of the field. Very low plant populations can ripen unevenly, whereas high populations can lead to smaller ears and low grain weight.

Varieties. There are very many cereal varieties now on the market and new ones are introduced every year. However, there are only a few outstanding varieties of each cereal and these are described in the annual *Recommended Lists* from the National Institute of Agricultural Botany (NIAB) for the United Kingdom—No. 8. Cereals. The UK has been divided into five main regions depending on climate and geographic factors.

Improvements in varieties over the last 30 years have greatly contributed to the increasing yield of cereals. Choice of cereal variety can affect both yield, quality, input requirements such as fungicide and growth regulators, and subsequent returns.

Time of drilling. Time of drilling will be affected by many factors including crop, variety, possible pest, disease and weed problems, soil type, machinery available, weather conditions and previous cropping. In winter cereals, sowing in late September will normally give a higher yield than drilling at the end of October. Drilling starts, in most parts of the country, with winter barley and first wheats from the middle of September. Second and third wheats which could be affected by take-all, or fields with serious grass weed problems, should be drilled last. Cereal sowing will normally be delayed following root crops. The latest safe date for drilling winter cereals will depend on the variety and its vernalization requirement. The NIAB gives recommendations on the latest safe date for sowing. Spring cereals (**mainly** spring wheat) can be sown during the winter. However, the majority of spring cereals are sown in the spring from January onwards. Time of drilling will depend more on suitable soil conditions than on calendar date.

Seed type and dressing. The most commonly grown category of seed sown is C2—second generation commercial seed. First generation C1 seed is only sown if it is being grown for seed on contract, or for home-saved seed, or if a very new variety is being tried. Most seed is of Higher Voluntary Standard (HVS) rather than the lower EC Minimum Standard (Seed Production, page 106).

Unless growing organic crops, seed (bought or home-saved) should be treated against fungal diseases such as bunt, leaf stripe or fusarium (Table 72). The EC has now banned the use of mercury-based products, which means that more expensive fungicides such as carboxin or guazatine should be used. Depending on the previous crop, pests such as wheat bulb fly and wireworm may be a potential threat (Table 71). Insecticides can be included in a seed dressing (a dual-purpose dressing).

Fertilizers and crop protection chemicals. It is now common practice to use a sprayer and fertilizer spreader on the cereal crop several times during the growing season. This is best achieved by following the same wheelings each time. These "tramlines" are normally introduced at drilling. The loss of land caused by tramlines is small and is more than compensated for by the ease and accuracy of spraying—provided the lines are initially properly positioned. Tramlines also allow easier crop inspection.

Fertilizer (mainly nitrogen) and pesticide (herbicides, fungicides and insecticides) applications, as well as the use of growth regulators, increased dramatically during the 1970s and early 1980s. Prophylactic/routine treatments were commonly used. With the changing returns from cereals, many farmers are now looking more closely at the inputs, and so a more integrated approach is being adopted.

Nitrogen rates depend on the type of cereal, yield potential and market, previous crop and the use of organic manures. Currently in the United Kingdom, the soil nitrogen index is based on these factors. In other parts of Europe, soil mineral nitrogen is analysed in the top 90 cm of soil, and fertilizer recommendations are made accordingly. This soil test is being studied in this country. Other researchers are looking at nitrogen models to help predict soil mineral nitrogen levels and subsequent recommendations.

Average nitrogen fertilizer use in the United Kingdom will vary according to the crop, i.e. winter wheat 190 kg nitrogen/ha, winter barley 150 kg and other cereals 100 kg. To avoid leaching, except in very exceptional circumstances, all nitrogen is now applied in the spring as split dressings between February and May (depending on crop and quality requirements). Phosphorus and potassium index levels are assessed by soil analysis and are now fairly high on the majority of arable farms. On fields with high reserves, only maintenance dressings are required or, indeed, none at all. Applications can be made at any convenient time during the growing season. If soil levels are very low (index 0 or 1), the fertilizer should either be applied to the seedbed or combine drilled. Care must be taken with potassium applications if straw is removed from the field rather than incorporated. Barley straw, especially, contains fairly high levels of potassium and its removal could well mean that subsequent applications of potassium will need to be increased.

Table 19 from the Potash Development Association and ADAS shows phosphate and potash recommendations for all cereals with a projected 5, 7.5 and 10 tonnes/ha yield.

There can be a yield response in cereals

TABLE 19 *All cereals—phosphate and potash*

P or K index	0[a]	1	2	3	Over 3
			kg/ha		
Straw ploughed					
Yield level 5.0 t/ha					
Phosphate (P_2O_5)	90	40	40M	40M	Nil
Potash (K_2O)	80	55	30M[b]	Nil	Nil
Yield level 7.5 t/ha					
Phosphate (P_2O_5)	110	60	60M	60M	Nil
Potash (K_2O)	95	70	45M[b]	Nil	Nil
Yield level 10.0 t/ha					
Phosphate (P_2O_5)	130	80	80M	80M	Nil
Potash (K_2O)	110	90	60M[b]	Nil	Nil
Straw removed					
Yield level 5.0 t/ha					
Phosphate (P_2O_5)	90	40	40M	40M	Nil
Potash (K_2O) Wheat	110	80	80M[c]	Nil	Nil
Potash (K_2O) Barley	120	95	100M[c]	Nil	Nil
Yield level 7.5 t/ha					
Phosphate (P_2O_5)	110	60	60M	60M	Nil
Potash (K_2O) Wheat	130	105	80M[c]	30M[c]	Nil
Potash (K_2O) Barley	150	125	100M[c]	40M[c]	Nil
Yield level 10.0 t/ha					
Phosphate (P_2O_5)	130	80	80M	80M	Nil
Potash (K_2O) Wheat	170	145	120M[c]	55M[c]	Nil
Potash (K_2O) Barley	190	165	135M[c]	65M[c]	Nil

Ref: ADAS and Potash Development Association.

[a]   At index 0 large amounts of phosphate and potash are recommended to raise the soil index over a number of years.

[b]   Not needed on most clay soils.

[c]   A lesser amount may be used on most clay soils.

M   – this indicates a maintenance dressing intended to prevent depletion of soil reserves rather than to give a yield response.

from controlling weeds, pests and diseases. The response will depend on the problem and infestation level. On average a winter cereal crop will be treated with two herbicides, at least two fungicides, an insecticide and a growth regulator during the growing season (Table 20 Winter cereals—chemical calendar).

Plant growth regulators and lodging. Plant growth regulators are mainly used in winter cereals to reduce plant height and increase straw strength and so reduce lodging and brackling; leaning is not a problem.

Lodging is caused by a number of factors:

1. Weak straw varieties. The NIAB Recommended List shows that there is quite a difference in straw strength between crops and varieties.
2. Nitrogen use. Excess nitrogen applications, excess soil nitrogen mineralization and/or too early applications of nitrogen can produce lush growth and weak straw.
3. Stem-based diseases. Fungal diseases such as eyespot can induce lodging.
4. Weeds. Weed competition can weaken straw strength.
5. Time of sowing and seed rate. Early drilled

TABLE 20 *Winter cereals – chemical calendar*

Month	Crop growth stage*	Herbicides: growth regulators	Insecticides/ molluscicides†	Fungicides
September	Crop sowing	Couch and weed control, post harvest pre-drilling	Seed dressing Control of slugs	Seed dressing
October	Crop emergence	Grass weed control – blackgrass, wild oats etc. and broadleaved weeds	Control of aphids (BYDV)	
November	Leaf emergence			Mildew control winter barley
December	Tillering			
January			Control of wheat bulb fly	
February				
March	Stem extension			
April		Broadleaved weed control; growth regulator, control of wild oats		Eyespot and foliar diseases
May	Flag leaf emerging			Foliar diseases
June	Ear emergence		Control of aphids (winter wheat)	One or two treatments
July	Flowering			
August	Ripening Harvest	Couch control pre-harvest		

* Will be affected by sowing date, crop (wheat or barley) and climate.
† Tend to treat when necessary, not on routine basis.
Pesticide inputs are normally much higher in winter cereals than in spring cereals.

wheat and late drilled winter barley and winter rye are more prone to lodging, especially if drilled at a high seed rate.

6. Soil type. Cereal crops grown on shallow, droughty soils are less affected by lodging than on deeper, more fertile soils.

7. Weather. The amount, timing and intensity of rainfall affects lodging.

Lodging at early ear emergence causes the greatest yield loss (up to 40% reduction has been recorded). Other problems induced by lodging can include the production of secondary tillers and uneven ripening, poorer grain quality, especially if the ears start sprouting, increased weed competition, delayed harvest, increased combining time and drying requirements. Growth regulators tend to be used routinely on fertile soils where high yielding quality cereal crops are being grown, and there is a history of lodging (page 99).

Pests. Cereals can be attacked by several pests (Table 71), some of which cause little damage (e.g. leaf miners); others can cause total crop loss if not controlled. The main pests in cereals include aphids, wireworm, slugs, leatherjackets, wheat bulb fly, frit fly, cereal cyst nematode, birds and rabbits. Many pests are specific to some, but not all, cereals; wheat bulb fly does not affect winter oats or spring cereals sown after the middle of March. The previous crop can also have a significant effect on the cereal. Pests following grass include wireworm, leatherjacket and frit fly. Some pest problems only occur in certain areas of the country or on particular soil types, e.g. wheat bulb fly is an eastern counties problem, and slugs are associated with heavier soils. Treatment for pests is not routine every year. Research organizations are trying to improve pest forecasting systems. Cultural control methods such as time of drilling, seedbed conditions and varietal resistance should be used before resorting to chemical control.

Diseases. There are many diseases which can affect cereals from those which are seed-borne (e.g. loose smut), to stem-based problems (e.g. eyespot), to foliar diseases (e.g. mildew and rust). The majority are caused by fungi, although there are a number of important virus diseases such as barley yellow dwarf virus (BYDV). Some diseases are specific to particular cereals; others can attack most cereals. Routine fungicide programmes are normally used in most crops. Varietal resistance, weather conditions, disease incidence and husbandry will affect the fungicide programme used. The most cost effective timing for fungicides will vary with the cereal crop. In wheat, for instance, it is essential to keep the flag leaf and ear clean, whereas in winter barley treatment at early stem extension (growth stage 30–32) is recommended. Care must be taken with fungicide choice as there are diseases which have now developed resistance to some commonly used fungicides (e.g. eyespot resistance to benzimidazoles and mildew to the triazoles).

Weeds. Weed problems in cereals vary, depending on rotations (both past and present), soil types, cultivation systems and area of the country. Continuous autumn sowing of cereals has encouraged autumn-germinating weeds such as blackgrass, especially on the heavier soils. Mixed rotations with autumn and spring-sown crops have a more varied weed flora. Non-ploughing techniques, particularly for autumn cereals, have favoured the bromes.

The most important grass weed problems in winter cereals are wild oats, common couch, blackgrass (not Scotland or south-west England), meadowgrasses (particularly in grass cereal rotations) and the bromes. Common broadleaved weeds in winter cereals include chickweed, cleavers, mayweed, speedwell and field pansy. In spring cereals, the polygonums (e.g. knotgrass and redshank) and common hemp nettle (in some areas) are more important. Broadleaved weeds are cheaply and easily controlled in the cereal crop, but this is certainly not the case with grass weeds.

With the increasing numbers of different pesticides and formulations, many farmers are relying on independent advisers/consultants and distributor representatives to help with their spray programmes. Many organizations run cereal/arable crop trial centres where it is possible to see new chemicals and husbandry techniques in practice. Organizations such as the Arable Research Centres are farmer-funded. Information from these centres is normally only available to members.

WHEAT

The present annual production of wheat in the United Kingdom is about 14 million tonnes of which 5 million tonnes are used for flour milling. However, this is over 80% of the total requirements for flour compared with less than 40% 20 years ago. The change to more home-grown usage is mainly because of higher prices received for quality wheat varieties and different methods of bread making. The 5 million tonnes of wheat for milling is made up of 3.2 million tonnes for bread making, 0.7 tonne for biscuits and 1.1 million tonnes for other uses.

Of the remainder, some is used for seed, breakfast foods and distilling whisky, but the majority goes into compound feeding stuffs or for export.

All milling wheat should satisfy the following general standards:

(1) Be free of pest infestation, discoloured grains, objectionable smells, ergot and other injurious materials.
(2) Not overheated during drying or storage.
(3) Moisture content of 15% or less.
(4) Maximum impurities less than 2% by weight.
(5) Pesticide residues within limits prescribed by legislation.
(6) Specific weight—usually at least 76 kg/hl required.

Intervention buying to cope with surplus production can be important in some years when marketing wheat (Standard Requirements for Intervention, Table 17). However, with the exclusion of feed wheat from Intervention this is unlikely to be so important in the future.

In the production and use of flour, quality requirements can be divided into two groups, milling and baking.

Milling quality

This refers to the ease of separation of white flour from the germ and bran (pericarp and outer layers of the seed). It is a varietal characteristic and can be improved by breeding. In the milling process, the grain passes between fluted rollers which expose the endosperm and scrape off the bran before sieving. The endosperm is ground into flour by smooth rollers and in this process the good quality endosperms (hard wheats) break along the cell walls into smooth-faced particles which slide over each other without difficulty and so can be easily sieved. However, the poor quality endosperms (soft wheats) produce broken cells with jagged edges which cling together in clumps. This makes sieving slow and difficult; some of the cell contents are also lost. "Milling value" is a measure of the yield, grade and colour of flour obtained from sound wheat. Varieties favoured for bread making are normally hard wheats. Soft wheats are used for biscuit making.

Baking quality

Wheat is the only cereal which produces flour which is suitable for bread making because the dough produced has elastic properties. This is due to the gluten (hydrated insoluble protein) present. The amount and quality of the gluten is a varietal characteristic, but it can also be affected by soil fertility and climatic conditions.

About 20% of the protein is in the wheat germ and is not important in the baking process. However, the other 80% (in the endosperm) is very important. It can be measured in various ways. For example, protein quantity is assessed by the **Kjeldahl** test (ground grain is digested in sulphuric acid to release nitrogen which is measured and converted by an agreed factor to give the protein percentage) which is a rapid automatic system taking only 20 minutes. Protein quantity can also be measured by a colour (dye) process and by a **near infra-red reflectance** (NIR) technique which is a very rapid dry method. In addition, this measures moisture content as well as hardness. Protein quality is best assessed by a mini-baking test, but this takes at least two hours and so is mainly used for testing varieties for bread, biscuit or cake making. The **Zeleny** test is used for Intervention buying. A white flour is mixed with lactic acid and isopropyl alcohol and the resulting sedimentation is measured; the greater the amount of sediment the better the quality. The **SDS** method is similar but faster. Another commonly used method is to wash the gluten out of the ground grain and assess it for colour, elasticity and toughness (strength); this takes about 20 minutes. The protein in overheated (more than 60°C) wheat grain is spoiled (denatured) for bread making.

Bread-making quality

To provide large, soft, finely textured loaves, the baker requires flour and dough with a large amount of good quality gluten (at least 10% of protein in 14% m.c. grain). This is produced by strong-textured wheats. A good dough will produce a loaf about twice its volume. The small amount of alpha-amylase enzyme present in sound wheat is desirable for changing some starch to sugars for feeding the yeast in bread making. However, wet and germinated wheat contains excessive amounts of alpha-amylase and so excessive amounts of sugars and dextrines are produced. This results in loaves with a very sticky texture and dark brown crusts. Alpha-amylase activity can be measured by the **Hagberg FN** (falling number) test. The number is the time in seconds for a plunger to fall through a slurry of ground grain and water, plus 60 seconds for heating time before the plunger is dropped. A high FN indicates a low (and desirable) alpha-amylase level. On the NIAB list, varieties are grouped as low <180, medium 180–220, high 220–260, and very high >260. This test takes about 20 minutes; it is very important in acceptance of wheat for

bread making (minimum 250) and Intervention buying (minimum 220).

A **dough machinability** test may also be required for Intervention buying; sticky doughs are likely to be rejected.

If the amino acid cystine is not present in the flour, as may happen if the crop suffered from sulphur deficiency, then the quality of the bread would be poor, even although the grain passed all the usual tests.

The amount of water absorption by flour is also important in bread making. It is a varietal characteristic and depends on the protein content and amount of starch granules damaged by milling. A high water uptake by damaged granules of hard wheats means more loaves, which keep fresher longer, from each sack of flour.

Biscuit-making quality

A very elastic type of gluten is **not** required for biscuits, which should remain about the same size as the dough. Weak-textured wheat varieties are suitable for biscuit-making flour.

Milling for other uses

As well as flour for bread making and biscuits, there is a large demand for use in cakes, confectionery, soups and household flour. Market specifications will depend on the end product.

The Flour Milling and Baking Research Association advises on the suitability of varieties and samples for various baking processes.

The poorer quality grain and the by-products from white flour production, i.e. bran (skin of grain) and various inseparable mixtures of bran and flour (e.g. wheatings), are fed to pigs, poultry and other stock.

Table 21 shows typical wheat quality requirements.

Wheat straw

This is used mainly for bedding, but some varieties produce good quality thatching straw

TABLE 21 *Typical wheat quality requirements*

Market	Maximum moisture content %	Minimum specific weight (kg/hl)	Maximum impurities (% by wt.)	Protein content % (14% m.c.)	Minimum Hagberg FN (sec)
Breadmaking	15*	76	2	>11	250
Biscuitmaking	15*	–	2	<10	140
Export milling	15	76	2	10.5	180–230
Export feed	15	73	2	–	–
Feed	15	72	2	–	–

*If grain is to be stored for more than four months, maximum moisture content should be 14%.
For Intervention Standards see Table 17.

which is very valuable. Some straw is still used for covering potato and sugar beet clamps.

Soils and climate. Wheat is a deep-rooted plant which grows well on rich and heavy soils and in the drier eastern and southern parts of this country. Winter wheat will withstand most frosty conditions, but it can be killed out by waterlogged soils. pH should be higher than 5.5.

Place in rotation. When the soil fertility is good, wheat is the best cereal to grow as its yields are higher and its returns generally better than the other cereals. It is commonly taken for one or two years after grassland, potatoes, sugar beet, beans or oilseed rape. It has also been grown continuously on many farms as eyespot and grass weeds can now be controlled. After four or five years, the so-called "take-all barrier" is passed and yields remain fairly constant. However, many farmers now grow as many first wheats as possible (rather than continuous) because of their higher yields.

Seedbeds. A fairly rough autumn seedbed prevents "soil-capping" in a mild, wet winter and protects the base of the plants from cold frosty winds. In a difficult autumn, winter wheat may be successfully planted in a wet sticky seedbed and usually it still produces a satisfactory crop. Spring wheat should only be planted in a good seedbed. Soil-acting herbicides require fine seedbed conditions.

Time of sowing. Winter wheat can be sown from September to March. Early sowings are usually preferable, but they may not be better than late sowings in some favourable autumns. However, sowing late may be impossible in some years. The NIAB list can be checked for "latest safe sowing dates".

Spring wheat can be sown from late autumn to April the following year. An increasing number of farmers are now planting spring wheat after late harvested root crops.

Method of sowing. Ideally, the crop should be sown steadily at a uniform depth (2.5 cm, depending on the moisture level in the soil).

Methods used:

(1) Combine drill with fertilizer in 15–18 cm rows.
(2) Grain-only drill in 10–18 cm rows (10 cm preferable).
(3) Broadcasting using the fertilizer spinner. This can be very satisfactory (especially if drilling is impossible) provided a good covering of the seed is achieved.

Varieties (NIAB Leaflet No. 8. Cereals). Winter wheats normally yield very much better than spring wheats, but the latter are usually of very good milling and baking quality. About 95% of wheat drilled are winter varieties. Most winter varieties have to pass through periods of cold weather (vernalization) and increasing day length (photoperiod) before the ears will develop normally.

Varieties differ in their susceptibility to disease, e.g. yellow rust and mildew, and it is advisable to grow at least three varieties selected to reduce the risk of spread of these diseases (see Diversification Scheme in the NIAB cereal leaflet). It is now well-established that varieties differ considerably in their management requirements in such matters as the best time to sow; fertilizer timing and treatments for yield and quality; disease susceptibility and control; risk of herbicide damage; straw stiffness and response to growth regulators; risk of sprouting in the ear; ease of combining and saleability.

Recommended varieties:

	Winter	Spring
1. Good quality milling and breadmaking	Mercia	Alexandria,Tonic, Axona, Canon
2. Fair breadmaking	Haven, Hussar	Baldus
3. Biscuit quality and others	Riband, Hunter Admiral Beaver	

New varieties are continually being developed for higher yields, quality, disease resistance, standing ability (including dwarfing genes), and also special hybrid seed.

Seed rates. The usual range is 150–250 kg/ha for winter wheat, and 170–220 kg/ha for spring wheat. A number of factors will affect the seed rate:

(1) Seedbed conditions (higher rates in very dry, cloddy and stony soils).
(2) Seed vigour. This can be tested. Some seed is better at establishing in poor conditions.
(3) Weather and time of sowing. The rate could be 30–50% higher in November than in September.
(4) Plant population. A target of 250–350 plants/m² in the autumn may be reduced to 150–300 plants/m² in early spring. In very good conditions, 100 plants/m² can give satisfactory yields, possibly with less lodging and disease problems than in thick crops (e.g. 400 plants/m²). Yield/ha is determined by the number of plants/ha x number of ears/plant x number of grains/ear x average grain weight (this last can be influenced by

variety, growing conditions, disease control and possibly by growth regulators). Low plant populations tend to produce more ears/plant (average about two) and possibly more grains/ear (about 30–50, with a few to over 90) and a heavier grain weight.

The following is a simple way to calculate seed rate, knowing the thousand-grain weight (g)—TGW.

$$\text{Seed rate (kg/ha)} = \frac{\text{no. plants/m}^2 \text{ required} \times \text{TGW}}{\% \text{ establishment expected}}$$

Example: $\dfrac{350 \times 45}{70} = 225 \text{ kg/ha.}$

Fertilizers. To return those nutrients removed from the soil, 8 kg/ha of P_2O_5 and K_2O should be applied for each tonne of grain/ha expected yield (e.g. at 7.5 tonnes/ha, 60 kg each of phosphate and potash is needed), if the straw is incorporated. If the straw is removed, the potassium rate should be increased to 11.5 kg/tonne grain expected yield (e.g. at 7.5 tonnes/ha, 90 kg of potash is needed, as well as 60 kg of phosphate).

The nitrogen rates depend on soil type, expected yield, previous cropping and the possible use of organic manures. The nitrogen recommendation for a 7.5 tonnes/ha crop of winter wheat following cereals is:

Soil type	Fertilizer N kg/ha
Sandy or shallow soils	185
Other mineral soils	160
Organic soils	100

These rates should be increased by 20 kg/ha for every extra tonne/ha yield expected. (Spring wheat, sown in the spring, requires 35 kg/ha less nitrogen than the above rates.) Wheat grown after a break crop normally yields more than second wheats or continuous wheats. However, more fertilizer is not always required as the available nitrogen appears to be used more efficiently.

There is a difference of opinion about the timing for spring nitrogen. It is important that adequate nitrogen is available when the crop growth rate is rapid at the beginning of stem extension—GS 30–32. Some researchers have correlated nitrogen timing and yield with ear development (Fig. 39b). In practice, this technique has not always proved to be feasible. If more than 100 kg/ha is to be applied, it will normally be split either two or three times. The first application of 40 kg should be applied during tillering, no earlier than March. Too early an application of nitrogen will encourage tillering and is more liable to leach. If the crop is late drilled, backward, thin or suffering from pest attack, then the first application should be in February to encourage tillering. The main dressing of nitrogen should be applied at GS 31, usually from the middle of April. Timing for spring wheat, drilled in the autumn, should be as for winter wheat. Spring-sown spring wheat should have the nitrogen split between the seedbed and the three-leaf stage. The earlier the drilling, the more nitrogen should be applied post-emergence and vice versa.

To increase protein by up to 1%, for milling wheat, extra nitrogen (30–50 kg/ha) is usually applied with the main dressing (low risk of leaching) or applied about late May—flag leaf emerging (GS 37) or later. If this is applied as a solid nitrogen fertilizer it may lie unused on a dry soil surface. A liquid application of 20% urea is often applied at the milky-ripe stage. Late foliar applications of nitrogen have no effect on yield.

Spring grazing of winter wheat. This is rarely practised now but, if there is a need for it and the crop is well forward and the soil is dry, it can provide useful grazing for sheep or cattle in late March. It should be grazed only once and as uniformly as possible in blocks. It should then be top-dressed with extra nitrogen (50 kg/ha), unless a reason for grazing was to reduce the risk of lodging. Yields are likely to be reduced by this grazing. It should not be grazed when

the young ear inside the base of the plant comes above ground level. Sheep can graze to ground level, but cattle only to 1 cm off the ground. Grazing should cease if the plants are being pulled out of the ground. Grazed crops have shorter straw at harvest, and sometimes less disease.

Seed dressings. Those liquids or powders containing chemicals such as flutriafol, TBZ or carboxin are useful for controlling diseases such as bunt, fusarium, septoria and loose smut. Pests such as wireworms and wheat bulb fly can be controlled using gamma-HCH.

Pests. In the autumn, slugs can be a serious problem in trashy, cloddy seedbeds, especially when the crop is late-drilled. The slugs can hollow out the seed and shred the young seedling leaves. Methiocarb should be applied if there is a high risk (Table 72). Early drilled wheat can be affected by aphid-transmitted barley yellow dwarf virus (BYDV). The optimum time for control is late October/early November with a suitable aphicide. In the new year, in susceptible fields, wheat bulb fly may require controlling either in January at egg hatch or later, when the first "dead heart" symptoms are seen. Aphids can cause direct damage to wheat, particularly when the grain is in the watery/milky-ripe stage (Table 71).

Diseases. Take-all disease can be a serious problem when growing wheat continuously until the "take-all barrier" has been passed.

The recognition of other cereal pests and diseases and their control are shown in Tables 71 and 72.

Weeds. In winter cereals, autumn-germinating grass weeds such as blackgrass and meadow grass should be controlled in the autumn with a residual herbicide such as isoproturon. If broadleaved weeds are controlled in the autumn they may require further treatment in the spring. Spring treatments, except for wild oats and cleavers, are normally applied by the first node detectable stage GS 31 (Chapter 11).

Plant growth regulators. Plant growth regulators containing chlormequat ("CCC") plus choline chloride (e.g. "Arotex Extra" and "5C

Cycocel") have been used by cereal growers for over 30 years. These chemicals can reduce lodging in cereals and allow yields to be increased by the use of higher fertilizer rates which otherwise would cause lodging and its attendant losses. These growth regulators shorten the internodes and strengthen the stem walls. The reduction in internode length varies (7–20 cm) between varieties. To prevent lodging in winter wheat it is important that they are applied at, and/or before, stem extension (GS 30) so that the lower internodes are shortened. In high risk situations, the treatment should be split between tillering and stem extension stage. Another plant growth regulator, ethylene-releasing ethephon, e.g. 2-chloroethyl-phosphonic acid ("Cerone") or plus mepiquat-chloride ("Terpal"), may also be applied at GS 37 (flagleaf just visible) to GS 45 (boot swollen). This should assist further in reducing lodging and increasing grain size and yield. These are more expensive products and tend only to be used in wheat if chlormequat cannot be applied.

Harvesting. Winter wheat ripens before spring wheat. The crop is harvested in August and September. Indications of ripeness for harvesting are:

(1) Binder (mainly of historical interest).
   Straw is yellowish with all greenness gone. Grain in cheesy condition, firm but not hard.
(2) Combine (7–10 days later than the binder). Straw: turning whitish and the nodes shrivelled. The grain is easily rubbed from the ears and is hard and dry.

Yield

Yield (tonnes/ha)	Average	Very good
Grain	7	10
Straw	3	4

Durum wheat

This is a different species of wheat which, if properly grown, has high protein and hard, amber-coloured vitreous endosperm. On milling it breaks into fairly uniform large fragments called semolina, which is made into a range of high quality pasta products by extruding a stiff dough (semolina mixed with warm water) through dies of various shapes. This high quality market (up to a 50% premium) requires grain which is a light amber colour and meets minimum standards, i.e. 12.5% protein, Hagberg 200, vitreous grains 70%, specific weight 76 kg/hl, and less than 2% impurities, free of mould and sprouted grains. Some durum wheat is used for breakfast cereals. It is an advantage to have a contract with a company and to follow its husbandry advice. Durum wheat receives EC support.

Soils and climate. Durum wheat will grow on a wide range of arable soils, but is probably best suited to medium quality land. It tolerates dry conditions better than ordinary wheat.

It is a Mediterranean crop and the new varieties which have been developed do reasonably well in a good season in the southern and eastern wheat-growing areas in England.

Seedbeds and sowing. Seedbed requirements and methods of sowing are similar to the wheat crop.

Time of sowing. As durum wheat does not require vernalization, it is sown in either October or March.

Varieties. They are all bearded and include *Ambral, Arcour, Cando, Escodur, Flodur.*

Seed rates. 180–200 kg/ha (350–400 seeds/m²). The target should be 300 plants/m², with a minimum of 100 after winter.

Fertilizers. A soil index of 2 for both phosphate and potash is ideal. 40 kg each of phosphate and potash should be applied in the seedbed. Nitrogen requirement and application will be as for milling wheat, i.e. up to 225 kg/ha.

Pests. It will be necessary to check for wheat bulb fly, slugs and grain aphids (Table 71).

Diseases. Durum wheat is very susceptible to eyespot, most foliar diseases and ergot.

Weed control. There are only a limited number of products which are approved for use in durum wheat (Chapter 11).

Harvesting. An autumn-sown crop is usually

combined in August. If possible, it should be given priority (about 20–22% m.c. for combining) to obtain good vitreous grains and a high Hagberg number. If it is left too late, the endosperm is white and there is a risk of sprouting. Grain should be stored at 15% m.c.

Spring-sown Durum wheat is normally harvested in September.

Yield. This is not so high yielding as ordinary wheat, averaging between 15–20% less. The range is from 3 to 6 tonnes/ha.

BARLEY

Barley is the second most important arable crop in the United Kingdom. 60% is now winter-sown. The grain is used mainly for:

(1) Feeding to pigs (ground), cows and intensively-fed beef (rolled). Some feed barley is also exported.
(2) Malting. The best quality grain (about 2 million tonnes each year) is sold for malting purposes. This grain should be plump and sound with a high germination percentage (about 95%) but a low nitrogen (protein) percentage—less than 1.75% nitrogen. Some varieties are more suitable than others for malting (NIAB Leaflet No. 8. Cereals). The final use for the malt will affect the grain nitrogen requirements. A premium, which varies from year to year, is paid for good malting barley (Table 24).

In the malting process the grain is soaked, then sprouted on a floor to produce an enzyme (diastase, maltase), dried in a kiln and the rootlets removed to leave the malt grains.

In the brewing process the malt is crushed and then soaked in warm water in mash tuns. The enzyme changes the starch into sugars which then dissolve in the liquor before being drained off. The remainder of the malt (brewers' grains) is a valuable cattle food.

The sugary liquor (wort) is boiled with hops to give it a bitter flavour and keeping quality, and to destroy the enzymes. The strained wort is then fermented with yeast which converts the sugars to alcohol and so produces beer.

To make whisky, hops are not added; the fermented wort is distilled to produce a more concentrated alcoholic liquid called malt whisky.

Some malt is used in other products, e.g. malt vinegar and breakfast cereals.

Barley straw is used mainly for litter, but an increasing amount is fed to both cattle and sheep.

Soils and climate. Barley can be grown on arable land throughout the UK, provided the pH of the soil is about 6.5; it is more affected by a low pH than wheat. Barley is a shallow-rooted crop which grows better than other cereals on thin chalk and limestone soils. However, it will grow on a wide range of soils provided they are well drained. On organic and very fertile soils, it may lodge, especially in a wet season, and the grain is unlikely to be of malting quality.

The crop is not exacting as regards climate, but little winter barley is grown in the colder parts of the United Kingdom.

Place in rotation. Barley, unlike wheat, is usually grown when the fertility is not very

TABLE 22 *Typical barley quality requirements*

Market	Maximum moisture content %	Minimum specific weight (kg/hl)	Maximum admixture %	Maximum nitrogen range %	Minimum germination %
Malting	16	–	2	1.75	95+
Export malting	15.5	64	–	1.6–1.8	95
Export feed	15	64	2	–	–
Feed	16	60–62	2	–	–

For Intervention Standards see Table 17.

high. On many farms, provided that the soil conditions are right, it has been grown continuously on the same fields and produced reasonable yields. As an early harvested crop, winter barley is commonly grown before oilseed rape.

Seedbeds. Generally a finer seedbed than wheat is required. Shallow sowing at 3–5 cm is important.

Time of sowing. Winter barley can be sown from early September (e.g. in Scotland) to early November (e.g. in South Devon). It is important to have the plants well established before the winter. Spring barley can be sown from January to early April. A good seedbed is more important than early drilling.

The sowing of barley follows the same lines as for wheat.

Varieties (NIAB Leaflet No. 8. Cereals)

	Winter	Spring
Malting quality (two-row)	Halcyon, Pipkin Bronze, Maris Otter, Puffin	Blenheim, Alexis, Chariot, Derkado, Triumph
Feeding quality (two-row) (six-row)	Pastoral, Fighter, Marinka Gaulois, Manitou	Hart, Chad
BYMV resistant	Target, Willow, Firefly	
Cyst nematode resistant		Decor

Modern two-row varieties of winter barley are frost-hardy, and yield better than spring barley—especially in areas which suffer from drought in the summer. However, many spring varieties have very good malting quality compared with the winter varieties. To obtain best results with winter barley, the crop must be sown early enough to develop a good root system and become well tillered before winter.

Seed rates. Winter barley 150–180 kg/ha; spring barley 125–180 kg/ha. The target is for an established 250/300 plants/m², which should result from sowing 300–400 seeds/m².

Pests. The main pest problem which requires control, particularly in early-drilled winter barley, is to reduce barley yellow dwarf virus transmitted by aphids (Table 71).

Diseases. The barley crop can be affected by a large number of seedborne diseases such as leaf stripe, fusarium, smuts and snow rot. Foliar diseases can also be prevalent in both winter and spring barley, such as net blotch, mildew and rhynchosporium. A serious problem where a large amount of barley has been grown is barley yellow mosaic virus. For control of these diseases see Table 72.

Weeds. Autumn weed control is desirable for both annual broadleaved and grass weeds, using suitable herbicides. Long runs of winter cereals tend to encourage grass weeds such as blackgrass and barren brome, and control can be very expensive.

Fertilizer requirements

(1) Nitrogen	Feed	Malting
Winter barley	40–160 kg according to fertility split between February and early April	40–120 kg according to fertility apply by end of March
Spring barley	up to 125 kg applied by early May	up to 125 kg applied by end of March

The nitrogen can all be applied to the seedbed with late-sown spring crops, whereas with the early-sown crop it should be split to avoid leaching.
(2) The phosphate and potash requirements are similar to wheat.

The very early ripening time of winter barley makes it easier to control perennial weeds by pre-harvest glyphosate. This is especially the case with onion couch which senesces earlier than ordinary couch (Chapter 11).

Plant growth regulators. To control lodging in winter barley, chlormequat at GS 30 can be used. This is not as effective as ethephon formulations (Terpal or Cerone) at GS 32–49. This should reduce lodging, necking and brackling. The treatments will be similar for spring barley.

Harvesting. Winter barley is ready for combining from mid-July to early August. Spring barley combining normally starts mid-August to early September.

The crop is ready for harvesting when the straw has turned whitish and the ears are hanging downwards parallel to the straw (although this depends on the variety). In a crop containing late tillers, harvesting should start when most of the crop is ready. Malting crops are usually left to become as dead ripe as possible before harvesting. Harvesting of feed barley is often started before the ideal stage. This is especially the case if a large area has to be harvested and the weather is uncertain. The ears of some varieties and over-ripe crops can break off very easily, resulting in serious losses.

Yield

Yield (tonnes/ha)	Average	Good	Excellent
Grain	5.5	7	9
Straw	3	4	5

OATS

In the 1940s about half of the cereals in the United Kingdom were oats; now it is only about 3% of the area. Of the 500,000 tonnes grown, about one-third is for milling for human consumption (usually under contract). The decline of the horse and the higher yields of wheat, as well as the Intervention support for wheat and barley, have been the main factors affecting the declining (until now) oat hectarage. However, the increasing demand for fibrous foods is now helping sales.

Oats for human consumption should have a high specific weight, 50 kg/hl, and a moisture content of 14% or less. There are no standards for oil and protein content.

Oats are mainly used for feeding to livestock. They are particularly good for horses and are also valuable for cattle and sheep. However, they are not very suitable for pigs because of their high husk (fibre) content. The best quality oats may be sold for making oatmeal which is used for bread-making, oatcakes, porridge and breakfast foods. Recently, there has been some renewed interest in the crop as a possible "break" from continuous barley, because oats are resistant to several of the diseases which affect barley, especially eyespot and take-all. There is also a renewed interest in the "naked" oats species which has a very high feeding value (better than maize) because the husk threshes off the grain. The normal oat grain contains about 20% by weight of husk (the lemma and palea) but it is easily separated from the valuable groat in naked varieties. New varieties of naked oats have been developed at the Welsh Plant Breeding Station, but they are lower yielding than normal oats.

Oat straw is very variable in quality; the best quality from leafy varieties is similar in feeding value to medium quality hay. Some short, stiff-strawed varieties are no better than barley straw when grown in the warmer, drier parts of the country.

Soils and climate. Oats will grow on most types of soil; they can withstand moderately acid conditions where wheat and barley would fail. The pH should be 5 or over although, if too much lime is present, manganese deficiency (grey-speck) may reduce yields.

Oats do best in the cooler and wetter northern and western parts of the country, but even in these areas they have been replaced by barley on many farms.

Place in rotation. Oats can be taken at almost any stage in a rotation of crops. They are a useful take-all break. If grown too often, cyst nematodes may cause a crop failure.

Seedbed and methods of sowing. These are similar to wheat and barley.

Time of sowing. Winter oats, late September–October.

Spring oats, February–March.

Varieties (NIAB Leaflet No. 8. Cereals. Winter varieties are not so frost-hardy as winter wheat or barley. They usually yield better, especially in the drier districts, and are less likely to be damaged by frit fly than spring oats. Winter varieties are mainly grown south of the Humber.

Winter varieties:	*Image, Aintree, Solva, Craig, Mirabel.* All except Aintree and Mirabel are resistant to stem nematode.
Spring varieties:	*Keeper, Dula, Rollo, Melys, Valiant.* Keeper and Rollo are resistant to cyst nematode.
Naked oat varieties:	*Kynon* (winter), *Rhiannon* (spring).

Seed rates. 190–250 kg/ha for both winter and spring oats. The target plant population is 300 plants/m².

Fertilizers. The phosphate and potash requirements are similar to wheat.

Between 80–120 kg/ha nitrogen is usually considered optimum at nitrogen index 0. This should be applied at stem extension stage in winter oats and on the seedbed for spring oats.

Spring grazing. This may be desirable if there is a risk of lodging, because grazing results in a shorter straw.

Pests. The main pest problem is aphids transmitting BYDV. The cereal cyst nematode is encouraged by close cropping with oats (Table 71).

Diseases. Mildew, crown rust and oat mosaic virus are the major diseases of oats. Disease-resistant varieties should be chosen. Winter oats may require a fungicide at GS 31 (Table 72).

Weeds. Grass weeds are very difficult to control in the oat crop. Few chemicals are recommended. It is preferable to grow oats where there is no blackgrass or wild oat problem. Oats are, however, a very competitive crop against weeds.

Growth regulators. Chlormequat formulations are useful in preventing lodging if applied at GS 32.

Harvesting. Winter oats are usually harvested just before winter wheat. Likewise, spring oats (depending on the district) are normally harvested before spring-sown spring wheat. Oats, like barley, are liable to shedding when ripe. The grain is not very dense and requires 40% more storage space than wheat.

There is still a very small proportion of the oat crop cut with the binder. This will be found in northern districts and in the south-west. It is usually cut when the straw is still green or just turning yellow. This early cutting can give good quality straw, and there is less shedding of grain.

Yield

Yield (tonnes/ha)	Average	Good
Grain	5	7
Straw	3	5

RYE

Rye is grown on a small scale in this country for grain or very early grazing (page 185). The grain is used mainly for making rye crispbread and there has been an increased demand for rye in soft grainbread. It is not widely used for feeding to livestock.

Rye should only be grown on contract and, for milling, a Hagberg Falling Number of 120 is required.

The long, tough straw is very good for thatching and bedding, but not for feeding.

Soils and climate. Rye will grow on poor, light acid soils and in dry districts where other cereals may fail. It is mainly grown in such conditions for grain because, on good soils, although the output may be higher, it does not yield or sell so well as other cereals.

It is extremely frost-hardy and withstands much colder conditions than the other cereals.

Place in rotation. Rye can replace other cereals in a rotation, especially where the fertility is not too high. It can be grown continuously on poor soils with occasional breaks.

Seedbeds and methods of sowing. The seedbed requirements and methods of sowing are similar to those for wheat.

Time of sowing. For grain production, the crop should preferably be sown in September. (There are at present no spring rye varieties grown in this country.)

Seed rate. 150–190 kg/ha. Lower seed rates are used for the hybrid ryes.

Varieties. Rye, unlike the other cereals, is cross-fertilized and so varieties are difficult to maintain true to type and new seed should be bought in each year.

Amando, Halo, Sentinel, Merkator are examples of varieties. Hybrid rye varieties (e.g. *Luchs*) are also now available; the seed is expensive, but yields can be up to 15% higher.

Fertilizers. Depending on the yield potential of the soil, nitrogen requirements are in the range of 100–150 kg/ha. This should be applied as a top-dressing in April. The phosphate and potash requirements are low; 40–50 kg/ha of each is normally considered sufficient.

Spring grazing. The grain crop can provide useful grazing in late February/March, depending on soil conditions, and before the plants start to develop hollow stems. Grain yield will be lower, but it can be sold for seed.

Crop protection. A limited range of grass and broadleaved herbicides is available. As rye is tall and weak-strawed, a growth regulator is required, e.g. chlormequat (GS 31) and "Terpal" (GS 37). Hybrid rye varieties have shorter, stiffer straw.

Rye is less susceptible to take-all than wheat or barley. It can, however, be affected by eyespot, mildew and brown rust. A fungicide at GS 31 is

usually worth while. As rye is open-flowering, it is more susceptible than other cereals to ergot.

Harvesting. The crop is normally harvested before winter wheat. It is combined when the grain is hard and dry and the straw is turning from a greyish to white colour. The ears sprout very readily in a wet harvest season, so it is usually harvested at 20% moisture content.

Yield

	Average	Good
Grain	4 tonnes/ha	5 tonnes/ha
Straw	4 tonnes/ha	5 tonnes/ha

TRITICALE

This is a relatively new crop, produced by crossing tetraploid wheat with diploid rye and treating seedlings of the sterile F_1 plants so that their chromosome number is doubled and they become reasonably fertile. It is bearded and intermediate between wheat and rye in most of its characteristics. Unlike other countries, only the winter varieties are being grown in this country.

There is no EC support. Triticale is in direct competition with other feed grains.

Soil and climate. Triticale is probably best suited to marginal and lighter soils for use as a feed grain of similar value as wheat. It is very winter hardy and tolerant of drought.

Seedbeds and methods of sowing. The seedbed requirements and methods of sowing are similar to those for wheat.

Seed rate. 270–330 seeds/m² (125–160 kg/ha); minimum plants after winter 150/m².

Time of sowing. Mid-September to October.

Fertilizers. Nitrogen requirements are in the range of 120–175 kg/ha top-dressed at the end of March/early April. Depending on the soil indices, the seedbed requirements for phosphate are 60 kg/ha and potash 45 kg/ha, at index 2, for both plant nutrients.

Varieties. Lasko, Cumulus, Purdy, Protius, Alamo.

Crop protection. The growth regulators chlormequat, "Terpal" or "Cerone" will reduce lodging.

Triticale can be quite weak-strawed on less droughty soils.

Weed control. There are more restrictions on the use of some grass weed herbicides in triticale than with the wheat crop. Several broadleaved weed herbicides are recommended (Table 67).

Diseases. Triticale is susceptible to ergot and eyespot, and fairly resistant to cereal foliar diseases such as brown rust. A fungicide at GS 31 for control of eyespot and, where necessary, foliar diseases is recommended. Take-all does not affect triticale to any extent, and it is resistant to barley yellow mosaic virus.

Harvesting. The crop is harvested in August, as for wheat. Drying and storage advice is also the same as for wheat.

Yield. Very variable: 3–9 tonnes/ha.

MIXED CORN CROPS

Mixtures of cereals (dredge corn) are grown in some areas, particularly in the south-west. The most common type is a mixture of barley and oats in varying proportions, with a total seed rate of about 220 kg/ha. The yield of grain is usually better than if either crop were grown alone. Varieties must be chosen which ripen at the same time.

Sometimes cereals and peas or beans are mixed (mashlum). This type can be used for grain or silage.

Winter and spring mixtures can be used both for dredge corn and mashlum.

Growing and harvesting. The growing and the harvesting of the crops are similar to oats.

MAIZE FOR GRAIN

Maize is a tall annual grass plant with a strong, solid stem carrying large narrow leaves. The male flowers are produced on a tassel at the top of the plant, and the female some distance away on one or more spikes in the axils of the leaves. (This separation simplifies the production of hybrid seed.) After wind pollination of the filament-like styles (silks), the grain develops in rows on the female spike (cob) to produce the maize ear in its surrounding husk leaves.

Climate limits the production of grain maize in the United Kingdom. It is a sub-tropical plant and in this country it is confined to a small area in the south-east where the yield is not at all reliable. The situation is likely to remain this way until, or if, more cold-tolerant varieties are produced. There is a limited market for "corn-on-the-cob" or sweet corn as a vegetable. This latter is a special type of maize in which some of the sugar produced is not converted into starch and is harvested when the grain is in the milky stage.

The soil and husbandry is generally the same as for forage maize (page 157).

Varieties

	Grain	Sweetcorn
Early	Mona	Sunrise
		Sweet Nugget
Late	Magda	Cocktail

The plant population and thus the seed rate is lower for the grain crop compared with the forage crop, viz. 90,000 plants/ha, and a seed rate of 24–30 kg/ha, with the distance between the plants, precision-drilled at 12–15 cm.

Harvesting. Maize for grain is usually ready between mid-October and mid-November. Frosts will kill off the foliage and this will facilitate combining and help to dry the grain, which in the yellow hard condition has a moisture content of 35–45% at harvest.

Various types of machines are available for harvesting. Some are ordinary combines with header attachments which only remove the ears from the standing crop and then thresh off the grain; others thresh the whole crop and some deliver the complete or dehusked ears into a trailer to be shelled later.

Drying the grain can be a problem and is very expensive. It usually has to be dried in two or three stages in a continuous drier from, say, 40% to 15% moisture content. Floor-drying cannot be used for threshed grain, but the whole ears can be dried in this way.

Wet storage is possible in air-tight tower or butyl silos, but the grain can be difficult to unload and it deteriorates rapidly when taken out of the silo.

Propionic acid, applied at 18–20 litres/tonne through a special applicator, will preserve the wet grain very satisfactorily in heaps in existing buildings.

Yield

Grain:	4–6 tonnes/ha
Vegetable cobs:	up to 75,000/ha

CEREAL SEED PRODUCTION

Under EC regulations all seeds bought and sold by farmers must be officially certified, tested, sealed and labelled. There are no restrictions on farmers sowing seed grown on their own farms.

The Cereal Seed Regulations 1980 state that the only seed which may be bought or sold is seed which has been produced under the official Certification Scheme. Under this scheme, seed is produced on a generation system starting with *Pre-Basic seed* produced by the maintainer/breeder of the variety. It is identified by a white label with a purple stripe. This seed is used to produce *Basic seed* (white label), which in turn produces *Certified Seed 1st generation (C1)* (blue label), and this produces *Certified Seed 2nd generation (C2)* (red label). No further certification is possible. Minimum and Higher Voluntary Standards (M/HVS) are set out for each category of seed with regard to previous cropping, isolation, weeds (especially wild oats), diseases, impurities and germination. Minimum standards are set by the EC, whereas HVS is a UK scheme. The majority of seed sold is HVS. Contracts for growing seed are made with registered merchants.

Seed production is specialized work. Great care must be taken to avoid contamination with other seed. Herbicides must be used at the correct time; lodging should be avoided by cautious use of nitrogen fertilizer. When growing seed, input costs can be more than normal. Low seed rates are also used to reduce the high seed cost. Extra herbicide, rogueing and growth regulators may be required. The combine must be cleaned thoroughly, and threshing must be carried out carefully to avoid damaging the seed, which should be dried at a low temperature. Care is also necessary when cleaning to avoid admixture with other grains, and seed corn must be stored separately.

Premiums are incentives for seed growers. Autumn-sown cereal seed crops are taken off the farm after harvest and so they do not have to be stored for long. Contract cleaning and seed-dressing mobile services are available for farmers who grow their own seed but who may not have these facilities. As much care should be taken in growing and harvesting home-saved seed as for seed for certification (NIAB Seed Growers' Leaflet No. 1).

FURTHER READING

Cereal Husbandry, E. J. Wibberley, Farming Press.
Technology of Cereals, N. L. Kent, Pergamon Press.

6

COMBINABLE BREAK CROPS

Oilseed rape

Oilseed rape is now established as an important and profitable crop in the United Kingdom. 1992 Returns show that 442,000 hectares are now grown.

The small black seed contains 38–40% oil which is extracted by crushing and used for the manufacture of margarine and cooking fats. Oilseed processors will only accept seed which is low in both erucic acid and glucosinolate for human or animal consumption. The residue left after the oil has been removed can be included in animal rations at a higher rate now that the glucosinolate levels are below 25 micromoles per gramme of seed. These varieties are known as "double lows". All recommended varieties are now double lows and the arable area payment will not be paid on the old single low varieties. Set-Aside land can be used to grow high erucic acid (HEA) varieties, e.g. *Martina* or *Askari*; the oil from these can be used as a light industrial lubricant.

Soils and climate. The crop will grow in a wide range of soil and climatic conditions provided the land is well drained, pH over 6, and the soil and subsoil structure is good.

Rotation. Ideally, rape and other brassica crops should not be grown more than one year in five, in order to avoid a build-up of diseases such as club root, and pests. It is an alternative host for sugar beet nematode and this could affect the place taken by sugar beet in a rotation.

Winter varieties (NIAB Leaflet No. 9. Oilseed Crops). *Falcon, Libravo, Samourai* are fully recommended double low varieties with average yields. Samourai has good standing power, whilst Libravo shows above-average disease resistance.

Newer, higher yielding varieties include *Bristol* and *Envol*, both of which are early maturing. Bristol has a high glucosinolate level. New varieties are being bred to give better disease resistance—the early double low varieties were very prone to fungal diseases, especially stem canker (phoma).

Spring varieties. A reduction in price received for the produce and a move to area payments has resulted in an increased interest in spring varieties. The cost of growing these spring crops is less than that of the autumn-sown varieties and the gross margins can be comparable.

High yielding varieties include *Bingo, Comet, Forte, Tanto*; all are double low.

Time of sowing. Winter crops—mid-August to early September, usually following winter barley.

Spring crops—early March, if possible.

Seedbeds. The seed is very small and so fine, moist soil conditions are required. Direct drilling, into stubble, can be very successful on soils with good structure. Pans and large flat stones restrict the growth of the deep taproots.

Seed rates and plant population. The seed can be dressed for control of flea beetle, alternaria and phoma (Tables 71 and 72).

According to seedbed conditions, 4–8 kg/ha can be drilled in 12–18 cm rows, 1–3 cm deep, to produce a final population of 80–100 plants/m². The seed can be broadcast. After a hard winter, the population may drop to 30–40/m², but this can still give a good yield on fertile soils. Spring rape does not branch as much and the seed is smaller than winter rape. Therefore, 120–130 plants/m² should be established. If the crops are too dense, the important secondary inflorescences do not develop fully. Ideally, there should be six to eight medium-sized leaves (10–15 cm high) on each plant before winter. If sown too early, or excessive nitrogen is used, too much leaf develops in autumn and this can reduce yields.

Fertilizers. Good crops will take up large amounts of plant food, but generally the response to phosphate and potash is low, especially on good soils.

TABLE 23

	N	P₂O₅	K₂O
	kg/ha	kg/ha	kg/ha
Winter rape			
seedbed	40	40–75	40–75
top-dressing Feb./March	200–240*		
Spring rape	100–200	40–75	40–75

(i) Assume phosphate and potash indices are 2–3.
(ii) Nitrogen level depends on yield potential.

Winter oilseed rape can respond to autumn nitrogen. A top-dressing of 40 kg/ha should be applied if signs of nitrogen deficiency are seen in October/November. Good early ground cover is vital to prevent pigeon damage and to smother weeds.

In some crops, e.g. on light soils with a high pH, boron deficiency may occur (stunted growth, curled leaves with rough mid-ribs and some stem cracking). 20 kg of borax or 10 kg of "Solubor" per hectare can be applied to the seedbed (if a deficiency is anticipated) or sprayed on the leaves in early spring. Sulphur deficiency (which could be mistaken for nitrogen deficiency) may occur on light, sandy soils low in organic matter. If possible (under these conditions), superphosphate, sulphate of ammonia, or gypsum can be used as basic fertilizers. A sulphur spray can be applied on to the leaves.

Growth stage. A knowledge of the growth stages in the oilseed rape plant may be helpful in deciding on the best time to treat the crop for a specific problem (Table 24).

Weed control. Good crops will usually smother broadleaved weeds. Thin crops are more likely to become weedy. Herbicides are available to deal with most weed problems.

Pests and diseases. The considerable increase in the area of oilseed rape grown in recent years has, as expected, resulted in increasing problems with pests and diseases (Tables 71 and 72).

Pigeon damage. This can be very serious in some years, especially in areas with a high pigeon population, near woods, where only a small area of rape is grown, and where there is little else for the pigeons to eat in the winter.

TABLE 24   *Growth stages in oil-seed rape*

	Definition	Code
0	*Germination and emergence*	
1	*Leaf production*	
	Both cotyledons unfolded and green	1,0
	First true leaf	1,1
	Second true leaf	1,2
	Third true leaf	1,3
	Fourth true leaf	1,4
	Fifth true leaf	1,5
	About tenth true leaf	1,10
	About fifteenth true leaf	1,15
2	*Stem extension*	
	No internodes ("rosette")	2,0
	About five internodes	2,5
3	*Flower bud development*	
	Only leaf buds present	3,0
	Flower buds present but enclosed by leaves	3,1
	Flower buds visible from above ("green bud")	3,3
	Flower buds level with leaves	3,4
	Flower buds raised above leaves	3,5
	First flower stalks extending	3,6
	First flower buds yellow ("yellow bud")	3,7
4	*Flowering*	
	First flower opened	4,0
	10% all buds opened	4,1
	30% all buds opened	4,3
	50% all buds opened	4,5
5	*Pod development*	
	30% potential pods	5,3
	50% potential pods	5,5
	70% potential pods	5,7
	All potential pods	5,9

TABLE 24 (cont.) *Growth stages in oil-seed rape*

Definition	Code
6 *Seed development*	
Seeds expanding	6,1
Most seeds translucent but full size	6,2
Most seeds green	6,3
Most seeds green-brown mottled	6,4
Most seeds brown	6,5
Most seeds dark brown	6,6
Most seeds black but soft	6,7
Most seeds black and hard	6,8
All seeds black and hard	6,9
7 *Leaf senescence*	
8 *Stem senescence*	
Most stem green	8,1
Half stem green	8,5
Little stem green	8,9
9 *Pod senescence*	
Most pods green	9.1
Half pods green	9,5
Few pods green	9,9

Note: To estimate later leaf stages judge the number of lost leaves by their scars. Note that stages from bud to seed development should normally apply to the main stem and seed development stages should normally apply to the lowest third of the main inflorescence. Otherwise, branch position on the inflorescence should be stated. Senescence stages apply to the whole plant.

Although the crop is often eaten in late autumn and early winter, the greatest damage is caused by grazing in late winter and early spring when the new shoots and buds are developing. This leads to uneven flowering and ripening and creates problems for timing of spraying and harvesting, as well as loss of yield and a lower oil percentage. Many sound and visual scaring devices such as bangers, kites, balloons and scarecrows are available, and when used they should be moved around from time to time. Shooting is also helpful, particularly when the weather is severe and in the more vulnerable periods. A full crop is always likely to be less damaged than a thin patchy crop.

Rabbit damage. This is an increasing problem during winter where rabbit populations are getting larger. Electric fencing and gassing are possible solutions.

Harvesting. Where necessary, a pre-harvest (one to two weeks before) spray of glyphosate is effective in controlling any perennial weeds which are actively growing; it also desiccates the crop.

To reduce shedding losses in windy areas, the winter crop is usually swathed. This should be carried out when most of the middle pods turn yellow and the seed is a chocolate colour. The swathing must be done well and a vertical knife may have to be used for heavy and tangled crops. Swathing is usually a contractor operation. The cut crop should be left on a 20-cm stubble to dry and ripen, and combined 7–10 days later. Sheltered crops and spring crops can be combined direct when the seed is black. At this stage losses can readily occur by the ripe pods shattering. Combining in dull weather and in evenings reduces shatter losses. When ripening is uneven, shattering from the ripe pods can be reduced by desiccating the crop with diquat about three days after the swathing stage (seed chocolate-black). This also desiccates any green weed growth before combining 4–7 days later.

Drying and storage. The moisture content of the seed at harvest time will be in the range 8–15% for swathed crops and 10–25% for direct-combined crops. The contract price is usually based on 9% moisture content and so it should be dried to a slightly lower percentage. Damp rape seed must be dried as soon as possible, either by a continuous drier (not above 60°C, and then cooled quickly) or by bulk drying on the floor or in bins. The undried grain should not be piled more than 1.25 m deep, and the drying ducts covered with hessian.

Normally, dry cold air is used for drying, but some heat may be required to lower the relative humidity to 70% so that the seed dries to 8% moisture content.

The bulk capacity of the seed is 1.4 m³ per tonne.

Yield. Winter rape 2–4 tonnes/ha, spring rape 2–3 tonnes/ha.

When rape seed is being conveyed in trailers or lorries, great care should be taken to block all holes through which it might escape.

Linseed and Flax

Linseed and flax are different varieties of the same plant which has been grown in this country since Roman times. Its importance has fluctuated over the centuries.

Figure 41 shows the difference between the two varieties. Linseed is short-strawed with capsules (bolls) to give a higher yield of seed than flax which is long-strawed and grown for its fibre.

Linseed. The seed contains about 40% oil and 24% crude protein. The drying oil produced from linseed is mainly used for making paints, putty, varnishes, oil cloth, linoleum, printer's ink etc. Over the last 20 years the introduction of plastics and latex paints, and the use of other oils, has reduced its importance. However, with a move towards more "natural" products, the use of linseed oil has increased recently. Better dyeing techniques have meant that linoleum, for instance, can be made bright and attractive to compete with modern PVC floor coverings. Unlike PVC, it does not give off poisonous gas when burnt. It is not used in products for human consumption because of its high linolenic acid content, but the by-product linseed cake is a highly-valued, protein-rich animal foodstuff.

The crop is supported by an area payment, but it lies outside the Set-Aside/Arable Area Payment Scheme. The deficiency payment on linseed is based on the difference between the world price and the EC minimum price as well as on the national yield. It is, therefore, variable. Approximately 40,000 hectares of linseed are required to meet the present demand in this country. However, the linseed area in the United Kingdom is now 140,000 hectares, and so much of the oil is exported.

Soils and climate. Linseed can be grown in any of the arable areas of the United Kingdom, but moisture-retentive soils are preferable because of the small root system of the plant. The pH should be about 6.5.

Rotation. Because of the comparatively small area grown, pests and diseases are not serious problems. Nevertheless, linseed should not, in theory, be grown more often than one year in five, although current practice is to crop it more frequently to maximize the area payment. It is a good cereal break crop.

Varieties. Modern linseed varieties such as *Antares, Atlante, Barbara, Blue Chip* are a great improvement on older varieties, especially with regard to yield, shorter and less fibrous straw, earlier and more uniform ripening, less shattering of seed, and disease resistance (NIAB Leaflet No. 9. Oilseed Crops).

Seed and sowing

The seed should be dressed with thiabendazole + thiram, or prochloraz, to control

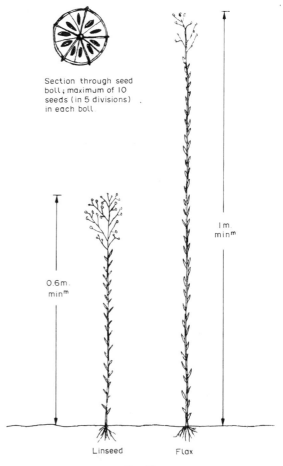

Section through seed boll; maximum of 10 seeds (in 5 divisions) in each boll

0.6 m. min^m

1 m. min^m

Linseed   Flax

Fig. 41.

damping-off and sown at 700–750 seeds/m² (80–90 kg/ha). It should be drilled at 1–2 cm deep in 9–12 cm rows into a fine, firm and moist seedbed. The recommended time of sowing is at the end of March/early April. Linseed is susceptible to severe frost and the frost damage can sometimes be confused with potash deficiency.

Fertilizers

N	P₂O₅	K₂O
kg/ha	kg/ha	kg/ha
100	50	50

Assuming phosphate and potash indices of 2–3.

There is very little response to phosphate and potash on fertile soils. Zinc deficiency may occur on over-limed soils; this can check growth and cause white spots on leaves. A foliar application of zinc sulphate may have to be made.

Weed control. The crop has a very "open" habit of growth and cannot compete well with weeds. Perennial weeds should be controlled in the previous year. Trifluralin in the seedbed will control annual meadow grass and some broadleaved weeds. The use of trifluralin has *off-label approval* for the chemical. Products given approval for use on oilseed rape have an off-label approval for use on linseed.

Disease control. Resistant varieties can be used for rust and wilt; approved seed dressings for the control of seedling blight, foot and stem rots.

Pests. Seed dressing should be effective against the flea beetle and leatherjacket (Table 71). Pigeons and rabbits can damage the young crop extensively in the spring.

Flowering. Most varieties have pale blue flowers which appear in early July. The plant flowers early in the morning, but only for a few hours. The flowering period for the crop may last several weeks. It is normally self-pollinated.

Harvesting. Linseed is ready for harvesting when the whole plant is dry, the stems are yellow-brown and the leaves have fallen from the base. The seeds should rattle in the bolls, be plump and brown and, for direct combining, show the first signs of shattering. Linseed straw is very tough and wiry and so the knife must be kept sharp. The crop may be desiccated with diquat before combining, especially where there are weeds. Glyphosate can also be used when there is couch present. The secret of successful desiccation is achieving penetration below the canopy of capsules so that the chemical can dry out the stems. Higher than normal volumes of water should be used and "crop tilters" are useful as long as they do not damage the crop.

Plant breeders are attempting to produce varieties which have their capsules near the top of the crop. The cutter bar of the combine can then be set high to keep as much of the fibrous stem as possible out of the machine. This will reduce blockages and tangles. The long straw can then be ploughed under after harvest and allowed to rot. Combining should only be carried out when the crop is dry.

Drying and storage. The seed must be carefully dried to about 8% moisture content for safe storage. It is harvested at about 12–16%.

Linseed is a small slippery seed which can easily fall through tiny holes; this must be watched carefully at all stages, i.e. at sowing, combining, in trailers, and in the drier and storage buildings.

Yield. A satisfactory yield for linseed is 2.5–3 tonnes/he. Price per tonne is about £80 plus approximately a £500/ha deficiency payment (1993 figures).

Flax. This is grown mainly for the fibres in the stem which are present as long bundles around a woody core, and which are used for linen and other products. Recent interest in the crop in Northern Ireland and Scotland is based on a new method of **retting** (a controlled rotting process which breaks down the pectin binding the fibres to the woody core). The crop is pulled from the ground by a special machine using rubber belts. It used to be tied into sheaves (beets) to be retted in field dams or heated water tanks, followed by drying in the field or over a drier. The new process is to spray the crop with glyphosate (up to 4 litres + 1 litre wetter/ha) about 10 days after the mid-flowering period (about late July) and leave for four weeks to desiccate and two weeks or more to ret. To be effective, continuous moist (humid) conditions are necessary and so it is very unreliable compared with tank retting. In September, when the moisture content is ideally about 16%, the crop is pulled and left for a few hours for the roots to dry. It is then baled in large round bales and later taken to the mill for scutching (the separation of fibres from the remainder of the stem). The pulling and baling is usually done by a contractor, costing about £250/ha. The yield of baled straw is about 6–8 tonnes/ha. The scutched fibre may sell for £500–£1500/tonne—depending on quality. However, there is a surplus of flax being grown in the EC at present. Most is grown in Belgium, North Germany and France and many growers in these countries have stockpiles of baled flax waiting to be processed.

When the crop is scutched, the straw is separated into the following fractions:

Long fibre (13%)—used for linen, yarns, ropes;
Short fibre or tow (16%)—blended with synthetic fibres for a range of materials, e.g. mail bags, tarpaulins, hessian;
Shive (40%)—fibre insulating board;
Seed (13%)—up to 1.5 tonnes/ha linseed;
Chaff and dust (18%).

Some by-products are used for papermaking.
Varieties. Varieties of flax, which are usually supplied by the contracting mill, include *Ariane, Hera, Regina, Belinka, Weira.*

The growing of the crop is similar to linseed except that the target plant population is 1800 plants/m² (125–150 kg/ha).

Sunflowers

Sunflower plants grow successfully in gardens and coverts for game birds in southern England but, when grown as a seed crop, birds are likely to cause serious losses in the seedbed as well as by eating the ripening seeds. Grey mould (botrytis) can cause serious damage to the leaves, the stems (resulting in lodging) and flower heads in a cool, damp season. As yet, there are no satisfactory methods of dealing with these problems, but resistant varieties and fungicides may overcome botrytis. A number of farmers in East Anglia have grown reasonable crops on a small scale, but harvesting and handling the seed afterwards is not easy. Earlier ripening varieties are needed for this country.

Modern sunflower varieties are potentially high-yielding. Short-stemmed hybrid plants are produced in a similar way to maize hybrids, by using a male sterility and restoring genes technique.

The decorticated (dehusked) seed contains over 40% of valuable, high-quality edible oil which is used in food manufacture, margarine and cooking oils.

A well-drained soil with a pH over 5.5 is essential, and it should not be grown in the same field more often than one year in four. 50/100 kg each of nitrogen, phosphate and potash should cover the nutrient requirements (soil phosphate and potash indices at 2 and 3). Too much nitrogen or very rich soil will cause lodging.

The seed is sown mid-April at 5–6 kg/ha, precision-drilled 25–70 mm deep, 15–20 cm apart, in 75-cm rows.

Trifluralin has off-label approval for broad-leaved weed control, although inter-row cultivations and cross harrowing should help to keep the crop clean.

The crop is ready for harvesting in late September/October when all the leaves have turned yellow. The heads dry out very slowly and desiccation should help. The seed and rubbish may have a moisture content of up to 30% and this has to be dried to 10% with a maximum temperature 46°C. Drying is not easy and the material is very bulky.

A yield of 2.5 – 3.5 tonnes/ha is possible.

Evening primrose

Evening primrose is a yellow-flowered biennial garden plant which is being grown on a very limited scale as a possible farm crop. The seed contains about 20% oil which is a very rich source of gamma-linolenic acid (GLA) and this is highly valued for many pharmaceutical uses. However, the market for both evening primrose and borage has been damaged by cheaper sources of GLA—notably the discovery that blackcurrant pips (a by-product of the fruit drink industry) contain large quantities.

Established plants will grow successfully on most well-drained arable soils in the southern half of England. Establishment is difficult. The tiny seed germinates unevenly and requires a very good, moist seedbed. It is best drilled in late July/early August in 40–60 cm rows, 8–12 mm deep at about 1–3 kg/ha. Ideally, this should give an established 30–50 plants/m² population (5–10/m² is the acceptable minimum). Better crops may be obtained by transplanting (30–50 plants/m²) from soil blocks in April. The plant branches freely and grows to about 1–5 m high.

Varieties include *Epinal* and *Constable*. The genetic make-up of the plant is such that varietal improvement is likely to be very difficult.

Fertilizer requirements are not exacting, with 25 kg nitrogen and 50 kg each of phosphate and potash (at index 2–3). In the spring, a top-dressing of 50–150 kg/ha can be contemplated.

Some oilseed rape herbicides have been given off-label approval, e.g. benazolin + clopyralid and trifluralin for broadleaved weeds. The cereal and pea herbicide pendimethalin has off-label approval for annual grass and broadleaved weed control.

Harvesting takes place in September/October. Ripening is very uneven. Desiccation and combining direct appears to be preferable to swathing. The seed is very small but not slippery. The moisture content at harvest is about 20% and this should be dried to 10% for safe storage.

The yield appears to fluctuate between 300–700 kg/ha.

Borage

Borage is an indigenous annual plant which has been growing wild or in gardens for many centuries. It was developed as a crop plant for its seed, with an oil content of about 30%, which is very rich in gamma-linolenic acid (GLA).

Very light soils and acid soils should be avoided for borage, which prefers a good loam soil in areas with a fairly low rainfall (especially at harvest time). Evidence suggests that it can be grown continuously in the same field. Seed rate is at 15–20 kg/ha sown in late March/April, in 12–50 cm rows about 20–25 mm deep. Seedbed and fertilizer requirements are similar to those of spring barley.

It is a vigorous crop growing to 1 m high, which can smother most weeds. No herbicides are yet approved for borage. The crop flowers (blue) appear from June onwards and throughout the summer.

Seed shedding is a major problem as the crop comes to harvest. It is usually swathed when the first seeds start shedding at the end of July. Weather permitting, the row is combined about a week later. The seed at harvest is about 20–25% moisture content and has to be dried to 10% for safe storage.

The yield is very variable—150–750 kg/ha.

Combinable pulses

Pulse crops are legumes which have edible seeds. The main ones grown in the United Kingdom are the various types of beans and peas. Bacteria found in nodules on the roots of these crops can fix nitrogen and so they should not require nitrogen fertilizers. Additionally, the crop following generally benefits from nitrogen left in the soil: this could amount to about 60 kg/ha.

Field beans and dry harvested peas are useful, protein-rich grains for blending with cereals for feeding farm stock. However, yields can be variable, and harvesting difficult. About half the combinable peas grown are for human consumption in the canned or dried state.

The smaller (tick) beans and maple peas are popular for pigeon feeding, and high prices are paid for good quality grain.

Field beans usually grow well on heavy soils

and loams, whereas peas prefer the medium and lighter soils.

Beans or peas provide a useful break between cereal crops, but they should not be grown in successive years because of fungus diseases and pests.

The introduction of improved varieties and better chemical control of weeds and pests, such as aphids, has resulted in a greater area being grown. The Arable Area Payments for legume crops under the Set-Aside regime are higher than those for cereals. This may mean that a larger hectarage of these crops will be grown as long as such factors as soil type, climate and the farming system allows (Set-Aside, page 70).

Harvesting, storage and drying facilities must also be of a high standard for these crops.

Irrigation is beneficial on light soils and in dry seasons, especially when the pods are setting and filling.

Field beans

This crop has been grown in the United Kingdom for many centuries, although it has fluctuated in popularity. It occupies about 70% of the total area of all types of beans grown in these islands. It is a good break crop for heavy land arable farms. It is, for instance, widely grown on the clays and chalky clays of East Anglia. Field beans do well on these soil types as well as the heavy loams, provided they have a good structure, are well drained with a pH over 6.5. Organic soils and high nitrogen content usually result in too much straw, lodging and low grain yields.

The varieties grown can be grouped as:

(1) Winter bean (large seed). *Bourdon* is a high yielding variety with medium early maturity and moderate straw strength. *Boxer* and *Punch* are also widely grown. *Banner* is an older variety which is becoming outclassed.
(2) Spring beans
    (a) Tick beans (small seeds). *Maris Bead* and *Barker* produce very small round seeds and are recommended for the lucrative

pigeon trade. *Frinebo, Gobo and Victor* are very high yielding varieties.
    (b) Horse beans (medium seed). *Pistache* is a small horse bean with early maturity.

Recent plant breeding work has led to the development of spring bean varieties with no tannin in the grain. They are white-flowered and the beans can be used at higher inclusion rates in animal feeds. The varieties *Albatross* and *Toret* are tannin-free but have a lower yield than the other recommended varieties of spring beans.

New spring varieties being bred have a more determinate type of growth which gives them more uniform ripening.

In general, compared with spring beans, winter beans yield about 0.3 tonne/ha more, ripen up to a month earlier, are more susceptible to chocolate spot disease and are lower in protein—23–27% compared with 27–30% for spring beans (NIAB Leaflet No. 10. Field Beans and Field Peas).

Manures and fertilizers. No nitrogen is required, but beans usually benefit from well-rotted farm-yard manure (25–50 tonnes/ha), or slurry (up to 40,000 l/ha).

Depending on phosphate and potash indices, 0–75 kg P_2O_5 and 0–110 kg K_2O_5 per hectare is generally advised; an average of 50 kg of each may be combine-drilled.

Seedbed. This is similar to the cereal crop but ideally deeper.

Time of sowing. Winter beans are sown October to mid-November. If sown too early in a mild autumn, the resultant soft growth is easily damaged by hard frosts and it can be attacked by an early infestation of chocolate spot disease. This early spread of the disease can subsequently be very difficult to control. Spring beans are sown as soon as soil conditions permit after early February.

Seed rates. Seed size varies considerably.

Winter beans: About 200 kg/ha (range 180–250 kg), i.e. about 25/m² to give an established population of about 20 plants/m². After a hard winter, a minimum of 10/m² (well distributed) is needed to give a good yield. Too thick a crop can give a poor pod set and therefore a disappointing yield.

Spring beans: Tick types—about 200 kg/ha (range 180–220 kg) to give a population of 30–40 plants/m².

Horse types: About 230–250 kg/ha to give a population of 40–50 plants/m².

The higher plant populations for spring beans are necessary because they do not branch out as much as winter beans.

Seed certified free of ascochyta is advised, and it should be dressed against damping-off (Table 72).

Sowing methods. These can vary. Winter beans may be ploughed-in (10–15 cm deep) in November, or broadcast on the ploughed surface and covered by harrowing. Normally, the spring crop is sown with the ordinary corn drill (15–18 cm rows), but care is required with the larger-seeded varieties to avoid "bridging in the drill" or damage to the seed.

Weed control. Perennial weeds, e.g. couch, thistles, field bindweed, must be destroyed before the crop is sown. Simazine has been used for many years to deal with, fairly cheaply, most annual weeds, but cleavers, polygonum species, wild oats and volunteer cereals are not controlled satisfactorily. This soil-acting herbicide is depth selective for beans, provided they are planted at least 8 cm deep; if not, the bean leaves become blackened round the edges and the plants may die.

The range of herbicides now available to the bean grower is fairly large, and simazine is often replaced (in the presence of problem weeds) by a chemical with a broader spectrum of control.

Products such as terbuthylazine + terbutryn and trietazine + terbutryn give pre-emergence control of broadleaved weeds and annual grasses. Propyzamide controls some perennial grasses as well, in addition to annual grasses and broadleaved weeds. There are several products used post-emergence to control wild oats and volunteer cereals, e.g. alloxydim-sodium and sethoxydim (Table 69).

Pollination. Honey bees (two or three hives to each hectare) are usually necessary, especially with seed crops, to obtain good pollination of the beans. Some new hybrid varieties being developed are self-fertilizing and so are less dependent on bees than the common open-pollinated varieties. Systemic aphicides which may be required to control blackfly should, if possible, be applied before flowering (i.e. in June) to give better control with less risk of harm to the bees. These chemicals, which are normally only required for spring beans, may be applied as sprays (e.g. pirimicarb) or granules (e.g. phorate or disulfoton).

Pests and diseases (Tables 71 and 72).

Harvesting. Winter beans are usually ready for harvesting in August–September; spring beans in September–October. They ripen unevenly, lower pods first. Most crops are now harvested by combine when the leaves have withered and nearly all the pods are ripe and the seeds ready to shatter. It is best to combine in dull weather or in mornings and evenings to reduce losses at the cutter-bar. Desiccation of any green growth with diquat can make combining easier.

Drying and storage. Combined beans may require drying before storage. These large seeds must be dried carefully in continuous driers, preferably (to avoid cracking) in two stages if the moisture content is over 20%. They can be dried successfully on floor driers, but the air may escape too easily and some beans near the floor may not be dried if the ducts are widely spaced.

The dried beans should be stored (1.25 m³/tonne) at 14% moisture content.

Yield. Winter beans 2–5 tonnes/ha. Spring beans 2–6 tonnes/ha.

Dry harvested peas

The pattern of pea production in the United Kingdom has changed considerably in recent years. Previously, most of the crops grown were used for human consumption and either harvested dry or vined (harvested fresh). Due to high subsidy payments from the EC for protein feed, the area of peas nearly doubled over the last decade, and the increased production was harvested dry for the feed compound trade. The compounding subsidy has been removed and peas, for both dry harvesting and vining, now receive Arable Area Payments under the present Set-Aside regime (page 70).

As a protein source for animal feed, peas are in competition with imported soya and fishmeal, and the market fluctuates in response to world prices for these commodities.

The present production for feed is about 240,000 tonnes (grown on 60,000 hectares), but some estimates suggest that up to a million tonnes could be used in future. Peas are preferred to field beans for feed.

All the modern varieties grown for human and animal consumption have white flowers, but some older varieties grown for stockfeed and racing pigeons are purple-flowered. All the varieties are dwarf types, with some being semi-leafless, i.e. the leaflets are changed to tendrils which intertwine with neighbouring plants and so the crop stands better. The semi-leafless plants have large stipules and are very efficient producers of pods, and easy to harvest. Their open type of foliage is less likely to be attacked by some diseases, although good weed control is essential. Other varieties are tare-leaved which have a fine foliage and an upright habit of growth.

The main types of production at present are for:

(i) *Human consumption*. These sell at a premium and are:
  (a) sold dry in packets or loose (dark green colour desirable);
  (b) soaked, dyed and canned as processed peas, pease pudding or soups;
  (c) blanched, cooked in oil, baked and sold as crispy "snackpeas".
(ii) *Compound feeds*. These are from high-yielding varieties, are preferably not brown and can include rejects from human consumption crops, but they must not be mouldy.
(iii) *Seed*. These are grown on special contract;

approximately 8% of the pea area is required for seed production.
(iv) *Racing pigeons*. These peas must be clean and disease-free.

Types of peas

Marrowfat. This is the main type used for human consumption, although their high yield makes them suitable for stockfeed. The seeds are large, blue-green and dimpled. Because of their size they are not easy to dry. *Princess* is the highest yielding marrowfat on the NIAB Leaflet No. 10. Field Beans and Field Peas. It is a semi-leafless variety with good standing ability and an early harvest. *Maro* is an old variety of moderate yield and excellent quality, but it lodges easily and can be difficult to harvest. *Progreta* is a very popular tare-leaved mutant selected from Maro; it yields better than Maro and is easier to harvest. *Bunting* has large, even-sized seed of good canning quality. *Guido* is a high yielding variety which yields better than Maro and has a larger pea size. It is of good quality for human consumption.

White peas. These have round, smooth, white-yellow seed and yield well. They are mainly used for feed, but there can be a market for pease pudding, soups and other products. Examples of varieties include: *Rex, Bohatyr, Countess, Baroness*. Rex is a very high yielding white pea which, like Bohatyr, stands well and holds its pods well off the ground. This means that they can be harvested relatively easily. Baroness is one of the highest yielding peas available. It stands well and is moderately early to harvest.

Small Blues. These have small, round, smooth, blue-green seed. They are very early with moderately good yields. There is limited use for canning as small processed peas. Examples include: *Orb, Conquest, Echo*. Orb and Conquest are very early to harvest and the latter variety is of particularly good quality.

Large Blues. These have large, round, smooth, blue-green seed. There is limited use for packet sales and they are mainly marketed for feed. These varieties are suitable for fertile soils. An example is *Solara*, which is the most widely grown variety of this type.

Maple peas. These have purple flowers and round, brown-speckled seed. These varieties are used for forage, feed and racing pigeons. *Minerva* is an example of a maple pea.

Climate. Combinable peas require good weather for harvesting (ideally, less than 50 mm rain in July and in August). Rain at harvest cannot only make combining difficult, but it may also lead to staining of the peas. Only high vigour seed of early varieties should be sown in northern areas. Although peas are deep-rooted, the crop will benefit from irrigation in dry situations, particularly at flowering time (which increases the number of peas) and at pod-swelling stage (which increases the size of the peas).

Soils. Peas grow best on well-drained loams and lighter soils. The pH should be about 6–6.5; if it is too high, manganese deficiency is likely. Lodging can be a problem on organic and very fertile soils and so suitable varieties should be selected.

Rotation. Peas, beans, vetches and oilseed rape ideally should be grouped as the same crop in the context of rotations. They should not be grown more often than one year in five to avoid a build-up of persistent soilborne pests and diseases. However, with a move to the shorter runs of cereals and a limited number of break crops available to the farmer, very often these crops are sown no more than two years apart. Great care should be taken to monitor the crops carefully for a build-up of problems and these should be treated where necessary.

Seedbed. It should be possible to drill the peas about 5 cm deep in a fairly loose tilth. On light easy-working soils it is possible to sow into well-ploughed land with minimum or no previous cultivation. Rolling may be necessary to push stones into the surface and to firm very loose soils, but it should not be done on wet, heavy land. Peas are very sensitive to waterlogging and compaction and every care must be taken to avoid these conditions.

Seed rates. The amount of seed to sow can be calculated from the following formula:

$$\text{Seed rate} = \frac{\text{Target population} \times 1000 \text{ seed weight} \times 100}{\% \text{ germination} \times (100 - \text{expected field loss})}$$

The target population generally accepted is:

Marrowfats 65 plants/m^2.
Large Blues and Whites 70 plants/m^2.
Small Blues 95 plants/m^2.

The figures for number of seeds/kg and percentage germination can be obtained from the seed merchant. The "field loss" figure usually varies between 10 and 25, depending on seed size, time of sowing, soil, seedbed and weather conditions.

The average seed rate is 230 kg/ha, but this can vary from 150 to 300 kg/ha.

Seed dressing. Downy mildew, ascochyta, and damping-off can be controlled by seed dressings (Table 72).

Time of sowing. As soon as possible after mid-February. Only seed of high vigour should be sown early.

In order to achieve quick germination, emergence and ground cover, ideally peas should be sown at a time when soil temperatures are increasing. However, there is an over-riding need to sow as early as possible and so every opportunity must be taken. No yield benefit has been found from autumn sowing of dry harvested peas.

Sowing. When herbicides are to be used the crop is normally sown in narrow rows (15–20 cm) with a conventional cereal drill. Depth of drilling is usually 4–5 cm, but very early sown crops should be drilled closer to the surface to encourage rapid emergence. Very late drilled crops will often be drilled a little deeper into moisture.

Fertilizers. No nitrogen is required and, depending on soil fertility, 0–50 kg/ha P$_2$O$_5$ and

0–150 kg/ha K_2O are recommended, depending on the level of fertility. The fertilizer is normally broadcast in the seedbed. Manganese deficiency is associated with alkaline or organic soils; in severe cases it can cause inter-veinal yellowing of the leaves, but the more usual effect is "Marsh Spot"—brown necrotic areas in the seed. This is especially damaging in peas for human consumption and for seed

purposes. Where symptoms appear, or where trouble is expected, the crop should be sprayed with manganese sulphate at or before flowering and again a week later (Table 72).

Weed control. It is most important that weeds are controlled in the pea crop because they can cause serious yield losses, encourage disease, make harvesting difficult and spoil the product. Peas are a fairly open crop, especially the semi-

TABLE 25   *Growth stages in peas*

Definitions and codes for stages of development of the pea

These definitions and codes refer to the main stem of an individual plant. Definitions and codes for nodel development refer to all cultivars; only nodes where a stipule and leaf stalk develop are recorded; descriptions in brackets refer to conventional leaved cultivars only.

Code	Definition	Description

Germination and emergence

000	Dry seed	
001	Imbibed seed	
002	Radicle apparent	
003	Plumule and radicle apparent	
004	Emergence	

Vegetative stage refers to main stem and recorded node. Two small-scale leaves appear first and the nodes where these occur are not recorded.

101	First node	(leaf fully unfolded, with one pair leaflets, no tendrils present)
102	Second node	(leaf fully unfolded, with one pair leaflets, simple tendril)
103	Third node	(leaf fully unfolded, with one pair leaflets, complex tendril)
:	:	
10x	x node	(leaf fully unfolded, with more than one pair of leaflets, complex tendril found on later nodes)
:	:	
:	Last recorded node	(any number of nodes on the main stem with fully unfolded leaves according to cultivar)

Reproductive stage refers to main stem, and first flowers or pods apparent.

201	Enclosed buds	small flower buds enclosed in terminal shoot
202	Visible buds	flower buds visible outside terminal shoot
203	First open flower	
204	Pod set	a small immature pod
205	Fat pod	
206	Pod swell	pods swollen, but still with small immature seeds
207	Pod fill	green seeds fill the pod cavity
208	Green wrinkled pod	
209	Yellow wrinkled pod	seed "rubbery"
210	Dry seed	pods dry and brown, seed dry and hard

Senescence stage refers to lower, middle and upper pods on whole plant.

301   Desiccant application stage. Lower pods dry and brown, seed dry, middle pods yellow and wrinkled, seed "rubbery", upper pods green and wrinkled

302   Pre-harvest stage. Lower and middle pods dry and brown, seed dry, upper pods yellow and wrinkled, seed "rubbery"

303   Dry harvest stage. All pods dry and brown, seed dry

leafless types, and offer little competition and so herbicides must be very efficient. Inter-row cultivations to control weeds are seldom used now because of the change to narrow rows and chemical control. If possible, couch and other perennial weeds should be killed pre-harvest with glyphosate in the previous cereal crop. Most crops are sown fairly early and weeds may be at a suitable stage for spraying when temperatures are still too low for good post-emergence control. Therefore, about two-thirds of pea crops will receive a pre-emergence residual spray. Residuals are not effective on organic soils and some may cause crop damage on sandy soils. Most herbicides for peas have different application rates for different soil types but, whatever the soil type, results will be poor if seedbeds are very dry or cloddy.

Pests and diseases. Peas can be damaged by many pests and diseases (Tables 71 and 72).

Growth stage. A knowledge of the growth stages in the pea plant will be helpful in deciding on the best time to treat the crop for a specific problem (Table 25).

Harvesting combinable peas. This crop is dependent on good weather at harvest time. Crops grown for human consumption may be rejected, or premiums reduced, if the produce is stained, bleached, rotted or sprouted. Peas sold loose or in packets should be dark green, but a light, even colour is acceptable for processing. If the crop dies off evenly and is weed-free, it can be combined direct (using lifting fingers), when the peas should go through the combine without being damaged (the moisture content should be about 25%). Delay can result in shelling out, splitting and spoilage losses. If the crop is very weedy or ripening unevenly, it will be helpful to desiccate it with diquat when the foliage is turning yellow and the lower pods are fully ripe, to be followed by combining 7–10 days later. Alternatively, glyphosate may be applied 7–14 days before harvest to control perennial weeds.

Desiccation may cause bleaching and reduce the value for some markets, but the peas can still be sold for animal feed. Windrowing is an alternative to desiccation, although it is only recommended for drier areas. The product, from windrowed crops, may be better in good conditions, but severe losses can occur in wet weather. For storage and sale, the peas should be dried to 14% moisture content. This is more satisfactory and safer in floor-ventilated bins. If continuous driers are used, the temperature should not exceed 43°C for human consumption and seed. If the peas are very wet, it may be necessary to dry in two stages with a two-day interval between drying.

Yield. The yield of dry-harvested peas ranges from 3 to 7 tonnes/ha, with an average of 4 tonnes.

PEA AND BEAN SEED PRODUCTION

Peas are self-pollinating and so require only an isolation gap of 2 m. Beans, on the other hand, cross-pollinate and need a 100-m isolation.

Both crops should not be grown any closer than one year in three. Because they are both large-seeded crops it is relatively easy to clean out most weed seeds. Wild oats are the main problem.

Growers can help pollination in the bean crop by using honey bees at one hive per hectare (NIAB Seed Growers' Leaflets Nos. 6 and 7).

Lupins

Lupins have never been an important crop in this country except where it has been used for green manuring on acid, sandy soils and, to a limited extent, for sheep folding. The main interest now is centred on grain production. The natural lupin seed has 30–40% protein (dehusked grain can be compared to soya bean in this respect) and 10–12% edible oil. Modern varieties have been bred for low alkaloid content (low toxicity) and are called "sweet" lupins. The most interesting species are the blue, yellow and white-flowered types. The blue lupin can yield well on deep fertile soils, but it is susceptible to fusarium wilt. It is also very late ripening and so is not suitable for this country. The yellow lupin is grown for seed in north-east Europe, but its best use is for green manuring when reclaiming very acid, sandy soils. The *pearl (tarwi)* lupin has possibilities because of its better disease resistance; it also produces 13–14% good quality oil and 40% protein, but it is late to harvest and toxic, although this can be bred out. The white lupin is the most promising for grain production in this country. A yield of 3 tonnes/ha would appear to be a realistic objective, but it is very dependent on rapid growing conditions during flowering and podding (mid-June to mid-August).

White lupins are only suited to the area south and east of a line from the Humber to the Bristol Channel. Within this area it would be advisable to select fields which face south, are fertile and well drained, with a good soil structure and a pH below 7.5. Very light and very heavy soils should be avoided.

Varieties. Lublanc, Kalina, Vladimir have done well in trials in this country.

Fertilizers. 40–50 kg/ha of phosphate and pot-ash is recommended at indices of 2 and 3. Nitrogen is not needed; it encourages tall stem growth, delays ripening and discourages nodulation.

Time of sowing. White lupins are reasonably frost-hardy and are usually sown about the end of March. However, cold seedbed conditions promote early flowering and reduced podding.

Sowing. It is recommended that the seed is dressed with a rhizobium inoculum for nodule formation, although this may make drilling more difficult. Thiabendazole + thiram is also recommended as a seed dressing for control of fusarium.

A good seedbed is necessary for the seed which should be sown with a corn drill in 15–20 cm rows, 3–4 cm deep, at approximately 90 seeds/m² (185–250 kg/ha depending on seed size). This should give an established 50–75 plants/m².

Weed control. This is very important; lupins are not very competitive. Perennial weeds should be controlled in the previous year. The stale seedbed technique can be used prior to sowing. Terbuthylazine + terbutrun applied pre-emergence can be used for broadleaved weed control and diclofop-methyl post-emergence for grass weed control.

Harvesting. This normally does not take place until the end of September and some varieties will not be ready for another month. Desiccation is helpful if ripening is uneven or weeds are present. Diquat should be used when the pods are brown, the seed coat a grey-green and the endosperm yellow. Combining direct should be possible 7–10 days later, although white lupins have very thick pod walls which are slow to dry.

Drying and storage. This is similar to dry harvested peas, but the grain is usually much wetter. It is stored at 14–15% moisture content.

Yield. An average crop should yield 2.25 tonnes/ha with a grain analysis of 33–53% protein and 9–13% oil.

Dried 'navy' beans

These are the dried beans used for tinned baked beans. About 80,000 tonnes/year are imported and when imports are scarce or expensive there is a renewed interest in growing them in the United Kingdom. They were grown in the 1970s in some favourable areas in the south-east but, on the whole, results have been a failure, if not very disappointing. Very good quality (only obtainable in excellent weather conditions) is important, otherwise they are unsaleable and are of doubtful value for stock feed unless heat-treated.

They are grown in the same way as green beans. Trifluralin and/or bentazone will deal with most annual weeds. If black nightshade is not controlled, the berries contaminate and stain the beans and make them unsaleable.

Varieties. These include *Edmond, Adrian, Lime-light* (a butter bean).

Fertilizers. Navy beans will respond to nitrogen as they do not normally develop nitrogen-fixing nodules. 100–130 kg/ha are recommended.

40/50 kg/ha of phosphate and potash should be given at index of 2–3.

Time of sowing. About mid-May.

Seed rate. 90 kg/ha should give about 40 plants/m² grown in 40-cm rows.

Harvesting. The crop is usually ready for harvesting in September when all the leaves have fallen off the plants, and when the moisture content of the seed is 20–25%; if under 18%, the seeds crack, and if over 30%, they are crushed in the combine. A low drum speed must be used. Desiccation with diquat will remove green rubbish before combining. They should be dried carefully—as for dried peas.

Yield. This is very variable (1–3 tonnes/ha); prices fluctuate wildly.

Soya beans

Soya bean seed contains about 40% protein which is used for food manufacture, industry and feedingstuffs, as well as 17–20% oil which is also used for human consumption and industry. It is the most important vegetable oil in the world.

The varieties now available are not really suitable for this country and the possibilities of its becoming a farm crop are at present only of academic interest. Types originating in Japan, which are insensitive to day length, and grown in a climate similar to the south of England, may prove to be useful.

FURTHER READING

Linseed Law, J. Turner, BASF.
Oilseed Rape, J.T. Wardle and others, Farming Press.

7

ROOT CROPS

POTATOES

The production of potatoes in the United Kingdom is, at present, strictly controlled by a quota (hectarage) system administered by the Potato Marketing Board. A heavy excess levy (and possibly a fine) is imposed if the area grown by a farmer exceeds the quota. Quotas may be purchased, or leased, from other growers. The Board also controls the marketing of the crop annually by setting and supervising quality standards and riddle sizes. Additionally, it organizes a support buying scheme to help maintain reasonable prices in difficult seasons.

Potatoes are tubers in which starch is stored and, depending mainly on the variety, they vary in **shape** (e.g. round, oval, kidney, irregular); **skin colour** (mainly cream and/or red); flesh colour (mainly cream or lemon) and depth of eyes (where sprouts develop).

In this country, potatoes are used mainly for human consumption, but in a glut year some subsidized lots may be used for stock-feeding. In several EC countries, some of the crop is used for producing starch, glue, alcohol, etc. The consumption of potatoes in the United Kingdom, at 112 kg/head, is fairly constant from year to year and so the prices obtained may vary considerably according to the supply. They are eaten as boiled or baked potatoes, chips, crisps, etc. An increasing proportion of the crop (about 30%) is processed as frozen chips, dehydrated instant mash and canned "new" potatoes.

Good quality ware potato tubers:

(i) are not shrivelled;

(ii) are not damaged, frosted or diseased;

(iii) are free of **greening** (an indication that a toxic substance, solanine, is present); **secondary-growth irregularities** such as knobs, cracks and glassiness; **sprout growth** which spoils quality and increases preparation costs;

(iv) are of good shape and size (40–80 mm);

(v) have clean skins with shallow eyes for easy peeling;

(vi) have not been damaged by pests such as wireworms, slugs and cutworms;

(vii) have flesh which does not blacken before or after cooking;

(viii) do not break down when being boiled or fried.

Potatoes must have special qualities to be suitable for processing:

(i) For crisps, chips and dehydration the tubers must have a high dry matter (starch) content and a low sugar content (too much sugar produces dark brown crisps and chips).

(ii) For canning, small (20–40 mm), waxy-fleshed, low dry matter tubers are required which do not break down in the cans.

In the United Kingdom, varieties are classified as first and second earlies and maincrops.

The difference in times of maturity between varieties is mainly because the earlies are "long day" potatoes, i.e. they can produce a crop during the long days of early and mid-summer, whereas the late varieties are neutral or "short day" types and will not produce full crops unless allowed to grow on until the shorter days in early autumn. Early and some second early potatoes can be considered as a "high risk/high profit" crop. Yields are lower than maincrops, but prices are often very high, especially very early in the season. Hitting the market at the right time before the price begins to fall is the secret of profitable early potato production. The main competition at this time of the year comes from imported potatoes from southern Europe and North Africa. With the inclusion of Spain and Portugal in the EC, and the development of their agriculture, it is likely that the returns from early potatoes will be harder to maintain in the future.

Many of the maincrop growers enter into contracts with processors or supermarkets. Maincrop growers have less opportunity to "play the market". Older maincrop varieties such as Russet Burbank are back in favour and grown on contract for the French fries trade.

The following are the more important (or promising) varieties:

First earlies: *Arran Comet* (d), *Maris Bard*, *Home Guard*, *Pentland Javelin* (r), *Rocket* (r).
Second earlies: *Wilja*, *Marfona*, *Estima* (d), *Maris Peer* (preferred variety for canning or high quality pre-packs), *Nadine* (r), *Stroma*.
Maincrop: *Ailsa*, *Cara* (r), also *Red Cara* (r) (both late-maturing), *Costella* (r), *Desiree* (d), *King Edward*, *Kingston* (r), *Maris Piper* (r) (versatile variety suitable for ware and chipping), *Navan* (r), *Pentland Dell* (processing), *Pentland Squire* (d) (ware, crisps and chips), *Record* (crisps), *Sante* (r) and *Stemster* (r).

Note: (d) is a deep-rooting (drought-resistant) variety. (r) is resistant to pathotype Ro 1 (formerly A) cyst nematode (eelworm).

It should be noted that all healthy plants of the same variety are alike because they are reproduced vegetatively.

Seed rates

The decision on the most profitable seed rates for potatoes is complicated by factors such as cost of the seed and expected price for the crop; variety and size of the seed must also be considered carefully. The figures given in Table 26 are for normal-sized seed (30–60 mm), but if healthy small seed (20–30 mm) is used the rate can be reduced by at least 25%. In some countries it is common practice to cut larger seeds into pieces by hand or machine before planting. The cut pieces are more likely to rot than whole seed, but they may be treated with thiabendazole.

Only certified seed can be bought or sold. Growers can plant uncertified home-grown seed, but this is risky if the crop is from poor stock and is not protected by suitable insecticides against aphids, which spread leaf-roll and mosaic virus diseases. Some varieties such as Cara and Pentland Javelin have a high resistance to virus diseases and so are better suited to growing-on as homegrown seed. A

TABLE 26

	Yield of tubers (t/ha)	Seed rate (t/ha)	Time of planting	Time of harvesting
Earlies	10–30	4–5	Feb–Mar	late May/July
Maincrop	35–65	3–4	April	Aug/Oct
Seed	25–40	4–5	April	Sept/Oct
Canning	10–20	5–7	Mar–June	June/Oct

reliable tuber-testing service should be used to check any doubtful stocks of seed before planting.

Most seed crops of potatoes are grown in Northern Ireland, Scotland or at a high altitude in England. Disease and aphid levels are lower in these areas.

Sprouting (chitting) tubers before planting

About half of the potato crops in the United Kingdom are grown from sprouted seed. This is a necessity with earlies to obtain high early market prices. Sprouting should be started in the winter before planting so that single sprouts develop on each tuber (an apical dominance effect). It is desirable for maincrops because, on average, it increases yield by 3–5 tonnes per hectare. It also allows more flexibility of planting time, and reduces the risk of yield losses if blight occurs early. Seed crops also benefit from sprouting. Rogues may be removed before planting; the crop bulks earlier and so the haulms can be destroyed early to check the spread of virus diseases by aphids. To obtain high yields from maincrops and seed crops, several sprouts should be encouraged to grow on each tuber (multi-sprouting).

The **physiological age** of seed tubers can have a marked effect on how most varieties develop. Physiologically old seed is produced from early planted seed crops which are lifted early in warm conditions and, when in store, have heat units added to them by keeping the temperature at levels above 4°C. Such seed produces earlier crops with fewer tubers, but they are more susceptible to drought. Physiologically young seed, produced and stored in cooler conditions, is better for maincrops which grow on to high yields.

The rate of growth of the sprouts is controlled by temperature and is usually quickest about 16°C, but varieties differ considerably. The strength of the sprout is controlled by light. Short, sturdy sprouts are formed in well-lit buildings such as glasshouses or barns fitted with warm-white fluorescent tubes. Storage buildings must be frost-proof and well ventilated to avoid high humidity and condensation. Sprouting outside under clear polythene sheeting can be carried out in sheltered places.

Chitting trays, each holding about 15 kg (60–80 per tonne), are still the most popular containers for sprouting, especially for earlies. Higher-capacity white plastic trays, or wire crates and pallet boxes holding at least 500 kg, can speed up the loading of high output maincrop planters.

Well-developed sprouts (2–3 cm) give maximum yield advantage. However, they are easily damaged by mechanical planting and so mini-chitted seed, with sprouts less than 5 mm, is used with automatic machines on at least one-third of sprouted maincrops. Mini-chitting is achieved by temperature manipulation.

If well-sprouted tubers are taken from a warm store and planted into cold ground, some varieties are likely to emerge very late due to "coiled sprout" development, or "little potato" may develop when no shoots appear and the old tuber is converted into a few new tubers.

Soils and climate. Earlies do best on light, well-drained soils in areas free of late frosts, but some frost protection is possible with irrigation. Covering with polythene sheeting can hasten crop establishment. This is removed when the crops grow bigger.

Maincrops are best suited to deep, fertile, loam soils because high yields are very important, especially if prices are low. Potatoes will grow in acid soils (over pH 5.5); common scab can cause problems on high pH soils in a dry season. Irrigation, if it is possible, should be considered.

Place in rotation. Potatoes can be taken at various stages, but usually after cereals. To control cyst nematodes, maincrops should not be grown more often than two years in eight, seed crops one year in five to seven years, but earlies may be grown continuously if lifted before the nematodes develop viable cysts (i.e. before 21 June).

Seedbed preparation. Except on light soils, early ploughing is necessary to allow for frost action. In spring, deep cultivations, harrowing, discing and/or power-driven rotary cultivators are used, as required, to produce a fine deep tilth without losing too much moisture. The land is then ridged up using only the weathered soil available. Bringing up unweathered soil will result in smearing of the ridge sides.

Some soils are very stony, which are likely to cause harvesting problems. These can be collected mechanically and removed, or mechanically separated into windrows between the ridges. If sufficient tilth has not been formed over winter by frost action, it may be necessary to form mini-ridges to start with and, after planting, build these up to cover the potatoes as the soil dries out. Building up the ridges in several stages also allows the soil to warm up faster and, in addition, this method can hasten the growth of earlies.

Planting. Nearly all potato crops in the United Kingdom are planted mechanically with 2–7-row machines which are either fully or semi-automatic. Hand planting is still practised on some farms, mainly in Scotland.

The ridges must be well formed to protect the new tubers from blight spores and light which would cause greening. Some farmers on light, sandy soils are now planting potatoes in beds. There are usually three rows in beds 150–180 cm wide, i.e. the width of two ridges. This is preferable for irrigation (there is less run-off), and it produces more tubers in the 40–60 mm size and less greening. Modern machinery can deal with the extra throughput of soil (it may,

in fact, reduce damage), but it is only feasible in fairly dry conditions on light soils.

Spacing of tubers (sets). Row width for earlies and seed potatoes is 60–70 cm; for maincrop, the commonest width is 90 cm. There is less yield loss and fewer problems in producing good ridges, keeping them free of clods, and fewer "greened" and blighted tubers with the wider rows. The spacing between the sets will depend on the seed rate and the size of the seed. Table 27 gives an indication of optimum seed rate and set numbers per hectare. This assumes that the seed size is 35–55 mm, the rows are 90 cm apart and the seed price about twice that likely to be obtained for the ware.

On highly fertile soils, especially if irrigated, and where higher yields (over 70 tonnes/ha) are expected, some of the tubers may be too large (over 80 mm). In these conditions, the seed rate should be increased by 10% to give more tubers of desirable size. Some varieties such as Cara produce fewer tubers than average and so they are often too large, hence the higher seed rates. Other varieties such as King Edward and Maris Piper produce a large number of tubers and they may not grow big enough (under 40 mm) and so a lower seed rate is acceptable.

Manures and fertilizers. Potato crops benefit from organic manures such as farmyard manure and slurry. Manure can be liberally applied when available at, say, 30–50 tonnes/ha of farmyard manure or slurry equivalent. Normally this is ploughed-in during the autumn.

Fertilizers are either broadcast during the spring cultivations or, more efficiently, placed in the ridges by the planter.

TABLE 27 *Optimum populations and seed rates*

Variety	Sets/50 kg	Tuber spacing 90-cm row	Sets/ha	Tonnes/ha
Wilja	630	31 cm	36,000	2.86
Desiree	650	25 cm	45,000	3.46
Pentland Dell	600	32 cm	35,000	2.96
Record	700	30 cm	35,500	2.70
King Edward	800	31 cm	36,000	2.25
Maris Piper	750	33 cm	34,000	2.26
Cara	750	17 cm	65,000	4.30

TABLE 28

	N kg/ha (assumed index 0)	P₂O₅ kg/ha	K₂O kg/ha
		(assumed index 2)	
Earlies	180 (1)	250	120
Second earlies & maincrop	220	250	250
Second earlies & maincrop (organic soils)	130	250	250
Seed (burnt off early)	180	250	250

Notes: (i) Reduce by 30 kg/ha if lifting very early.
  (ii) Recommendations for phosphate and potash are for incorporation into seedbed before planting. If fertilizer is placed adjacent to the seed, amounts reduced by 25%.
  (iii) 25 tonnes/ha FYM could supply 35 kg N, 50 kg P_2O_5 and 100 K_2O.

Too much nitrogen applied to the seedbed usually delays tuber development which can reduce yields in some seasons. The tuber size is increased by nitrogen and potash, whilst phosphate increases the number of tubers (Table 28).

Weed control. Weeds, by tradition, have been controlled in good weather by cultivations, including ridging after planting. However, the frequent passage of rubber-tyred tractors tends to produce clods in the ridges and this hinders mechanical harvesting. Cultivations also damage the potato roots and stolons, and increase moisture loss from the soil. Consequently, chemical weed control, mainly by contact and soil-acting residual herbicides, has become normal practice on the majority of farms (Table 69).

Blight. This is the worst fungal disease which attacks the potato crop. It can seriously reduce yield by killing the foliage early; during periods of heavy rain the spores of the fungus can be washed into the soil and on to the tubers, so causing them to rot in the ground or during storage. The disease spreads rapidly in warm, moist conditions which, when they occur, are notified to growers by the Potato Marketing Board, MAFF, etc., as "Blight Period Warnings". Growers can then respond by an application of fungicides; a wide range is available. Most fungicides have a contact action and are effective if sprayed regularly with the leaves being well covered, e.g. the dithiocarbamates. Other fungicides have a systemic action which may be preferable in bad weather conditions, e.g. cymoxanil. Fentin compounds are suitable for killing spores which fall on the soil before they reach the tubers, but may scorch in hot weather. To reduce the risk of spores spreading from the leaves to the tubers, the haulms should be destroyed when about 70% have been killed by blight; this is especially important if heavy rain is expected (Table 72).

Aphids. Increased physical damage to the crop and the spread of virus by aphids in some hot summers have necessitated the use of soil-acting granular insecticides such as phorate or disulfoton, or sprays on more than half the potato crops in the country (Table 71).

Irrigation. In dry seasons, very profitable returns can be obtained from irrigating the potato crop. Quality may be improved by the tubers being more uniform in size and having less common scab, but the dry matter per cent may be lowered if an excessive amount of water is used. If water is applied to potatoes which have almost died due to drought, secondary growth may develop as knobs and cracks; this will obviously spoil the crop.

Earlies may be protected from frost damage by keeping the soil moist on those warm sunny days which are followed by radiation frosts at night. In more severe cases, they can be protected by spraying with about 2–3 mm of water per hour during the frosty period; the latent heat given out, as icicles form on the crop, prevents damage to the leaves. Irrigation may also be used to assist in breaking down clods when preparing spring seedbeds in a dry time.

Harvesting. Earlies are harvested when the crop is still growing. To make lifting (and later cultivations) easier, the tops should be destroyed with a flail-type machine. Earlies must be treated gently because the skins are very soft. Maincrops, especially if they are to be stored, should not be harvested until the tuber skins have hardened. This is usually about three weeks after the tops have died, or have been desiccated.

Early crops are often picked by hand, but most maincrops are lifted mechanically. One-row and two-row harvesters of various designs and capabilities are available to lift and load the crop into trailers. Very high rates of working are possible with some machines in stone-free and clod-free conditions, when it is reasonably dry.

Damage to tubers by harvesting operations can result in serious losses due to bad design and/or faulty operation of the machines. Damage can be extremely serious in very dry soil conditions when the tubers are somewhat dehydrated and more susceptible to damage, and the clods are very hard.

Storage. A high proportion of the maincrop has to be stored. Storage may only be for a short period for some of the crop, but in other cases the potatoes may have to be stored until May/June when the earlies come on to the market again. Obviously, badly damaged, diseased and rain-wet tubers will not store satisfactorily and should be sold at harvest time.

Most crops are now stored indoors, mainly in 150–500 tonne bulk stores. Care is required when filling these stores to ensure an even distribution over the floor and to avoid "soil-cones" which can prevent air circulation. If the heap is more than 2 m deep, some form of duct ventilation will be required to prevent over-heating. In the more sophisticated buildings, very accurate temperature control is possible by using refrigeration and recirculation ventilation. In most ordinary stores it is usual to put a deep (30–60 cm) layer of straw or nylon quilt on top of the heap to prevent greening and frost damage, and to collect condensation moisture, so keeping the tubers dry. Good ventilation, and allowing the heap to warm up to 15°C for 7–10 days after filling, helps to dry the tubers and to heal wounds. Some farmers apply "Storite" to prevent rotting in store.

Potatoes can also be stored in pallet boxes. These are expensive, but ventilation is usually very satisfactory and the tubers retain a good appearance.

For late storage (by whatever method), sprout growth has to be prevented in the spring. This can be done by keeping the heap cool (4–7°C) by cold air ventilation. This method is not satisfactory for crisping and chipping potatoes, because some starch changes to sugars (some reversal of this process is possible by warming the tubers before sale). Alternatively, sprout suppressants such as tecnazene can be applied when the crop is going into store; it is effective for up to five months when ventilation is restricted.

Grading. The stored tubers are riddled (graded or sorted) during the winter or early spring when the best potatoes (ware) are separated from chats (small tubers), diseased, damaged, over-sized and mis-shapen tubers. They are usually sold off in 25-kg paper bags.

Yield. See Table 26.

POTATO SEED PRODUCTION

Seed potato production has, traditionally, been mainly carried on in Scotland, Northern Ireland, and the hill areas of England and Wales. In these areas, the low temperatures and strong winds keep the aphid populations in check. This means that the severe virus diseases (leaf-roll and mosaics), which are spread from diseased to healthy plants by aphids, are less likely to

TABLE 29   *Tolerances: percentage of plants affected (final field inspection)*

Disease and defects	Basic seed				Certified seed CC
	VTSC	SE	E	AA	
Rogues and variations	0	0.05	0.05	0.1	0.2 } 0.5 0.3
Leaf roll	0	0.01 } 0.1		0.25	2.0
Severe mosaic	0	0.00			
Mild mosaic	0	0.05	0.5	2.0	10.0
Blackleg	0	0.25	0.5	1.0	4.0

Notes: VTSC = virus tested stem cutting; SE = super elite (generation 1–3); E = elite.
0.01% = 5 plants/ha; 1.00% = 500 plants/ha.

occur. All seed crops must be certified during the growing season to ensure that they are true to type and variety, and are as free as possible from virus and other diseases. The certification grades of seed potatoes are shown in Table 29.

Only Basic seed can be used for further seed production, although AA grade seed is sometimes used for ware production. Certified seed (CC) grade is healthy commercial seed, also for ware production. Farmers are, however, allowed to grow their own seed from CC grade. This is known as "once grown" seed, and great care should be taken to ensure that it is free from viral and fungal diseases. Crops grown from Basic seed should contain less than 4% virus infection, and Certified seed less than 10% severe virus. Growers should not expect virus infection in the seed they purchase to be limited to the tolerances set for crop inspection. This is because of possible unseen infection spread during the season of inspection.

Full details of Certification are obtainable from MAFF or other Departments of Agriculture. Fields used for seed production must also be certified free of cyst nematodes. In addition to the fee payable for Certification, Basic seed growers may have to pay a hectarage royalty on recently introduced varieties. Strict grading standards must be observed when selling seed potatoes and, ideally, the seed containers should be inspected and sealed on the farm before

dispatch to other growers. The seed may be treated with thiabendazole to help control tuber rots. Additionally, some growers have tuber samples tested for presence of viruses.

The top-grade seed, VTSC, is produced by rooting stem cuttings of virus-free plants in sterilized compost in isolated glasshouses. The resultant plants and their progeny are then grown on in isolation from other stocks. This method ensures freedom from viruses and tuber-borne diseases such as gangrene, skin spot and blackleg.

The use of stem cuttings produces about 30 times as many tubers as would be obtained by planting tubers in the normal way. Micropropagation techniques are now used for producing disease-free seed stocks very much more quickly than the VTSC method, although it is more expensive.

Ideally, seed potato crops should be planted with sprouted tubers and at a high seed rate to produce a good yield of seed-size tubers (35–55 mm). The haulm should be destroyed, chemically or otherwise, when most of the tubers are seed size. This reduces not only the risk of spread of virus diseases by aphids but also the spread of blight to tubers. It is an undesirable practice to let the crop grow to maturity so that additional income can be obtained from the large, ware-size tubers.

The crop should have been kept clean through the growing season, and this attention to detail must continue through to storage. Wounds should be allowed to heal and wet loads kept out of the store.

Seed potatoes should not be grown in the same field more often than one year in five years (one in seven is preferable), because of the possible carry over of groundkeepers from the previous crop. Seed crops may not be produced where wart disease has occurred.

SUGAR BEET

The sugar which is extracted from the crop supplies the United Kingdom with approximately 60% of its total sugar requirements.

About 200,000 hectares are grown on contract by 10,000 growers for British Sugar which has 10 factories in England, most of which are situated on the eastern side of the country. A contract price per tonne of washed beet containing a standard percentage of sugar (usually 16%) is determined annually by the EC. Sugar percentage, transport allowance, early and late delivery allowances and other levies all determine the price which the grower will receive.

There is a world surplus of sugar and to restrict output a quota system operates for the production of sugar in the EC and the price the grower receives. The annual beet contract for UK growers is based on quota A—the top price for, at present, approximately one million tonnes of sugar, and quota B—25% less received per tonne for, normally, 1000 tonnes of sugar. Both quotas are adjusted to a 16% sugar level. Any sugar surplus to the A and B quotas is C quota and is sold (usually at a much lower price) on the world market after carrying forward about 70,000 tonnes to the next year's production. This is an insurance if, for some reason, the crop is a poor one. Nationally the quota should be met; if not, it could mean a lower quota in subsequent years.

Each grower is allocated a quota level based on a proportional figure for A and B quotas. It is important that the contracted tonnage is met each year:

(i) To maximize the gross margin for the crop.
(ii) Failure to maintain it could mean a reduction of quota in subsequent years for the grower. The advice is to endeavour to exceed the quota by up to 10% to compensate for yield fluctuations outside the grower's control and which may occur in a year.

The average sugar content is about 17%.

Apart from its sugar, sugar beet has two useful by-products:

(1) Beet tops—a very succulent food, but which must be fed wilted; they are not always easy to utilize.
(2) Beet pulp—the residue of the roots after the sugar has been extracted; an excellent feed for stock.

Soils and climate. Sugar beet can be grown on most soils, **except** heavy clays which are usually too wet and sticky; thin chalk soils which do not retain sufficient moisture and on which harvesting is difficult, and stony soils on which cultivations and harvesting can be very difficult.

Sugar beet is a sun-loving crop which will not grow well when there is too much rain and cloud. Sugar is produced by the conversion of the energy of sunlight into the energy of the plant—sucrose in sugar beet. Consequently sunlight, and the amount of crop leaf area which can absorb it, is an important factor in determining the yield of sugar produced by the crop. The grower should see that as much sunlight as possible is intercepted and utilized by a well-developed crop (irrigation may help) in the long days of May and June. This underlines the importance of early drilling and establishment of the crop as well as a uniform distribution of the plants.

Place in rotation. In order to prevent the rapid build-up of rhizomania (Table 72), as a condition of the contract sugar beet may not be sown in a field which has grown any Beta species, e.g. fodder beet, mangels, red beet, in either of the two preceding years. This restriction also helps to prevent the build-up of the beet cyst nematode (Table 71) and weed beet (page 134), although most brassica crops as host crops can perpetuate the nematode. A three or four year interval is, in fact, preferable between sugar beet and any other closely related crop as well as the brassicae.

Ideally, sugar beet will follow a cereal. This allows an opportunity for stubble cleaning and still time for careful October/November ploughing to allow for subsequent essential weathering.

Seedbed. The importance of a good seedbed for sugar beet cannot be over-emphasized. The success or failure of the crop can, to a great extent, depend on the seedbed. It must be deep yet firm, fine and level (page 41).

In most seasons, except on heavy soils, the less work done on the seedbed in the spring the better the conservation of moisture. The one-pass cultivation/drilling technique with front-mounted equipment and rear-mounted drill has much to commend. Good early winter plough-ing followed by a correctly timed winter culti-vation (where possible) is, as always, a pre-requisite on the majority of soils. The exceptions are the light peats and sandy soils; they can be ploughed in the spring as weathering is unnecessary.

A well-worked understructure by deep loos-ening equipment with no soil inversion is, in some situations, preferable to the plough. The latter should be avoided when blowing could be a problem. Non-ploughing techniques may also help to prevent weed beet germinating in the beet crop. The weed can more easily be dealt with in other crops in the rotation.

Varieties. The NIAB Leaflet No. 5—Sugar Beet, and the factory field officer will help the grower to decide which varieties to use. There is a selection of varieties and they can be grouped according to the yield of roots, the sugar percentage, and their resistance to bolting, i.e. running to seed in the year of sowing. This is influenced by a cold spell in the three-to-four-leaf stage of the plant. Apart from perpetuating the weed beet, bolting is highly

undesirable as the roots become woody with a low sugar content. If there are more than 10% bolters in the crop, there is a risk that some of the bolted beet may finish up as part of the sample taken for determining the sugar percentage at the factory gate. Varieties with a high resistance to bolting should be used for early sowing and in colder districts.

Size of top, height of crown (which could affect harvesting losses), establishment in field trials and resistance to both downy and pow-dery mildew are other factors which have to be considered in deciding on the variety or varieties to grow.

All varieties are purchased from British Sugar, not from the seed merchant.

Manuring. The following recommendations can be made:

(1) Farmyard manure—25 tonnes of well-rotted manure per hectare should, ideally, be applied in the autumn. Only about 30% of the sugar beet grown in the country actually receives FYM.

(2) Lime—a soil pH of between 6.5 and 7.0 is necessary. Alkaline soils can suppress some plant foods and, on peat soils, which may have a low content of phosphorus, magnesium and manganese, a pH 6.0 to 6.5 is acceptable.

(3) Salt—500 kg/ha should be broadcast a few weeks before sowing (except on poorly-structured soils). This can give approxi-mately 500 kg extra sugar per hectare.

(4) Magnesium—magnesium deficiency has be-come more evident in recent years, parti-cularly on light, sandy soils. 500 kg/ha Keiserite (magnesium sulphate) should be applied if necessary. Magnesium limestone can also be used if available. This will help to maintain magnesium levels, as well as remedying any lime deficiency.

(5) Boron—may also be necessary, particularly (but not always) on well-limed sandy soils where it could be in short supply (Table 72). A boronated fertilizer is often recommended in this situation.

(6) Manganese—a deficiency of this trace element is found where the pH on peaty soils is above 6 and on sandy soils above 6.5 and also on old ploughed-up grassland. Manganous oxide incorporated with the pelleted seed material is a useful insurance against manganese deficiency, but it may also be necessary to spray manganese sulphate as soon as deficiency symptoms show on the leaves (Table 72). A severe deficiency can reduce sugar yield by up to 30%.

Other plant foods required can be summarized as in Table 30, assuming phasphate and potash indices at 2 to 3.

TABLE 30

	N	P_2O_5	K_2O
	kg/ha	kg/ha	kg/ha
With FYM and salt	100	50	90
Without FYM but including salt	125	60	125

It is important to use less fertilizer with farmyard manure, otherwise sugar content (and thus profitability) will be reduced.

Kainit can be used at 500–600 kg/ha instead of potash, salt and magnesium (except in severe cases of magnesium deficiency where Keiserite is still necessary). It should be applied some weeks before sowing the seed.

The phosphate and potash can be broadcast and worked into the seedbed at any time over the preceding winter months.

Except after a very wet autumn and winter which may leave nitrogen reserves on the low side, nitrogen in excess of 125 kg/ha is unnecessary. It may, in fact, depress the sugar percentage without increasing the overall sugar yield. Nitrogen (not urea) must be applied in the spring. Band application between the rows at drilling is ideal, but otherwise half the application should be broadcast straight after drilling, with the balance after the beet has emerged.

Seed and sowing. Genetical monogerm seed has now replaced the hitherto used "natural seed"—a cluster of seeds fused together which usually produced more than one plant when it germinated. Monogerm seed contains a single embryo from which only one plant will grow.

All monogerm seed is pelleted, i.e. it is coated with clay to produce pellets of a uniform size and density—3.50–4.75 mm. Unpelleted monogerm seed is rather lens-shaped and, as such, cannot easily be handled in the precision drill.

Seed rate. This depends on seed spacing and row width. The recommendation is that pelleted seed should be sown at 0.8–2.9 units/ha (approximately 5–18 kg/ha); each unit has 100,000 seeds and there are about 15,500 seeds per kilogram.

Time of sowing. Mid-March to mid-April.

Sugar beet seed will not germinate at soil temperatures below 3°C and only very slowly between 3° and 5°C. In most years 5°C is not reached until about 20 March and sowings before that date will normally result in poor establishment and many bolters. It is best completed by early April, after which the yield penalty gradually increases until, in May, about 0.6% is lost for each day's delay in sowing (Fig. 42).

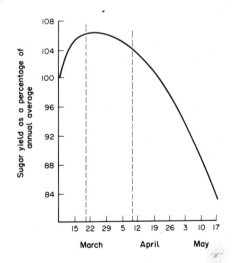

FIG. 42. Effect of drilling date on sugar yield; the average of 12 years of trials at Broom's Barn, West Suffolk. Maximum yield, with minimal bolting, is most unlikely to be achieved by drilling between 20 March and 10 April.

Drilling-to-a-stand. 90% of the sugar beet crop is now drilled-to-a-stand, i.e. the individual seeds are placed separately in the position required for the plant, using the precision drill. A very good seedbed and efficient pest and weed control (with herbicides) are essential for the system to be successful.

The number of plants which make up a crop of sugar beet is one of the most important factors in determining not only the yield, but also the sugar content of the roots.

The main aim of drilling-to-a-stand is to have a reasonably spaced, yet adequate, plant population. Ideally, this should average 75–80,000 plants/ha; when it is less than 65,000 there can be a serious reduction in final yield.

The seed spacing should be decided on the basis of the germination percentage of the seed (obtainable from the factory which supplies the seed), the possible losses before and after plant emergence (spray damage, inter-row cultivation, wireworm and other pests) and the row width. As far as is consistent with any inter-row work and the harvesting of the crop, the row width should be as narrow as possible, but at present it is unlikely to be less than 46 cm. If, as is more usually the case, a 50-cm row width is being used and the seed is spaced, for example, at 15 cm apart, with a plant establishment of 70%, this will give a plant population of 87,000/ha. With the probable losses mentioned, this will bring the population down to a very satisfactory figure. The seed spacing will, in fact, normally vary, depending on conditions, from 12 to 17 cm.

Apart from the saving in labour costs, another important advantage of drilling-to-a-stand is that, compared with what was known as conventional growing, the need (with larger areas) to space out the actual drilling dates (to make singling easier) no longer arises. Thus, the seed can be sown under the best possible conditions.

Evidence shows that, provided it is carried out successfully, and this means one mature plant for every two seeds sown, there is no difference in final yield comparing drilling-to-a-stand with other methods of growing, assuming that there is a high level of crop protection.

Shallow-sowing will aid germination; 18–25 mm is ideal with a well-prepared seedbed. A press wheel fitted immediately behind the seed coulter effectively compresses the soil over the seed and in dry weather this will help germination.

The bed system. The development of the bed system for growing beet has logically followed the now common practice of overall spraying of the crop for weed control and the use of multi-row harvesters. One of the most important advantages of the bed system is that all tractor wheelings are avoided where the beet is actually grown. The wheelings are concentrated on, at the most, 15% of the land surface (i.e. the area between the beds—suggested 60 cm) and, if necessary, any compaction there can be dealt with by soil-loosening tines following the tractor wheels.

No compaction should lead to better establishment and a consequent quicker crop cover and earlier interception of essential sunlight. Where applicable, straw can be placed between the beds to reduce wind erosion.

To compensate for the larger beet growing in the rows adjacent to the 60-cm gap (less competition), the plants should be closer-spaced than those in the inner rows. In this way a more uniform-sized crop should be obtained.

Harvesting of the beds (at least four rows at a time) could be a problem in some seasons, and heavier soils in the wetter West Midlands, for example, could be a limiting factor for the success of the system.

With the right conditions, beet grown in beds will almost always do better than conventionally-grown crops, both in yield and sugar percentage.

The prevention of wind damage. A recent survey by British Sugar showed that 20% (about 40,000 hectares) of the national sugar beet crop is at risk from wind damage. However, only about 16% is actually protected. Unless preventive measures are taken, it is not unusual for crops in vulnerable situations to be completely destroyed or damaged beyond an economic recovery. This necessitates a late re-drilling with a consequent serious loss of yield, apart from the extra cost involved.

It is on the lighter and/or organic soils that the beet is at greatest risk from the wind, from sowing until the 6-leaf stage.

Straw cover, as a wind barrier, can be provided to reduce the wind speed at ground level by creating a straw planted "hedge" about 110–120 cm apart between or across the beet rows, using a straw planting machine. Another type of barrier is a spring cereal (normally barley) drilled at about 60 kg/ha some three weeks **before** the beet is sown (the latter being drilled between the cereal rows) or, using an ordinary corn drill, the cereal can be sown at right angles to the intended direction of the beet rows. Before there is any competition, and when the beet has established itself sufficiently to withstand any damage, the cereal is sprayed out using, for example, alloxydim-sodium.

Instead of a straw cover the beet can be drilled direct into ploughed and furrow-pressed land. Compaction is avoided and the press helps to stabilize the surface centimetres of the soil. Direct drilling into the undisturbed soil, usually a stubble, using precision drills modified for the purpose, is sometimes used as a technique to prevent blowing.

Chemicals such as polyvinyl acetate, which bind the surface particles together, are commercially available. They are, however, expensive (more than £100/ha) and are not widely used. A barley cover crop is much cheaper.

Treatment during the growing period. Band spraying with herbicides at the time of drilling is no longer the usual method of controlling weeds **within** the row. When it was introduced it was cheaper than conventional overall spraying and also it had the advantage of leaving weeds for a time **between** the rows (to be removed later). These provided alternative food for pests and on light soils the weeds, acting as a barrier, helped to reduce wind damage.

However, band spraying is a slow operation and it has to be followed up by inter-row culti-

vations. Consequently, most growers now use a low-volume low-dose technique for overall application of the herbicide. Normally, a sequence of sprays is necessary, but application should always be carried out when the weeds are in the cotyledon to first true-leaf stage (page 237).

Because of the cost of chemicals, the sequential treatment is becoming more expensive, and there is a move back (albeit at present not very widespread) to band spraying.

Where overall application has not been carried out, steerage hoeing (in the early stages of plant establishment) will be necessary, perhaps twice, to assist in weed control. Further hoeing, after the beet has reached the 5–7-leaf stage, should be undertaken, but only as much as is necessary. Weeds have to be kept in check, but any inter-row cultivation does mean a loss of moisture. It is another point in favour of overall herbicide application.

Singling (*thinning*). Where there are problems of poor establishment, particularly on late, heavy soils, some growers (with sufficient labour) use a relatively high seed rate and precision drill at a 7.5–10 cm spacing for subsequent thinning at the 4–7 leaf stage. This last is done by hand, usually at piece-work rates. The final average distance apart of the plants in the row must depend on row widths, e.g. with rows of 50 cm, the plants are thinned to 22 cm apart, or with a row width of 55 cm, the plants are thinned to 20 cm apart. This, in theory, will give an ideal plant population of 75–80,000/ha.

Virus yellows. It may be necessary to spray the crop to kill the aphids which spread virus yellows (Table 72), or to apply the insecticide as granules spread on the soil.

Powdery mildew. A fungicide should be used if the disease becomes prevalent in late July/August (Table 72).

Irrigation. Although it is a deep-rooted crop, the beet plant will respond to supplementary water in the majority of seasons. This will obviously depend on summer rainfall as well as soil type.

If, by irrigation, the plant can develop a reasonable canopy of leaves early on in its life, it will be able to make more efficient use of sunlight for growth and sugar production. But the crop will also need water during June and July. In hot, dry and bright weather it may require up to 350 mm during these two months. Many soils (notably the sands), particularly in eastern England, are unable to supply this need. With irrigation on the farm and, as part of an overall irrigation plan, a water balance sheet will have to be kept to determine the soil moisture deficit and when it will be necessary to irrigate the sugar beet crop (page 38).

However, irrigation should not be necessary after August, even if there is a water shortage; by then the deep root should easily be able to cope.

Weed beet

Weed beet is a big problem for sugar beet growers. About 50,000 hectares of the total sugar beet area are affected. Weed beet is any unwanted beet within and between rows of sown beet and, of course, other crops. Unlike true beet, which is a biennial, it produces seed in one year. It originates from several sources, but mainly from naturally-occurring bolters in the commercial root crop, from contamination of seed (by cross pollination from rogue plants), from beet ground keepers either growing in crops following beet, or from old clamp and loading sites, and from seed shed from weed beet itself.

Once weed beet becomes established it is self-perpetuating (although it can remain dormant for many years) and on average it produces 2000 seeds in the year. However, only about 50% survive, but nevertheless a light infestation (suggested 1000/ha) can mean one million weed beet/ha the following year.

Control measures. It is very important to prevent any weed beet seed returning to the soil and so, to start with, inter-row work should be carried out to clear any seedlings which are between the rows (at present there are no selective herbicides to control weed beet in sugar beet). Following this, as far as possible,

all bolters should be removed from the beet field, no later than the middle of July, i.e. before the seed has set. Rogueing, i.e. by hand pulling, is the most economical method if there are less than 1000 weed beet/ha. If rogueing is not possible, the bolters can be cut down mechanically. This is done usually twice, the first time 7–10 cm above crop leaf height and then, 14–20 days later (about the end of July), to prevent lateral shoots and late flowering plants from producing seed, cutting is repeated—this time at crop leaf height.

Instead of mechanical cutting, the "weed wiper" fitted to the front or rear of the tractor can be used. This consists of a boom containing a herbicide (at present glyphosate is recommended) which permeates through a nylon rope "wick" attached to the boom. As it goes through the crop the wick wipes the taller weed beet and bolters (taller, that is, than the beet plants) and so deposits the herbicide. Usually more than two treatments are necessary, starting at the beginning of July and then repeating it at 10–14 day intervals until the middle of August.

In addition to preventing the propagation of seed from bolters or weed beet which have already contaminated the crop, it is important that weed beet in crops **other** than sugar beet should also be controlled. Weed beet found growing on old clamp and loading sites, headlands and waste land must also be destroyed.

The effect of cultivations on weed beet is being examined to try and find out the ideal sequence of cultivations necessary to reduce the weed beet population more rapidly. Direct drilling, tine cultivation and ploughing are all being studied.

Finally, with a bad infestation in a field, it will be necessary for the grower to widen his rotation to six to seven years, rather than the more usual three to four year interval.

Harvesting. In theory, early to mid-November is the right time to harvest the beet crop. In most years it is still slightly increasing in weight, with the sugar percentage reaching a maximum at that time before beginning to fall off.

However, apart from the fact that in most

years conditions for harvesting deteriorate as autumn proceeds, if all the beet were delivered at this time there would be tremendous congestion at the factory. Therefore a permit delivery system is used, whereby each grower is given dated loading permits which operate from the end of September until about the end of January. It means that growers must be prepared to have their beet delivered to the factory at intervals throughout the period. It is important that the permits are used and are not lost through failure to have sufficient beet out of the ground for delivery to the factory. In the early part of the harvesting period (up to about mid-October), when the beet is still growing in the field, the aim should be to keep just ahead of the delivery permit (in case bad weather stops lifting). As far as possible, the poorest part of the crop should be harvested first. It will never be able to respond to open weather at the end of the year in the way that a full crop can. More beet should be clamped in November and, depending on soil conditions, all lifting should be completed by mid-December.

To minimize dirt tares, direct delivery from field to factory should be avoided, especially when the beet is harvested in wet conditions. If the beet can be clamped at least a week before delivery, and then reloaded using an elevator with a cleaning mechanism, dirt tare will be considerably reduced; it should not exceed 10%. Very dirty loads may not be accepted by the factory.

Topping. As far as possible, the topper should be set to remove the tops to the level of the lowest leaves or buds on the plant (Fig. 43). A top tare of between 2% and 5% is acceptable; below 2% probably means that the crop has been overtopped with an excessive loss of beet. Adjustment to the topping mechanism will normally be necessary throughout the period of harvesting. It should not be neglected.

There are two main types of topping mechanisms:

(1) The feeler-wheel topper. This unit consists of a feeler wheel which moves up and down

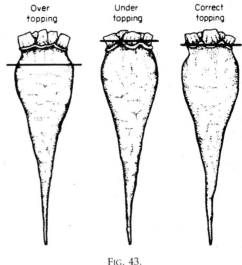

FIG. 43.

(floats) according to the height of the beet. This controls the position of the knife in relation to the crown of the beet.
(2) Flail topper. This consists of metal flails, which remove most of the leaf and stem, followed by scalper blades which, with the aid of guide bars, trim off the remaining leaves.

Following topping, the leaves are swept off the row prior to its being lifted.

"Top savers" which windrow the tops (away from the line of the harvester), leaving them in a clean condition for subsequent grazing or ensiling, are not widely used. Most tops are ploughed-in.

Harvesters. There are a number of different types of harvesters and harvesting systems. These range from the single-row trailed tanker-harvester to the multi-row (2–8 rows) self-propelled tanker, as well as the single and/or multi-row side-loading trailed machine. In addition, there is the two-stage harvesting system.

Depending on the distance to the clamp, the tanker harvester can be operated as a one or two man system. This can fit in well with the steady supply of beet needed for the permit delivery arrangements without seriously interrupting other work on the farm.

The two-stage system can harvest from two to eight rows at a time. A two-, rather than a three-stage system is now more widely used. The first stage tops and windrows the leaves with a separate topper unit. This is followed by lifting, cleaning and side-loading the beet as the second stage. To maintain the sugar percentage, the beet should be lifted as soon as possible after topping. This system is tending to be replaced by the multi-row self-propelled tanker harvester, although a two-stage system on the one tractor is quite commonly seen. This has the topper and lifting unit at the front of the tractor which trails a side-loader or tanker.

A recent survey by British Sugar indicates an average 8% loss of yield by machine harvesting, i.e. about 3.5 tonnes/ha per 45-tonne yield of crop per hectare. Over-topping and breakage of roots below the ground and leaving whole roots on the ground are the main reasons.

Storage. Beet stored on the farm should be done carefully, with the clamp adjacent to a hard road and accessible at both ends so that beet which has been clamped longest can be delivered first. Plenty of air circulation should be allowed. Ideally, use a ridge-type clamp for beet which is clamped early whilst the temperature is still high. However, as it gets colder this is less important and large square clamps, or heaps against a wall, can be built.

It is important to prevent the beet being damaged by frost, either in the ground or in store, as this has a very quick and deleterious effect on the sugar percentage. Badly frosted beet will be rejected by the factory as it is useless for sugar extraction. Depending on the district and incidence of frost, the clamp may have to be covered. Plastic sheeting, rather than straw, is now being used. It has the advantage that it can be removed quickly (with a rise in ambient temperature) and easily replaced when necessary.

Yield

Roots (washed) 40 tonnes/he.
Sugar at 17%   7.6 tonnes/ha.

Tops 25–35 tonnes/ha depending on the variety. A stock farmer may prefer a large topped variety, although fewer are grown now.

SUGAR BEET SEED PRODUCTION

All seed crops grown in the United Kingdom must be entered for seed production under the UK Seed Certification Scheme for Sugar Beet (implemented by the NIAB Beet Registration Scheme). They must be grown on contract with a seed company.

The seed crops are mainly concentrated in Lincolnshire, Cambridgeshire and Essex, but some are grown in parts of Gloucestershire, Oxfordshire and Northamptonshire. These are all reasonably dry areas for harvesting. (A great deal of seed now comes from southern Europe.)

Seed crops must be isolated (in zoned areas) at least 500 metres from other sugar beet seed crops and 1000 metres from mangels, fodder and garden beet. Sugar beet is a biennial plant and so seed is produced in the year following sowing.

There are two main methods of production:

(1) Grown *in situ*. The seed is sown at 8–11 kg/ha to produce about 370,000 plants/ha which are then left to grow on to produce seed. These crops may be drilled under a cover crop of barley (one row beet, two rows barley, one row beet, etc.). This is a useful way of checking downy mildew and virus yellows which can be a problem in the Eastern Counties. Sugar beet for seed may also be sown in July after a fallow or another crop, e.g. winter barley, arable silage, peas, early potatoes, etc. Isolated areas, such as the Cotswolds, usually produce healthy seed because of the absence of ordinary beet crops.

(2) With transplanted crops—stecklings (young plants). These are grown in seedbeds in summer and transplanted in the autumn or the following spring (at about 37,000 plants/ha). This method is now mainly used for elite or special seed production because

it returns about six times as much seed as that grown *in situ*.

Fertilizer requirements are similar to ordinary beet, but soil type is not so important because the roots are not harvested.

The seed is harvested in late August/early September. It is cut, left in windrows on a high stubble (30 cm) and combined 7–10 days later.

Yield. This is in the range 2–5 tonnes/ha.

Mangels and other beet are grown in a similar way.

FURTHER READING

Sugar Beet: A Growers' Guide, R. W. Jaggard, British Sugar.

The Potato W. G. Burton, Longmans.

The Potato Crop: The Scientific Basis for Improvement, P. M. Harris, Chapman and Hall.

8

FRESH HARVESTED CROPS

VEGETABLE PRODUCTION ON FARMS

Farm-scale production of vegetables such as broad and green beans, vining peas, Brussels sprouts, cabbages, carrots, onions, turnips, swedes, celery and cauliflowers is now well established on many farms.

These cropping enterprises have been made possible by the introduction of:

(1) precision drilling and fluid drills;
(2) graded and pelleted seed;
(3) varieties suitable for single-pass harvesting;
(4) safe and efficient herbicides, fungicides, insecticides and growth regulators;
(5) bed systems, plastic mulching, soil blocks and cell transplanters;
(6) new and improved machines for harvesting;
(7) better storage, grading, washing and packaging facilities.

These crops can also fill the role of break or "alternative" crops. Returns can be high, although the costs and risks are also high. There are no EC support schemes.

Limiting factors may be:

(1) unsuitable soils and sites;
(2) distance to market;
(3) finding suitable staff and supervision;
(4) finding a suitable water supply for washing;
(5) effluent disposal if washing is required;
(6) adverse weather conditions affecting both growing, harvesting and market demand.

It is very important to know the market requirements and to produce only what is needed when it is needed, and in suitable quality and quantity. It is usually advisable to arrange contracts with buyers, such as processors, pre-packers, supermarkets, chain stores or co-operatives. Alternatively, all or part of the produce may be sold direct to the public at the farm gate, by pick-your-own, or door-to-door delivery. The profitability of the enterprise can be enhanced by finding suitable outlets for "out-grades", or feeding the surplus to livestock on the farm. The yields on a farm scale may not be as high as for an intensive market garden. However, the scale of the enterprise compensates for this and, on a large farm with other enterprises, a bad year either for yield and/or price need not be a disaster.

Useful literature is produced by the advisory services and Horticultural Research International at Wellesbourne.

Harvested fresh peas

(1) *Vining peas*. These peas make up two-thirds of the domestic market trade. They are used for canning fresh ("garden peas"), quick-freezing or artificial drying and are normally grown under contract, or by co-operating groups. The contracting companies usually supply the variety they require and decide on times of sowing (February/May) based on a "heat unit" system. This is in order to spread the harvesting period to ensure a regular and constant supply to the processing factories.

Some examples of varieties are:

First earlies: *Sprite, Sparkle, Avola, Banff, Span.*

Second earlies: *Galaxie, Tessa.*

Early maincrop: *Scout, Tristar, Markado, Bikini. Waverex* and *Citadel* for the "petit pois" market.

Maincrop: *Perfection, Puget, Hurst Green Shaft, Markana.*

(2) *Fresh picking (pulling) peas.* These make up approximately 3% of the market. For sale in pods, e.g. *Feltham First, Onward* and for "mange tout" (whole pods edible), e.g. *Sugar Ann.*

Strict EC quality standards apply to the sale of picking peas.

The growing of vining and pulling peas is broadly similar to that of dried peas (page 115). However, the following points should be noted:

(i) Vining and pulling peas are best suited to the main arable areas and where there are local processing plants or markets.

(ii) Most vining pea growers are members of vining co-operatives, often linked to a specific processing factory. The factory must ensure a continuous and reliable supply of produce so that it can operate efficiently. Growers must be certain, therefore, that the harvest of their vining peas is spread out over several weeks, rather than all coming together in the same week.

This is achieved by: (a) growing different varieties with different maturity dates, and (b) staggering the **drilling** of the peas based on the calculation of accumulated heat units (AHUs). Thus, if the difference between one day and the next in East Anglia at harvest time is 11.5 AHUs, and the vining capacity of the co-operative is 40 hectares per day, each block of 40 hectares must be drilled 11.5 AHUs apart back in the spring. (This could mean delays in drilling of several days if the weather is cold.) This should result in the shared harvesters moving round the pea crops just at the right stage of ripeness. If there is a delay because of weather or breakdown, then the peas can often get too tough for freezing or canning. They can then be allowed to grow on for harvesting dry.

(iii) Vining peas are ready for **harvesting** when the crop is just starting to lose its green colour and the peas are still soft. The firmness of the peas is tested daily (near harvest time) with a tenderometer. This gives a guide to the best time to start cutting. The reading for freezing peas is about 100 and, for canning peas (which can be a little firmer), at 120. When ready, the crop must be cut as soon as possible. The pods are picked up and put through a special vining machine which gently separates the peas from the pods. The shelled peas are rushed to the processing plant for freezing, canning or drying. The haulms (vines) left by some of the harvesters may be made into silage or hay, or, more usually, incorporated into the soil to increase organic matter.

(iv) Pulling peas are harvested by removing the pods when the peas are in a fresh, sweet condition. This requires a large gang of casual workers or very expensive machinery and so the market is declining. However, a number of farms have set up "pick-your-own" enterprises which include fresh peas among their crops.

The yield of both vining and pulling peas is in the range of 3–8 tonnes/ha.

Broad beans

Broad beans belong to the same species, *Vicia faba*, as field beans. The husbandry, diseases and pests are similar.

Weed control—some varieties may be damaged by *simazine*. This should be checked, and also whether other products are approved for broad beans.

Types and varieties

(a) Processing. These types are sown from early April to mid-May as directed by the processors. For canning and freezing, white-flowered types are used, e.g. *Triple (three-fold) White, Eureka, Metissa, Listra*. The coloured-flowered types, e.g. *Primo, Pax (Felix), Statissa*, have excellent flavour but, because they stain on cooking and can be bruised when harvested, are used only for canning. *Beryl* is a small-seeded variety and *Filigreen* has green seeds.

(b) Fresh market and "pick-your-own". Some types are sown October/November and are harvested June/July, e.g. *Aquadulce, Claudia*. Other types are sown in early spring, examples of varieties being *Seville, Triple White, Fillbasket*.

Fertilizers. Table 31 gives a guide as to the plant nutrients necessary — assuming phosphate and potash indices at 2–3:

TABLE 31

Crop	N kg/ha	P_2O_5 kg/ha	K_2O kg/ha
Broad beans	0–60	150	100
Dwarf green French beans	130–170	100–120	50–60
Runner beans	120–180	50–90	120–180

TABLE 32

Crop	Row width cm	Plant/m² nos.	Seed rate kg/ha
Broad beans (processing)	40–50	16–20	200–220
Green beans	15(10–20)	27(25–30)	130(90–180)

TABLE 33

Crop	Dates of sowing	Dates of harvest	Yield (tonnes/ha)
Broad beans (processing)	April/ mid-May	mid-July/ end August	2–4
Green beans	mid-May/ end June	mid-August/ late September	4–9

Sowing. The large seeds are likely to be damaged in some drills. Precision, belt-fed and vacuum types are more suitable.

Seed rates. For processing types (Table 32). Fresh and pick-your-own types are sown about 7–10-cm apart in 50–76-cm rows at a rate of 125–160 kg/ha.

Harvesting. The fresh and pick-your-own crops are normally picked over two or three times. The fresh produce is usually sold in 8–10-kg boxes. The average yield is 4 tonnes/ha, although some late crops can yield as much as 8–10 tonnes/ha.

The processed crops are harvested according to the tenderometer readings (TR), e.g. for freezing, TR should read 110–140; for canning, the TR should read 130 and over. Yield, Table 33.

Green beans

Thirty years ago this crop was mainly grown in market gardens, but it is now an important farm crop chiefly grown for processing. This development has been encouraged by the expanding demand for convenience foods. The introduction of the one-pass harvester has also been an important development. This flicks the bean pods and leaves off the plant and separates the leaves in an air stream. The freshly-harvested pods are taken to the processing plant to be canned, frozen, or dried. They are not so perishable as peas or broad beans.

Efficient weed control has also helped to popularize the crop; trifluralin into the seedbed, and/or bentazone post-emergence will deal with the annual weeds. Perennial weeds should be dealt with before planting. If weeds are not controlled, the crop may be unsaleable.

Green beans prefer loams and the lighter types of soil. It is a late-sown, short-season crop and so requires a deep, free-draining, moisture-retentive soil, and a very good seedbed for fast, steady growth. It is also susceptible to frost and cold weather and so most crops are grown in the eastern counties from the Fens southwards.

Types and varieties. Tendercrop (introduced from North America) has grown well in this country. However, it has a poor flavour with rather long pods (15 cm), irregularly-shaped. (The unattractive shape was hidden by processing them as sliced beans.) It is being replaced by better quality types, which are, however, lower-yielding and more difficult to harvest. These are:

(a) Those with **short pods**, about 10 cm long, which can be processed whole. This is more attractive for canning or freezing, e.g. *Riviera, Stip, Gitana*.

(b) Those with **smooth, slender pods** (12 cm long) which are easier to grow than the short pod types. They are vigorous, erect plants, e.g. *Valja, Monaco, Rafasl*. A good proportion of these varieties are small enough to be processed whole; they can be separated at the factory with out-grades used for slicing, canning and freezing as transversely-cut beans. This group also includes *Midas* and *Kingdom Wax*; they have waxy yellow/golden pods popular on the Continent and in North America; they could be developed here for bean salads, etc.

(c) **Flat-podded beans** are being developed by crossing runner and green bean types to give the quality and pods of runner beans and growth habit of green beans.

Fertilizers

French beans do not normally develop nitrogen-fixing nodules, as is the case with other legumes. Nitrogen, therefore, has to be applied (Table 31). The phosphate and potash requirements should also be noted.

Dates of sowing. These are determined by the factory which instructs the grower with whom there is a contract.

Seeding. The seed is sown with a precision or force-feed drill at a seed rate calculated according to factors such as number of seeds/kg, plant population required, percentage germina-

tion, time of sowing and an estimate of likely seedbed losses.

Irrigation. This can be very beneficial in a dry season, especially when the pods are developing. Care is required, however, because it may spread some fungus diseases.

Diseases and pests

The crop is subject to attack by a number of pests. Halo blight and anthracnose can be a problem in some seasons.

Harvesting. The harvested bean pods must be whole, undamaged, separated (not in clusters) and free of stems, leaves, soil and stones. As the crop matures, the pods lengthen rapidly; they then enlarge as the seeds develop. The average length of the largest seed in each of ten randomly-selected pods is used as a guide to the time to harvest the crop. In practice, the length of the ten seeds is measured, and the crop is ready for freezing when the total length is 100 mm for large-seeded varieties and 80 mm for small-seeded varieties.

For canning, the respective lengths are 120 mm and 100 mm.

Brussels sprouts

Brussels sprouts can be grown on a commercial scale in most parts of the country, but chiefly the crop is seen in Bedfordshire and Lincolnshire.

Soils. The ideal soil is a well-drained medium to heavy loam with a pH of at least 6.5.

The growing and marketing of the crop should now be considered in two different ways:

(1) For the fresh market or successive picking.
(2) For quick freezing or single harvesting.

The fresh market

Transplanting. The plants are raised from seed in a seedbed with a seed rate of 0.5–2 kg for every hectare to be planted out eventually. The actual rate will depend on the seed count (1000 s/10 g), "field factor" (conditions which affect seedling emergence and growth) and plant population required for the crop.

The seed should be treated with a combined gamma HCH/thiram or captan seed dressing against the flea beetle, cabbage stem weevil and soil-borne fungi (Tables 71 and 72).

It is important that the seedbed used for transplants is free from disease, particularly club root. The pH should be 6.5–7.0. Phosphate is especially important; the equivalent of 100 kg P_2O_5 should be broadcast before sowing.

The seed is sown:

(1) Under cold frames either drilled or broadcast in February and March for transplanting late April–May, ready for picking mid-August to November,

or

(2) In an open seedbed drilled or broadcast in March and April for transplanting end May–June, ready for picking October to March.

Plant population is important. The plants are often planted on the square, although how they are arranged is not as important as actual density. For hand harvesting, sufficient access is necessary for the pickers. The hybrids need high densities, up to 30,000 plants/ha (2–3 plants/m²). The non-hybrids are better suited to wider spacing and hence a lower density. Traditionally, they are planted at 90 cm × 90 cm—12,000 plants/ha.

The transplants should be handled carefully and graded when pulled. Exposure to any wind and sun must be kept to a minimum, as should the time between lifting and replanting.

Machine transplanting is normal now and, operated properly, it does give a very satisfactory stand. However, to ensure a full and uniform stand, any gaps should, if possible, be "dibbled in" by hand within 10 days after the main plant. Cabbage root fly is a big problem for summer-planted crops, i.e. after the end of April. Granules applied at the time of planting or worked into the top centimetre of soil before space-drilling will give quite useful control of the larvae as they emerge from the eggs.

Space-drilling (Direct seeding). This is also often somewhat confusingly referred to as direct drilling. Depending on the variety, it is carried out from early April to mid-May for harvesting October to March.

Brussels sprouts may be space-drilled in rows at anything from 45 to 90 cm apart. The wider spaced crop is adopted where high yielding large sprouts are required, although now this is unusual. The seed rate is from 1 to 2 kg/ha, depending on seed count, germination percentage and required spacing. The seed is normally dropped in groups of three or four seeds (group sowing) at 25–35 mm apart or, alternatively, one seed may be placed every 10 cm apart. In both cases, surplus plants are removed when large enough to leave a final spacing of 90 cm.

Place in rotation. Brussels sprouts are fairly flexible in this respect. Provided the land is clean and fertility is reasonably high, the crop will fit well into the cereal rotation. The earliest crops are cleared by the end of September and so it is certainly possible to follow with winter wheat. Because of club root it is inadvisable to grow sprouts less than at least one year in three.

Manures and fertilizers. The plant foods required for transplanted and space-drilled crops are shown in Table 34. A phosphate and potash index of 2–3 is assumed.

TABLE 34

	N	P_2O_5	K_2O
	kg/ha*	kg/ha	kg/ha
With FYM	150	75	105
With slurry	190	75	75
With fertilizer only	200	60	150

* Base dressing.

Nitrogen top-dressing will be required to make a total nitrogen application up to at least 300 kg/ha depending upon the appearance of the crop. It is generally applied within one month of emergence or transplanting. The F_1 hybrids normally need a heavier top-dressing.

The need for potash should be noted.

Varieties. For the fresh market, a smooth, round dark green button, free from disease, is required. A 20–40 mm diameter range is considered necessary for a high yield.

The NIAB Vegetable Growers Leaflet No. 3. Brussels Sprouts gives details of the current recommended varieties for the fresh market and for freezing. A number of the varieties are suitable for either outlet, and where a continuity of supply is required, the examples in Table 35 can be used.

Most varieties are now F_1 hybrids. Compared with the open-pollinated forms, they produce a more uniform plant with a more even distribution of buttons along the stem length (ideal for machine harvesting) and, most importantly, particularly for freezing, evenness of maturity.

TABLE 35

Examples of varieties of Brussels sprouts

For the fresh market	
Early—up to mid-October	Adeline, Oliver
Mid-season—mid-October to end December	Lunet, Rampart
Late—after December	Cascade, Rasalon
For freezing	
Early to mid-season	Adeline, Cavalier, Lunet
Late	Sheriff, Cascade

Treatment during the growing period. A programme of herbicide treatment may be necessary to keep the crop clean of weeds. This could include a pre-drilling (or pre-planting) treatment or the use of residual herbicides, followed by post-emergence (or post-planting) treatments (page 237).

Control measures may be necessary against the cabbage root fly aphid and cabbage caterpillar.

Powdery mildew is at present the most serious disease likely to affect Brussels sprouts, but ring spot and light leaf spot could become an increasing problem in eastern England.

Harvesting. Picking usually starts in August and extends until March, according to the variety and season. For maximum yield the plants are picked over four to eight times (at intervals of about four weeks) during the season.

The EC Standard for Fresh Sprouts (1974) applies for produce sold fresh to the consumer, other than farm gate sales.

Fresh Brussels sprouts for sale on the domestic market must attain a quality requirement before they can be displayed, offered for sale, sold, delivered or marketed.

There are three classes: Class I for produce of good quality; Class II—reasonably good quality; Class III—poorer quality but still marketable. Class III may be suspended if there is a sufficient supply of sprouts in Classes I and II to meet consumer requirements.

A summary of the minimum requirement for all classes shows:

Size (compulsory for Class I and optional for Classes II and III). Minimum diameter—trimmed sprouts, 10 mm; untrimmed sprouts, 20 mm. In any package the difference between the diameter of the largest and the smallest sprout should not exceed 20 mm.

Appearance. Fresh.

Sound. Free from attack by insects or other parasites and free from all traces of disease, as well as free from insects or other parasites.

Clean. Free from soil and visible traces of chemicals. Not frozen.

Flavour. Free from foreign smell or taste.

The stalk of trimmed sprouts must be cut just beneath the outer leaves. The stalk of untrimmed sprouts must show a clean fracture at the base (without other parts of the plant adhering).

Yield. 17–22 tonnes/ha.

For freezing

The growing of the crop is similar to that for the fresh market with, for early varieties, February sowings (protected) for transplanting in May and harvesting late August and through September. Outdoor sowings start in March through to mid-April (depending on variety), for transplanting in May and early June, and harvesting from late September to early March.

Space-drilling is normally carried out from March to May.

The plant population for the frozen sprout market is higher than that for the fresh market. 45 cm x 55 cm will give a population of 40,000 plants/ha for early harvesting. Later varieties are more widely spaced, down to about 36,000 plants/ha.

Varieties (Table 35).

Stopping the plants. By removing the growing point (or terminal bud) from the plant or by destroying the bud with a sharp tap using a rubber hammer, sprout growth is stimulated and this will produce a 5–10% higher yield, especially for the earlier varieties. It will also enable harvesting to be carried out at an earlier date.

Time of stopping depends on the intended harvest date. It should be about four weeks before August harvesting and 10 weeks prior to harvesting in December. There is no point in stopping after the end of October.

Harvesting. Mobile machines with seated operators are increasingly used to cut the stalks, strip off the sprouts and at the same time separate them from the leaves. This system is taking over from cutting the stalks in the field for transport to the sprout-stripping machine. De-leafing, which leads to more efficient machine harvesting, can be carried out not more than two days prior to harvesting.

Occasionally, the earliest maturing varieties are picked over by hand before the whole plant is cut for stripping.

Yield. 12–15 tonnes/ha from a single-harvest crop.

Cabbages

Few cabbages are, these days, specifically grown for feeding to stock. Those used in this way normally cannot be graded for the domestic market, or are surplus to that market.

They can be grouped according to their traditional time of marketing, i.e. spring cabbage, summer and autumn cabbage, winter cabbage and savoys. However, new production

TABLE 36   *Cropping*

Type	Examples of varieties	Sown	Marketed	Average yield per ha
Spring cabbage	*Duncan* (F[1] hybrid)	In seedbed – July. Transplanted – September or sown from June onwards direct into field, three seeds/ 25 cm, chopped out to leave one plant/station in 40–cm row widths.	March to June	15 tonnes
Summer/ early autumn/ cabbage	*Golden Acre Winnigstadt*	Seed drilled direct, three seeds/30 cm, chopped out to leave one plant/station in 40–cm row widths.	July to October	35 tonnes
Late autumn/ winter cabbage (including Savoy cabbage)	*Late-Purple Flat Poll Langedijk 4* (*January King*)	In seedbed – April–May. Transplanted – June–July, or sown direct in rows 60 cm apart, three seeds/60 cm, chopped out to leave one plant/station.	November to March	35 tonnes

Note: Winter white cabbage can be stored quite successfully in clamps, but usually in bulk boxes placed in barns (from where they can be marketed direct) from December to April. It can be an expensive operation and does require careful management at harvest.

techniques and, where appropriate, storage, have extended the respective seasons.

The seed is either sown in specially prepared seedbeds or, as is becoming more popular now, drilled direct in to the cropping field.

Table 36 gives details of varieties, sowing and marketing.

Soil type for these cabbages is not exacting, provided it is well drained yet retains moisture.

The plant food requirement will vary according to the type of cabbage and the plant food indices (Table 37).

Where winter cabbage is intended solely for **stockfeeding**, a variety such as *Flagship, Marabel* or the savoy *January King* should be used. For this purpose moist, heavy soils are preferable for maximum yield, which can be up to 90 tonnes/ha, although this can make harvesting difficult.

Cabbage rootfly can be a problem, particularly with summer and winter cabbage.

TABLE 37

Type	N kg/ha	P_2O_5 kg/ha	K_2O kg/ha
Spring cabbage	50–100	60–180	60–250
Summer/autumn	175–275	60–180	60–250
Winter/savoys	175–250	60–180	60–250

Notes:
(i) A phosphate and potash index of 2–3 is assumed.
(ii) A lower amount is advised where the soil indices are higher.
(iii) Appropriate adjustments downwards should be made when farmyard manure (up to 40 tonnes/ha) is applied.
(iv) Nitrogen is applied at the bottom of the range where winter cabbage is to be stored.

Carrots

Carrot production for human consumption in this country is now a large and specialized enterprise for most growers. Very high capital investment in grading, washing and packing plants is involved (over £1200/ha of crop).

The crop is marketed in several ways. Some farmers grow and pack it on the farm, whereas others grow the carrots for packing to be carried out elsewhere. There are other growers with very little land who rent the carrot fields from other farmers

The fixed costs of running an efficient plant can be more than £60/tonne and, if the through-put is 75–90 tonnes/day, it will require about 400 hectares to keep it supplied through the season.

Carrots are not grown now for stockfeed but the very considerable quantities of rejects (up to 30%) from the packing lines may be fed to cattle.

Types and varieties. Carrots are grown for many different markets and quality requirements by the buyers can be very exacting. There are five main types of carrot (based on shape) with over 100 varieties (NIAB Vegetable Growers' Leaflet No. 5. Carrots). The cylindrical types such as *Nantes* and *Berlicum* are popular for pre-packing and some processing; small finger carrots for freezing are usually *Amsterdam Forcing* types; small *Chantenay* types are used for canning. Where large, uniformly-coloured carrots are required for slicing and dicing, *Autumn King* and *Berlicum* types are suitable. Varieties are also grouped by date of harvest: **first earlies** are harvested at the end of May, **second earlies** at the end of June and **maincrops** from July onwards.

Uniformity and freedom from damage and disease are very important. Some carrots are stored in buildings, but they lack the fresh appearance associated with newly-lifted carrots. Consequently, much of the crop is left in the ground during the winter and lifted as required. This can mean earthing-up or covering with straw or black polythene sheets to protect from frost. This may also help to stop new wasteful sprout growth in the spring. Some crops are left in the ground, risking frost damage. Lifting may continue until the crop becomes too woody or otherwise unsaleable, usually by early May.

Soils and climate. The climate in most arable areas of the country is suitable for carrots; the main limiting factor is soil type. Sandy soils and peats are the most favoured as they normally give good yields of well-shaped roots which may be harvested at any time and are easily washed. Most crops are grown in the eastern counties.

Rotation. One in five years is preferable to avoid disease.

Time of sowing and seedbed. To provide for supplies over a long period, it is necessary to sow at different times. First earlies are sown under polythene in October. Varieties grown in this way must have good resistance to bolting. Second earlies are also sown under polythene, but much later, i.e. in January or February. Maincrop carrots are sown from March to May, depending on the variety and intended harvest date.

To produce more than 70% of carrots of a desired size is very difficult, and every effort has to be made to prepare a really good uniform seedbed. This should be moist; carrot seed is difficult to germinate because of seed coat "inhibitors". These must be dissipated away from the seed in the soil solution.

Sowing and plant populations. The seed should be precision-drilled at a uniform depth "on the flat" or in a bed system. The seed rate (1–2.5 kg/ha) will depend on the seed quality, field conditions and market requirements. Where small carrots are required for freezing, there should be about 1100 plants/m²; large carrots for dicing require about 50/m²; 60–200/m² are recommended for earlies under plastic.

Manuring. Fertilizer recommendations can be summarized as follows (assume a phosphate and potash index of 2–3):

TABLE 38

	N	P_2O_5	K_2O
	kg/ha	kg/ha	kg/ha
On peats	60	125	125

Salt can be used for carrots (preferably ploughed in). 400 kg/ha (150 kg sodium) will replace 60 kg/ha potash.

Weed control. Any perennial weeds should, if possible, be controlled in the previous year. Triallate pre-drilled into the seedbed will control wild oats, blackgrass and annual meadow grass. Trifluralin can be used for a range of annual weeds. Most annual broadleaved weeds and annual meadow grass, post-emergence, can be controlled by chlorpropham, linuron, prometryne or metoxuron.

Pest and disease control. Carrot fly, cutworms, willow-carrot aphids and nematodes, and the diseases violet root rot and cavity spot, are the main problems in the growing of the crop (Tables 71 and 72).

Irrigation. This is often required in a dry period, but great care should be taken with water management to avoid splitting the roots.

Harvesting. The crop is harvested from June onwards according to market demand and size of roots. The roots are very easily damaged and so careful handling is required at all stages. Modern machines are designed to minimize damage. Those which lift by the leaves, and then top the roots, work well during the summer and up to the end of October (or when the tops become too weak) because of frost damage, for instance. The digger (share-elevator types) require the tops to be flailed-off first, and these machines have to be used from late October onwards. To maintain freshness, the time from lifting to dispatch to market should be as short as possible. Therefore, the whole operation has to be well planned.

It is normally desirable to have contracts for the various types of production.

Yield. This can vary from 20 tonnes/ha for early crops to well over 60 tonnes/ha for maincrops. However, rejects in grading can be 30% or more.

Swedes (for the domestic market)

The growing of swedes for human consumption follows the same lines as swedes for livestock feed (page 151). Farmers may grow the crop for the domestic market but, depending on the season, a proportion of the crop will very often be used for stockfeed.

However, the following points can be noted:

(1) Varieties such as *Ruta Otofte* are favoured as they are equally suitable for both outlets.
(2) Singling the crop (15 cm apart) is necessary on all but the most accurately-drilled stand.
(3) Although machine harvesting is increasingly being used for swedes, a sizeable proportion of the crop is still pulled by hand, particularly on the smaller farms in the west of the country.

Bulb onions

Bulb onions have increased in importance in recent years. It is a high value crop and, because of precision drilling, improved chemical weed, pest and disease control, and better methods of harvesting and storage, it is now a far more reliable crop than in the past.

About half of the total requirements (350,000 tonnes) of the domestic market is homegrown. There is therefore a good market for increased production, always provided that it is of good quality.

Soils and climate. Bulb onions can be grown on a wide range of mineral and peat soils provided they are well drained, have a good available-water capacity, a pH of 6.5 or more, and are friable enough to produce good seedbeds. Shallow soils, thin chalk, dry sands and sticky clays are not suitable.

They are grown in many parts of the United Kingdom, but do best in the eastern and south-eastern counties. If growth is stopped by drought, and a secondary growth follows after rain, the outer layers of the bulb may split open.

Environmental factors can also alter bulb size, skin colour and thickness, texture, flavour and storage quality.

They can now be supplied to markets throughout the year because of modern developments in storage and the production of autumn-sown crops

which, in mild areas, can be harvested from June onwards until the spring-sown crops are ready.

Rotation. Onions, and crops such as field beans, broad beans, oats and parsnips which carry the same pests, should not normally be grown more often than one year in five in the same field because of the risk of nematode, onion fly and white rot disease.

Fertilizer. Fertilizer recommendations are as follows:

TABLE 39

	N	P$_2$O$_5$	K$_2$O
	kg/ha	kg/ha	kg/ha
Average/good conditions	90	250	150

Notes:
(i) Assumed phosphate and potash index at 2–3.
(ii) A heavy use of phosphate for this shallow-rooted crop is often necessary.

Varieties. Spring-sown maincrops: *Durco, Karato, Maru, Targa.* NIAB Vegetable Growers' Leaflet No. 6. Bulb Onions.

Time of sowing

Over-wintered onions ideally should be sown between the first and third weeks in August. Bulbing starts about mid-May to be ready for harvest in July/August. Yields are no better than spring crops. However, there may be an early premium and they can help in providing an all-the-year-round supply.

Spring-drilled crops should be sown as soon as possible after mid-February; they begin to bulb in late July regardless of growing conditions.

Transplanted crops are planted out late March/early April.

Sowing. Over-wintered and spring-sown onions can be grown successfully on good soils by precision-drilling in shallow rows, 1–2 cm deep. The bed system is now used with the seed spaced in various ways.

The recommended seed rate for dressed graded seed is 4–6 kg/ha, or 28 kg/ha for pelleted seed. The target is 75 plants/m² for the spring-sown crop and 100/m² for the August-sown crop. Alternatively, for spring crops, multi-seeded transplants, grown in glasshouses, can be used to give an earlier (up to a month) and more reliable crop on late and difficult soils. The seed is sown in late January/early February in soil blocks or small-celled trays (about seven seeds in each cell). These are transplanted as sturdy plants in late March/early April, usually in four- or five-row beds, aiming for 10 units of six seedlings/m². The transplanted crops are very expensive to produce, but yields can be up to 50% higher than drilled crops.

Establishment from sets. It is possible to establish farm crops from sets and interest in this method has increased. The thumbnail size sets can be planted quickly with special machines and are not likely to be affected by soil capping. The sets are produced from high density crops (2500/m²) and stored at 10-15°C over winter to prevent bolting.

Weed control. Perennial weeds should, as far as possible, be controlled in the previous year. Tri-allate can be used in the seedbed for the control of wild oats, blackgrass and annual meadow grass. All the common annual broadleaved weeds and annual meadow grass can be controlled by one or more of the following:

Soil-acting residuals, chloridazon+propachlor and chlorpropham mixtures. Contact and residuals, aziprotryne and prometryne. Contact only, ioxynil and cyanazine.
Bulb onions are not very competitive and grass weeds can be a problem post-emergence. Alloxydim-sodium and diclofop-methyl can be considered.

Diseases and pests. The following can affect bulb onions:

Leaf blotch, downy mildew, onion white rot, neck rot, onion fly, cutworms and the bulb and stem nematode.

Harvesting. Bulb onions are ready for harvesting when about 80% of the tops have fallen over. This is usually from July to early September, depending on the time and method of planting. If left later, yields might increase slightly, but the outer skins are more likely to crack which could lead to disease loss in store.

Before harvesting, the bulk of the tops are cut off with a flail harvester, leaving only 80–120 mm of tops. The beds are then lifted, windrowed and left to dry. The bulbs are then elevated into a trailer and taken to the store where they can be stacked to a height of 4–5 metres. Warm air (30°C) is blown through ventilation ducts until the moisture is removed from the leaves and top of the stem. For the following three weeks, air at about 27°C is blown through from time to time to complete the drying of the neck tissue. The skins should be a golden colour. When fully dried, the stack temperature is slowly lowered to near freezing to control sprouting. Refrigeration will be needed for storage, at about minus 2°C, until July/August. To prevent condensation, the temperature should be raised for five days before the onions are sold. Temperature and humidity control are very important for the control of quality.

Onions are usually graded into various sizes before sale, e.g. over 60 mm, 45–60 mm, 30–45 mm and "picklers".

Yield. This is variable from 30 to 50 tonnes/ha.

FURTHER READING

Brussels Sprouts Facts, National Vegetable Research Station.
Cabbage Facts, National Vegetable Research Station.
Pea Growers' Handbook, PGRO.
Principles of Vegetable Crop Production, R. Fordham and A. G. Biggs, Collins.

9

FORAGE CROPS

FODDER BEET AND MANGELS

BOTH fodder beet and, now to a much lesser extent, mangels are grown for feeding to cattle and sheep in the more southerly parts of the country where they are better croppers than swedes and maincrop turnips. Traditionally, swedes and turnips have been grown in the cooler and wetter regions, but increasingly fodder beet is seen in these areas as well.

About 11,500 hectares are grown, which represent a fairly significant increase in the last five years. This renewal of interest is almost solely confined to fodder beet which, because of its higher dry matter content (Table 40), is now considered to be far more important than the mangel. The latter could perhaps be considered

preferable for feeding to young stock, although it would hardly be grown for that purpose alone. Fodder beet fits in well with the emphasis on growing and feeding more fodder crops for the dairy herd at the expense of at least part of the concentrate ration.

Fodder beet

This has been bred from selections from sugar beet and mangels. Thirty-five years ago it was quite popular as a feed for pigs.

Fodder beet should not be grown on heavy and/or poorly-drained soils, nor on stony soils. There could well be establishment and harvesting problems in these conditions.

TABLE 40   *Yield comparison of forage crops*

Crop	Fresh yield (tonnes/ha)	% DM	Dry matter yield (tonnes/ha)	ME* (MJ/ha)	DCP† (kg/ha)
Fodder beet	90	17.0	15.3	190,000	688
Mangels	120	11.0	13.2	164,400	726
Yellow turnips	70	8.6	6.0	67,200	438
Forage turnips	45	8.3 (tops)	3.7	41,440	267
Swedes	90	11.0	9.9	126,720	900
Kale	55	13.0	7.2	79,200	885
Forage rape	34	11.0	3.7	35,110	529
Cabbages	75	10.0	7.5	78,000	750
Fodder radish	40	8.5	3.4	37,200	650
Comparison					
Forage maize silage	50	25.0	12.5	135,000	750
Grass silage	40	25.0	10.0	93,000	1,070

*ME – Metabolizable energy.
†DCP – Digestible crude protein.

Varieties. Medium dry matter group (14–16% dry matter), e.g. *Kyros, Monoval*. High dry matter group (17–19% dry matter), e.g. *Trestel, Monofix*.

The latter group has a higher percentage of root growing in the ground and will generally carry more soil at harvest compared with the medium dry matter varieties. Most varieties are genetically monogerm, but some have larger tops than others and stock farmers may wish to consider utilizing them, either by grazing or conserving as silage. Depending on the variety, up to 30% of the value of the crop is in the tops.

The NIAB Leaflet No. 1—Fodder Crops will give more details of varieties.

Many aspects of the growing of fodder beet are similar to those of sugar beet, but there are some important differences as follows:

Seed rate per hectare. 140,000 seeds/ha. The seed is sold in packs. Graded (3.50–4.75 mm) pelleted seed is used, which should be precision-drilled if possible.

Time of sowing. Early to mid-April; sowing too early will cause bolting (some varieties are more susceptible). Sowing too late will reduce yields.

An increasing number of crops are now drilled-to-a-stand which will obviate any thinning out. Crops not grown in this way will need to be rough-singled. The drill width will depend on the harvesting system; it is normally at 50 cm with a 15–17 cm spacing in the row. The aim is to achieve at least a 65,000/ha plant population, but because of inevitable plant losses through pests and diseases, post-emergence herbicides and/or accidents with inter-row work if carried out, a target figure of 75,000 plants is not unrealistic.

Manures and fertilizers.

The phosphate and potash and/or agricultural salt can be applied several weeks before sowing the crop. Normally half the nitrogen will go on the seedbed and the balance at the 3–7 leaf stage.

A pH of 6.5–7.0 is necessary. Acid conditions cause stunted and mis-shapen roots. However,

TABLE 41

	N	P_2O_5	K_2O
	kg/ha	kg/ha	kg/ha
With FYM	100	50	50
With slurry	100	50	50
Fertilizer only	125–150(1)	75	150(2)

Notes:
(i) A phosphate and potash index of 2–3 is assumed.
(ii) 150 kg nitrogen can be considered if the tops are required for feeding, although excessive nitrogen could affect frost resistance with lower dry matter fodder beet.
(iii) If sodium, as agricultural salt, is used at 370–400 kg/ha, the potash can be reduced to 50 kg.

as with sugar beet, overliming can cause boron and manganese deficiency.

A pre-emergence herbicide should be considered almost as routine. It is important to keep the crop clean for at least the first six weeks; sole reliance on post-emergence sequences will not give optimum control. However, a post-emergence programme is desirable when necessary (page 237).

Weed beet can also be a problem. In fodder beet it tends not to be treated as seriously as in the sugar beet crop. Apart from the fact that the problem is perpetuated if the weed is allowed to grow unchecked (obviously a serious problem if sugar beet is also grown on the farm), it does compete with the fodder beet itself and with the following crops in the field. Control measures are the same as for sugar beet.

The seed should be dressed with gamma-HCH and thiram against footrot and the flea beetle (the latter more a problem in south-west England). The mangel fly can also be a problem (Table 71).

Virus yellows can be very serious in some years, particularly with an early attack (Table 72).

Powdery mildew, appearing as a white powdery growth on the leaves in early autumn, is not (unlike with sugar beet) generally considered serious enough to treat with a fungicide.

Harvesting. This is usually in October and November, although, depending on the season, the crop can continue to grow through the autumn. However, late harvesting generally increases the risk of more difficult conditions, particularly on heavier soils. Many growers will also harvest earlier at the expense of yield in order to follow with winter cereals (usually wheat), preferably sown before the end of October.

The crop is lifted either by specialist harvesters or modified sugar beet harvesters. Top savers can be used when necessary or, as is quite popular on the Continent, a forage harvester is used for topping the beet (although, unless the crop is very evenly grown, it is not done all that accurately). Topping must be carried out correctly when it is intended to store the beet for any length of time. The roots should be topped at the base of the leaf petioles. Over-topping soon results in mould growth; leaving too much green material causes high losses through increased respiration in the clamp.

Fodder beet can be fed fresh to stock, but normally it has to be stored. This must be done carefully and for outside storage an interior temperature of 3–6°C should be maintained. Heating in the clamp should be avoided, although the fodder beet must be adequately protected against frost. A straw cover or plastic sheeting can be used, but a strip along the top of the clamp should be left uncovered for ventilation, except under freezing conditions.

The storage capacity needed will vary according to the size of the beet; it is in the range 700–750 kg/m³.

Yield per hectare
80–100 tonnes—fresh crop
14.0–15.5 tonnes—dry matter

Mangels (mangolds)

As shown in Table 40, there is a significant difference in dry matter yield and energy when comparing mangels with fodder beet. Very few mangel crops are grown these days.

When grown, the husbandry is similar to fodder beet. Most of the varieties such as *Wintergold* (a low dry matter traditional variety) are multigerm, with the seed producing more than one plant when sown (*Peramono* as a monogerm variety is an exception). The seed, although graded and pelleted, will produce a relatively full row which may have to be thinned. The ultimate target population is 65,000 plants/ha.

Although the mangel grows with much of its root out of the ground, harvesting—except by hand—is difficult. The root "bleeds" very easily and so the tops (not the crowns) are either cut or twisted off. The tops are small and are rarely fed. The roots are not trimmed. Ideally, they should be left for a period in the field to "sweat out" in small heaps covered with leaves. Following this they can be clamped in the same way as fodder beet.

SWEDES AND TURNIPS

Although their popularity has declined in the past 30 years, swedes especially and maincrop turnips are still widely grown, particularly in the northern and western parts of the United Kingdom. At present, about 58,000 hectares are grown, not including the quick-growing forage turnip.

In appearance, the difference between the two crops is that swedes have smooth, ashy-green leaves which grow out from an extended stem or "neck" whilst turnips have hairy, grass-green leaves which arise almost directly from the root itself.

Both are valuable for cattle and sheep and, depending on the variety grown, they can be used as table vegetables. A limited market has developed for contract growing for freezing, and for turnips particularly this is quite a specialist crop.

The growing of swedes and main crop turnips is similar.

Climate and soil. The crops like a cool, moist climate without too much sunshine. Powdery mildew can be a problem in the warmer and drier parts of south and east England; varietal

TABLE 42 *Varieties of swedes and maincrop turnips*

Type	Examples of varieties	Average dry matter %	Remarks
Swedes (grouped according to skin colour and type of root)			
Purple Globe	Angela	8.8	Could be described as semi-tankard. Useful for grazing.
	Ruta Otofte*	9.6	
Green Globe	Melfort	11.1	Good winter hardiness.
Bronze Intermediate	Angus	10.9	Good winter hardiness.
Turnip			
Yellow-fleshed	Yellow Top Scotch; The Wallace	8.6	Slower to mature than other types; hardy and good keeper.

Notes:
(i) Varieties such as Seegold (Green Globe) should be used where club root is suspected.
(ii) *Suitable for the domestic market.
(iii) NIAB Leaflet No. 1—Fodder Crops will give more information on varieties.

resistance and fungicide treatment are not yet the complete answers.

With the exception of heavy clays, most soils are suitable.

Seedbed. A fine, firm and moist seedbed is necessary to get the plants quickly established. In very wet districts the crops can, with advantage, be sown on the ridge, although this is mainly seen in northern areas and in Scotland. It is not a widespread practice now, mainly because of precision seeding and overall chemical weed control.

Manures and fertilizers. The average plant food requirement is summarized in Table 43. If possible, 25–40 tonnes/ha farmyard manure should be applied in the autumn. This is especially important for improving the water-holding capacity of the lighter soils. Slurry can be used instead at about 35,000 litres/ha, although unlike FYM, it is not a soil conditioner.

The lower amount of nitrogen is recommended in wetter areas, although how much is used will depend on the nitrogen index. More

phosphate may be needed on the heavier soils, but this again will depend on the soil phosphate index; if the fertilizer is placed 5–10 cm below the seed, the phosphate can be reduced by half. Potash is important, but savings can be made by using organic manures.

TABLE 43

	N	P_2O_5	K_2O
	kg/ha	kg/ha	kg/ha
With FYM	50	50–75	50
With slurry	50	25–50	0
Fertilizer only	50–100	100–125	60–100

Lime is most essential. Soil pH should be above 6.0 as clubroot (finger and toe disease) can be prevalent under acid conditions. However, over-liming is equally serious as it can cause brown heart (raan) in swedes (Table 72).

Time of sowing. Mid-April to end June. Swedes can now be sown earlier than has been the case in the past, thus increasing the yield. Powdery mildew, which is more prone with earlier sowing, can be reduced by chemical treatment, e.g. tridemorph. In addition, an increasing number of swede varieties are showing reasonable resistance to mildew.

The yellow-fleshed turnip is normally sown in May and June.

Seed rate per hectare. 0.5–4.5 kg. The lower amount is used with the precision drill. The pelleted seed is sown 7–15 cm apart in the row at row widths of 17–35 cm. This is particularly applicable where overall chemical weed control is carried out and/or the crop is to be folded. The aim is for a high plant population of up to 100,000/ha.

With inter-row cultivation, or if the crop is to be lifted, the plants are spaced (or rough-singled) at 20–22 cm apart with row widths of 50–75 cm. A plant population of 60,000/ha is acceptable, although, if the roots are rather large and dry matter content is low, they will not keep well.

Both swedes and turnips can be broadcast at 3.5–4.5 kg/ha.

Dual-purpose seed dressings, containing gamma-HCH as a protection against flea beetle, and thiram for protection against soilborne fungi, should be used.

Treatment during the growing period. The stale seedbed technique, using a contact herbicide to kill the weeds at the time of drilling, followed, if necessary, by the use of a post-emergence herbicide such as propachlor applied when the crop has three to four leaves, should keep it clean of weeds. However, inter-row cultivations are still carried out by many growers, and this will follow the same lines as with most root crops.

Harvesting swedes. Swedes, like turnips, are very often grazed off in the field, but both crops can deteriorate under wet and frosty conditions.

Varieties with globe-shaped roots, rather than tankard-shaped, are less prone to damage when harvested by machine. In most districts they are lifted about November, before they are fully matured. It is therefore advisable to allow the roots to ripen off in the clamp to minimize scouring when later fed to stock. In mild districts, swedes may be left to mature in the field, and for this high dry matter, more winter-hardy types are recommended. Often in this case the swedes are only stored for two to three weeks, with the grower endeavouring to keep just ahead of any possible bad weather.

Because of the low dry matter content and probable soil contamination, the tops are usually wasted. However, they are quite high in protein.

Swedes are vulnerable to rotting under poor storage conditions. They should be handled carefully, not clamped in wet conditions, nor should they be too dry. Clamps should be no higher than two metres and adequately protected against frost.

Harvesting turnips. The main crop (yellow-fleshed) is normally harvested in October when the outer leaves begin to decay. They can be grazed *in situ*, but are often lifted and stored in one main clamp. Machines which merely top, tail and windrow the roots for subsequent collection are now being replaced by complete harvesters. On mainly livestock farms, harvesting is often contracted out.

Yield per hectare. Swedes, 70–110 tonnes/ha (10 tonnes/ha DM); turnips (yellow-fleshed) 60–80 tonnes/ha (9 tonnes/ha DM).

KALE

Kale is grown for feeding to livestock, usually in the autumn and winter months. It can be grazed either in the field or, cut (normally with a forage harvester) and carted off for feeding green. In this case a heavier yielding crop is needed. It is mostly grown in the south and south-west of England, although it is not as popular as it was 15 years ago. About 23,000 hectares are grown at present.

Climate and soil. Kale is an adaptable crop, although under very dry conditions it may be difficult to get well established. For grazing it is preferable to grow it under drier, lighter soil conditions, or on well-drained soils.

Place in rotation. On the livestock farm kale is often direct drilled into grass, either from which an early graze or silage cut has been taken. It can also follow a catch crop put in after cereals the previous autumn. Finger and toe (club root) (Table 72) can be a problem. It should not be grown in the same field more than one year in three.

Seedbed. A fine, firm, and clean seedbed is required. The crop can be direct-drilled and this has the important advantages of conserving moisture at sowing time and leaving a much firmer surface for grazing and good annual weed control.

Manures and fertilizers. Up to 50 tonnes/ha of farmyard manure can be applied in the autumn. It is especially important when a heavy-yielding crop is the aim. Slurry at about 35,000 litres/ha can be used instead and, again, this is normally an autumn application.

Other plant foods required can be summarized as in Table 44. Assume phosphate and potash index of 2.

TABLE 44

	N	P$_2$O$_5$	K$_2$O
	kg/ha	kg/ha	kg/ha
With FYM	125	60	60
With slurry	150	60	0
Fertilizer only	150	50	50

Less nitrogen is used if the crop follows grass.

The fertilizer is usually applied during final seedbed preparations. The nitrogen can be split and part top-dressed, but in most seasons it is unlikely to prove worth while.

A pH of 6.0–6.5 is recommended.

Varieties. NIAB Leaflet No. 1—Fodder Crops now lists recommended varieties of kale for use in the autumn (up to the end of December) and in the winter (January–April, although in the north of Scotland it is seldom possible to utilize kale after January).

The previously grouped types of kale, marrow-stem and thousand-head, have become outclassed by the hybrid kales which carry the more desirable features of the former types. Hybrids, such as *Maris Kestrel* and *Proteor*, perform well for autumn and winter feeding. Maris Kestrel is particularly useful for grazing, as it is shorter than the other kale varieties (dwarf-thousand-head kale excepted, which is not grown now to any extent). *Bittern*, a cross between marrow-stem kale and Brussels sprouts, and *Hereford* are outstanding varieties for both autumn and winter use.

Time of sowing. End of March–mid-July. Early sowing gives heavier crops, although sowing later may fit better into a cropping sequence.

Seed rate. Precision-drilled—0.5–2.0 kg/ha, the seed spaced at 2.5–10 cm apart in 35–75 cm rows. Drilled—3.0–5.0 kg/ha in 35–75 cm rows; the wider row when machine harvesting. Broadcast—5.0–7.5 kg/ha.

The seed should be dressed with gamma-HCH as a protection against flea beetle.

Treatment during the growing period. With drilled crops, steerage hoeing may have to be carried out a number of times from an early stage, but the herbicide desmetryne has eased the problem of keeping the crop weed-free.

If necessary, the crop should be dusted against flea beetles with gamma-HCH.

The drilled crop is seldom singled. When precision-drilled, the aim should be to produce a plant every 7.5–10 cm in the row. With an unthinned crop a higher proportion of leaf to stem is obtained, which produces more succulent plants. Sometimes the crop is harrowed across to thin it out.

Utilizing the crop. With cattle grazing in the field, using the electric fence cuts down the handling of the crop. But light or well-drained soils are essential, otherwise both stock and soil suffer. Wastage can also be high; it averages 15–30%. A 70% waste is by no means unknown. Cutting before grazing can reduce this waste.

To avoid wet and sticky conditions, which normally get worse towards the end of the year, the tendency these days is to start feeding kale to dairy cows much earlier in the autumn, although in a mild season it will continue to grow on well into early winter.

If strip-grazing is not possible, the forage harvester may be used to chop the crop coarsely and blow it into a trailer.

Very high intakes can lead to anaemia and fertility problems.

Yield. 35–75 tonnes/ha.

CATCH CROPPING
(FORAGE CATCH CROPS)

This is the practice of taking a quick-growing crop between two major crops. With the dairy herd and sheep flock it can be very profitable provided:

(1) It does not interfere with the following maincrop. This can happen when the main crop is planted late in a badly prepared seedbed.
(2) The catch crop is not grown when more attention should have been given to cleaning the land of weeds.
(3) It is grown cheaply, without expensive seed-bed preparation, and with limited use of fertilizers, except nitrogen.
(4) The fodder catch crop fits the farming system. In rotations where brassicae and sugar beet are grown, the catch crop can propagate club root and the beet nematode.

On the mixed farm, catch cropping is now widely practised. Direct drilling has helped make the practice more reliable. It has the advantage of conserving what moisture there is available in the soil at the time of the year. This helps to establish the catch crop more quickly. However, broadcasting gives a better cover and it should still be contemplated in the wetter areas.

Examples of catch cropping for forage crops.
(1) Maincrop—winter barley.
    Catch crop—for feeding until Christmas:
    (i) Forage turnips at 3 kg/ha, with 75 kg nitrogen/ha; or
    (ii) Forage rape at 6 kg/ha, with 75 kg nitrogen/ha.
    Maincrop—spring cereal or possibly kale.
    (For feeding after Christmas, more frost-resistant crops such as the English yellow turnip at 5 kg/ha should be grown with 60 kg nitrogen/ha.)
(2) Maincrop—cereal undersown.
    Catch crop—Italian ryegrass at 30–35 kg/ha with 60 kg nitrogen/ha.
    This grass can provide:
    (i) stubble grazing;
    (ii) first bite (the following spring), after having applied 75 kg nitrogen/ha;
    (iii) possibly another grazing or silage before planting.
    Maincrop—kale.
(3) Maincrop—second early potatoes (harvested in July) or a ley (ploughed in July).
    Catch crop—rye at 190 kg/ha with Italian ryegrass at 10 kg/ha and 50 kg nitrogen/ha to provide:
    (i) possibly an autumn graze (depending on the season);
    (ii) first bite (the following spring), after having applied 75 kg nitrogen/ha;
    (iii) silage, before planting.
    Maincrop—kale.
(4) Maincrop—early potatoes.
    Catch crop:
    (i) Forage turnips at 3 kg/ha with 75 kg nitrogen/ha; or
    (ii) Fodder radish at 8 kg/ha with 75 kg nitrogen/ha. Either forage turnips or fodder radish will provide grazing until the end of September before sowing.
    Maincrop—winter cereals.

Forage turnips (Dutch turnips)

These quick-growing, white turnips which originated in The Netherlands are often known as stubble turnips from the practice of growing them as a catch crop on the stubble following an early harvested cereal crop. However, in the north of England and in Scotland it is seldom possible to obtain a reasonable yield by stubble growing. In colder regions they are sown earlier and, indeed, in England, these turnips can be sown in April to provide a useful mid-summer feed, although they are not often used in this way. As far as possible, sowing should be completed by the end of July. In most seasons it is hardly worth-while to sow later than mid-August. Direct-drilling can be considered; it helps to conserve moisture at what is usually a dry time of the year.

Seed rate. 2–5 kg/ha drilled in rows 18–25 cm; 6–9 kg/ha broadcast. The seed should be dressed with gamma-HCH as a protection against flea beetle.

Varieties. On the current NIAB Leaflet No.1—Fodder Crops, 10 varieties of quick-growing white turnips are described. It should be noted that it is the tops (the proportion of the root above the ground) rather than the bulbs which are scored for dry matter content (mean 9.5%). The leaf is the important feature and the higher the leaf-to-bulb ratio the better. As a result, with most varieties, the bulbs (the so-called roots) can easily be knocked out of the ground when being grazed, and so are wasted. This has been overcome with the breeding of turnip hybrids, such as *Appin* and *Tyfon*, with little bulb but with fibrous roots to give better ground anchorage. These two varieties have a slightly lower dry matter content and not such good club root resistance.

Powdery mildew and alternaria can also be a problem with some of the varieties. Bolting can occur with a number of varieties (e.g. *Appin* and *Tyfon*) and these should not be used when sowing early, i.e. in April.

Fertilizer. The emphasis with forage turnips should be on nitrogen, when up to 100 kg can be broadcast when sown. Phosphate and potash will not always be necessary, although this will depend on soil indices.

Weed control. Should not be necessary.

Utilization. The crop is ready for grazing three months after drilling.

Yield. 50–60 tonnes/ha—fresh crop; 5 tonnes dry matter, produced 100 days after sowing by end of July. Up to 100 tonnes from crops drilled in spring and early summer.

Forage rape

This is a quick-growing palatable crop which is ready for grazing after 12 weeks. It generally lacks winter hardiness and should be used before the end of the year.

There are two main types:

(1) Giant, with well-developed leaves and stems and an average height of 85 cm and a D value of about 62. Most varieties show reasonable resistance to powdery mildew, but are susceptible to club root. *Hobson* and *Barsica* are examples of giant varieties.

(2) Dwarf—shorter, more prostrate habit of growth; at present *Caron* is recommended by the NIAB Leaflet No. 1—Fodder Crops, and it should perhaps only be considered where club root is a problem and/or with conditions of high fertility. Like all dwarf types (*Nevin* is a classic example but is now outclassed), *Caron* has a relatively low yield of digestible organic matter. It is also very susceptible to mildew.

Seed and sowing. The seed is either drilled at 4.5 kg/ha or 10 kg/ha broadcast from the end of April until mid-August, usually following in this last case an early harvested grain crop. Direct drilling is often carried out. All seed should be dressed with gamma-HCH as a protection against flea beetle.

Fertilizer. Up to 100 kg nitrogen can be used. Phosphate and potash are not always necessary, but this will depend on soil indices.

Yield. 24–34 tonnes/ha (4 tonnes dry matter).

Rape kale and hungry gap kale

These are not true kales but are related to, and similar to, rape. Both are particularly susceptible to club root and mildew.

They are grown on a small scale only in the southern part of the United Kingdom, and can be used for a limited period after Christmas. However, they do produce some useful regrowth in March and April (they are sometimes referred to as a two-feed crop)—the so-called "hungry gap" before spring grass.

Seeding and growing requirements are similar to forage rape.

Fodder radish

This is grown for forage purposes, green manuring (page 63) and game cover. By producing quick ground cover it can also be considered as a protection against wheat bulb fly. It is very quick growing (some varieties grow a metre high) and is suited to most soil conditions. *Nerys* and *Slobolt* are two commonly-grown varieties. Fodder radish is normally sown in July for forage at 8 kg/ha drilled, or 13 kg broadcast. For green manuring it may be sown earlier, with a heavier seed rate of 17 kg. Up to 75 kg nitrogen can be applied with the seed. Phosphate and potash should not be necessary.

It should be grazed off before flowering and whilst it is still palatable, being normally utilized within 8–12 weeks of sowing.

Fodder radish is very susceptible to frost but resistant to club root and mildew.

Forage peas

Forage peas is a relatively new crop in the United Kingdom, which was introduced in 1975, following development in Germany.

It can be grown as far north as southern Scotland and, although ground conditions may limit utilization, soil type is not exacting. The pH should be at 6.0–6.5.

It is a flexible crop sown at any time between late March and August for utilization 12–14 weeks later.

There are a number of varieties of forage peas but as yet there is no NIAB Recommended List. Examples are *Nadja*, *Progasil*, *Magnus*. Most varieties have pink/purple flowers, although Progasil (a British variety) is white-flowered; Magnus is the first semi-leafless variety.

The crop is sown at 80–100 kg/ha (depending on seed size), using the ordinary corn drill and at 3–4 cm depth. Depending on soil indices, as a guide, 80 kg each of phosphate and potash should be broadcast with 30 kg nitrogen prior to sowing the crop. Seedbed nitrogen will help to get it away more quickly.

Weed control can be a problem in the early stages of growth (MCPB can be considered) but, once the crop has developed, any weeds should be smothered.

Forage peas are probably best used as silage, sown alone or preferably part of an arable silage mixture when it gives a higher dry matter yield. It should be cut when the crop is still green and with the lower pods formed but with swelling not completed. Wilting and an additive will generally be necessary when ensiling. Apart from silage, the crop can also provide useful grazing when grass is short in late summer and autumn (also as an emergency-grown crop in the summer following, for example, a hard winter which may have had an adverse effect on the ryegrass leys). Because of a reasonably high tannin content, forage peas are less likely to cause bloat than most other legumes.

FORAGE MAIZE

A small amount of maize grown for forage is cut and fed green, but mainly it is ensiled and, as such, it makes an extremely palatable, high-energy (but low protein) silage with a high D (digestibility) value (page 165).

Approximately 57,000 hectares are now being grown in England and Wales.

Climate. The crop will not grow until the soil temperature at 10 cm has reached 10°C. Following sowing it needs an accumulated temperature (expressed as Ontario units) of approximately 2500 to reach the 30% dry matter necessary for a satisfactory silage. There are, therefore, climatic limitations for this crop and it is best grown in England only as far north as Cheshire and South Yorkshire.

Soils. Ideally, a rich, deep well-drained loam is best. Light soils are reasonable if they do not dry out. Thin chalk soils should be avoided and heavy clays are usually very slow to warm up in the spring, besides leading to possible difficult harvesting in a wet autumn. A soil pH of 6.0 or over is necessary.

Place in rotation. When maize became established as a forage crop some 25 years ago, it was considered possible to grow it continuously without any problems. It is now realized that maize can carry take-all on its roots, and so it is not a good break crop. Fusarium stalk-rot and smut are more troublesome in fields where maize is grown in close succession. However, it is not affected by eyespot and the usual foliar diseases of cereal crops.

Occasionally, two successive maize crops can be advised in order to control couch grass using atrazine. At the high rate necessary to kill couch, atrazine persists for two years in the soil, and crops other than maize immediately succeeding the first maize crop (which had atrazine) could be seriously affected.

Seedbed. Good ploughing in the autumn, down to 20–25 cm, is ideal, leaving a level surface for frost. Soil loosening is advised if any compaction is apparent. Deep cultivation should be avoided in the spring.

Manuring. Maize is not very responsive to nitrogen, phosphate and potash, although it can remove considerable amounts of potash from the soil (approximately 13 kg potash per tonne dry matter produced). This should be borne in mind for the following crop.

As maize is not sown until the end of April, it can be a useful crop to take farmyard manure or slurry.

The plant food requirements can be summarized as in Table 45, phosphate and potash indices at 2.

TABLE 45

	N	P_2O_5	K_2O
	kg/ha	kg/ha	kg/ha
With FYM or slurry	60	0	0
Fertilizer only	100	40	100

All the fertilizer should go on the seedbed. Combine drilling will delay germination and could well reduce plant population. Nitrogen top-dressing will cause leaf scorch.

Sowing. Soil temperature should be adequate in the south of England at the end of April/early May. Precision-drilling is necessary, with the seeds at 7.5–10 cm apart at 5–7 cm depth in rows 55–75 cm apart. (The row width should suit the harvester). A 20-row headland should be allowed to facilitate harvesting.

Seed rate. A final plant population of 11–12 plants/m^2 (110,000–120,000 plants/ha) should be regarded as optimum. Allowing for up to 10% establishment loss, approximately 125,000 seeds/ha should be sown (40–50 kg/ha). It is preferable to state seed rate as number of seeds/ha since seed size and shape can show variation between varieties.

Varieties. An important development in the past 10 years has been the introduction of earlier ripening varieties. This should mean harvesting about 14–20 days earlier than later ripening varieties. Although it depends on the season, it should be possible to have an easier, late September harvest compared with probable awkward conditions in the following six weeks.

Examples of varieties are *Bastille, Labrador, Leader, LG 2080,* NIAB Leaflet No. 7—Forage Maize.

Pests and diseases. Tables 71 and 72.

Weed control. Weeds can ruin a maize crop because of their faster growth rate in May and June. It is very uncompetitive in the early stages of growth and so pre-emergence control is more satisfactory than post-emergence.

Harvesting. Maize silage (page 200).

Yield. 35–45 tonnes/ha (average 13,500 kg DM/ha).

BRASSICAE SEED PRODUCTION

Kales, swedes, turnips, rapes. These are biennial plants which produce seed in the year following sowing. Swedes and turnips sown in August and September are usually hardy enough to overwinter. Special precautions have to be taken to avoid cross-fertilization. They are harvested in late summer–usually combined direct or from a windrow (NIAB Seed Growers' Leaflet Nos. 2 and 3).

FURTHER READING

Alternative Forage Crops, P. Jarvis, Milk Marketing Board.
Forage Conservation and Feeding, Ed. Frank Raymond, Farming Press.

10

GRASSLAND

GRASS is the most important crop in the United Kingdom. It occupies about two-thirds of the total area of crops and grass grown, although in the drier, sunny eastern areas only about one-third is in grass. Moreover, it is well to remember that grass is a crop, not something which just grows in a field!

CLASSIFICATION OF GRASSLAND

In the British Isles the total area of grassland, including rough grazings, is just over 11 million hectares. It can be broadly divided into two groups.

Uncultivated grasslands

These represent about 44% of the total area of grassland. They consist of:

(1) *Rough mountain and hill grazings*. The plants making up this type of grassland are not of great value. They consist mainly of fescues, bents, nardus and molinia grasses, as well as cotton grass, heather and gorse. Only sheep and beef cattle at very low stocking rates are possible. In some areas where the soil is extremely acid, reafforestation is being successfully carried out.

(2) *Lowland heaths*. Sheep's fescue is very often the dominant grass on the prevalent acid soils. These heaths are found in south and east England and some of them have been reclaimed.

(3) *Downs of south England*. Apart from herbs, grasses such as sheep's fescue and erect brome are found on these chalk and limestone soils.

(4) *Fen area in east and south-east England*. Any unreclaimed areas are mostly poorly drained and are dominated by water-loving plants such as molinia, rushes and sedges. Fens, when drained, are associated with intensive arable cropping (page 29).

(5) *Maritime swards*. These consist of salt marshes and coastal dune areas with marram and cord grass.

Cultivated grasslands

These represent about 56% of the total area of grassland. They consist of:

(1) *Permanent grassland*. This is grass over five years old and it represents 41% of the total area of grassland. As a group, permanent pastures cover a wide range, the quality depending chiefly on the amount of perennial ryegrass in the sward. A **first-grade** permanent pasture contains more than 30% perennial ryegrass with a small amount of agrostis; **second-grade** pasture has between 20% and 29% ryegrass and more agrostis and **third-grade** pasture tends to contain very much more agrostis species with less than 20% ryegrass in the sward. **Very poor** permanent pastures will be dominated by agrostis, and on poorly drained soils most of the sward will be made up of agrostis, rushes and sedges. Many of these permanent pastures can be improved (page 193).

(2) *Leys*. These are temporary swards which have been sown to grass or grass/clover for a limited period of up to five years. Leys represent about 15% of the total area of grassland. In

most cases, depending on management, they will produce more than permanent pasture due to the more productive plants which make up the sward. But generally this sward will not stand up so well to treading as the permanent pasture.

PLANTS MAKING UP THE LEY

These can be grouped as follows:

Grasses.
Legumes: (a) clovers,
            (b) lucerne and sainfoin.
Herbs.

The grasses and legumes can be divided into varieties. Farmers' Leaflet No.4—Recommended Varieties of Grass and Herbage Legumes (published annually by the National Institute of Agricultural Botany) gives details of the grasses and legumes used by farmers. The NIAB now classifies grass varieties as early, intermediate and late according to when the variety reaches 67D (page 165).

Early heading within a species is broadly associated with early spring growth and a less densely tillered plant with an erect habit of growth. These varieties are not very persistent. In the past, they were referred to as hay types or strains.

Late heading is associated with densely tillered plants which generally means good persistency. They have a prostrate growth habit and are more suitable for grazing. These varieties used to be called pasture or grazing types or strains.

Intermediate varieties tend to combine the advantages of both the previous groups of varieties, viz. relatively early spring growth but late flowering and good persistency. They were previously referred to as dual-purpose types or strains.

Varieties come from both British and Continental breeders. There is little difference in characteristics except that, especially with the ryegrasses, those from the Continent tend to be slightly more winter-hardy.

There are still some old, so-called commercial varieties (local varieties) used, although not now to any great extent. They are usually short-lived types more suited for conservation, e.g. Scots Timothy and Scottish Italian ryegrass.

TABLE 46 *Recognition of grasses in the vegetative stage*

	Short duration ryegrasses	Perennial ryegrass	Meadow fescue	Cocksfoot	Timothy
Leaf sheath	Definitely split Pink at base Rolled in shoot*	Split or entire Pink at base Folded in shoot	Split Pink at base Rolled in shoot	Entire at first, later split Folded in shoot	Split Pale at base Rolled in shoot
Blade	Broad Margin smooth Dark green	Narrow Margin smooth Dark green	Narrow Margin rough Lighter green	Broad Margin rough Light green	Broad Margin smooth Light green
Lower side of blade	Shiny	Shiny	Shiny	Dull	Dull
Ligule	Blunt	Short and blunt	Small, blunt, greenish-white	Long and transparent	Prominent and membranous
Auricles	Medium size and spreading	Small, clasping the stem	Small, narrow and spreading	Absent	Absent
General	Veins indistinct when held to light. Not hairy	Not hairy	Veins appear as white lines when held to light. Not hairy	Not hairy	Base of shoots may be swollen. Not hairy

*Identification of hybrid ryegrass is more difficult; it can show a mixture of vegetative features between Italian and perennial ryegrass. Round tillers tend to predominate, although this is not always the case.

GRASS IDENTIFICATION

Before it is possible to recognize plants in a grass field (Table 46), it is necessary to know something about the parts which make up the plant.

Vegetative (leafy) parts

Stem

(1) The *flowering stem or culm*. This grows erect and produces the flower. Most stems of annual grasses are culms.
(2) The *vegetative stem*. This does not produce a flower, and has not such an erect habit of growth as the culm. Perennial grasses have both flowering and vegetative stems.

Leaves. They are arranged on two alternate rows on the stem and are attached to the stem at a node. Each leaf consists of two parts (Fig. 44):

(1) The *sheath* which is attached to the stem.
(2) The *blade* which diverges from the stem.

The leaf sheath encloses the *buds* and younger leaves. Its edges may be joined (entire) or they

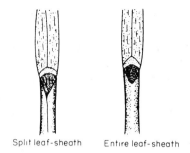

Split leaf-sheath   Entire leaf-sheath

FIG. 45. Parts of the grass leaf.

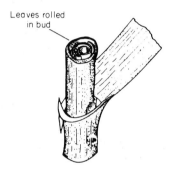

Leaves rolled in bud

FIG. 46. Parts of the grass leaf.

Leaf folded in the bud

FIG. 47. Parts of the grass leaf.

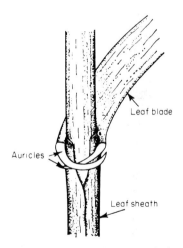

Leaf blade

Auricles

Leaf sheath

FIG. 44. Parts of the grass leaf.

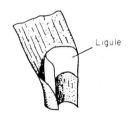

Ligule

FIG. 48. Parts of the grass leaf.

may overlap each other (split) (Fig. 45). If the leaves are rolled in the leaf sheath, the *shoots* will be round (Fig. 46), but when folded the shoots will be flattened (Fig. 47).

At the junction between the leaf blade and leaf sheath is the *ligule*. This is an outgrowth from the inner lining of the sheath (Fig. 48).

The *auricles* may also be seen on some grasses where the blade joins the sheath. They are a pair of clawlike outgrowths (see Fig. 44).

In some species, the leaf blade will show distinct veins when held against the light and, according to the variety, the underside may be shiny or dull.

Other features of the leaves are more variable, and are not very reliable; they can vary with age.

Inflorescence—the flower head of the grass

The inflorescence consists of a number of branches called *spikelets* which carry the flowers. There are two types of inflorescence:

(1) The *spike*—the spikelets are attached to the main stem without a stalk (Fig. 49).

(2) The *panicle*—the spikelets are attached to the main stem with a stalk (Fig. 50). In some grasses the spikelets are attached to the main stem with very short stalks to form a dense type of inflorescence termed *spike-like* (Fig. 51).

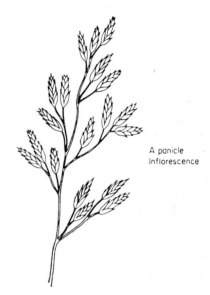

A panicle
inflorescence

FIG. 50. Grass inflorescence.

A spike
inflorescence

FIG. 49. Grass inflorescence.

A spike-like
panicle
inflorescence

FIG. 51. Grass inflorescence.

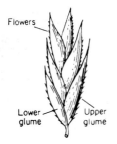

FIG. 52. The spikelet.

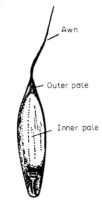

FIG. 53. The pales.

The spikelet is normally made up of an *axis*, bearing at its base the *upper and lower glumes* (Fig. 52). Most grasses have two glumes.

Above the glumes, and arranged in the same way, are the *outer* and *inner pales*. In some species these pales may carry *awns* which are usually extensions from the pales (Fig. 53). Within the pales is the *flower*. The flower consists of three parts (Fig. 54):

(1) The male organs—three *stamens*.
(2) The female organ—the rounded *ovary* from which arise the feathery *stigmas*.
(3) A pair of *lodicules*—at the base of the ovary. They are indirectly concerned with the fertilization process, which is basically the same in all species of plants (Fig. 22).

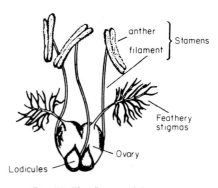

FIG. 54. The flower of the grass.

TABLE 47   *How to recognize the important legumes*

	Leaves, etc.	Stipule	General	Species
Mucronate tip	Centre leaflet with prominent stalk Leaflets serrated at tip	Broad, serrated and sharply pointed	May be hairy	Lucerne
	6–12 pairs leaflets, plus a terminal one	Thin, finely pointed		Sainfoin
No mucronate tip	Trifoliate, dark green with white half-moon markings on upper surface	Membranous with greenish purple veins Pointed	Stems 30–60 cm high. Slightly hairy Hairy	Red clover
	Trifoliate, serrated edge with or without markings on upper surface	Small and pointed	Not hairy	White clover

IDENTIFICATION OF THE LEGUMES

Leaves. With the exception of the first leaves (which may be simple), all leaves are compound. In some species the midrib is extended slightly to form a *mucronate* tip. Other features on the leaf may be serrated margins, presence or absence of marks, colour and hairiness (Figs. 55 and 56). The leaves are arranged alternately on the stem, and they can consist of the *stalk* which bears two or more leaflets according to the species (Fig. 57).

Stipules. These are attached to the base of the leaf stalk. They vary in shape and colour (Fig. 55).

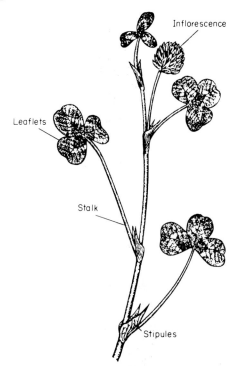

FIG. 57. Parts of the legume.

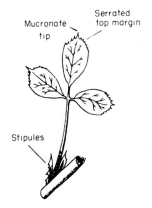

FIG. 55. Parts of the legume.

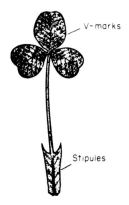

FIG. 56. Parts of the legume.

Flower. The flowers are brightly coloured and, being arranged on a central axis, form an indefinite type of inflorescence (page 12).

Some terms used

Before discussing grasses and grassland management it is important to understand what is meant by the following terms:

Seeding year. The year in which the seed mixture is sown.

First harvest year. The first year after the seeding year, and thus the second and third etc. harvest years.

Undersowing. Sowing the seed mixture with another crop (known as a cover crop). It is usually a cereal crop.

Note—*cover cropping.* A new term which describes the practice of growing a crop to reduce the amount of nitrogen leached from the soil over the winter months (page 64).

Oversowing. A term not often used now. It is the practice of sowing seed on to unprepared ground. The technique is mainly used to patch up a grass field (around troughs, field gates, etc.) where stock may, for example, have been outwintered.

Direct sowing or seeding. Sowing on bare ground without a cover crop.

Direct reseeding strictly means sowing without a cover crop, and putting the field straight back to grass, the previous crop having been grass. Very often it is used in the same way as direct sowing.

Note—direct sowing and direct reseeding should not be confused with direct drilling (page 41).

"Sod-seeding". A term sometimes used in grassland husbandry. It is the same as direct-drilling.

"Stitching-in". A method of improving a sward (page 193).

HERBAGE DIGESTIBILITY

One of the main characteristics which determines the feeding value of herbage plants is its digestibility, i.e. the amount of the plant which is actually digested by the animal. In the past, the digestibility of foods was only determined by animal feeding experiments. But, with the development of the laboratory *in vitro* technique which simulates rumen digestion, it is now possible for the digestibility of any species of herbage plant to be assessed very much more easily without using animals.

The digestibility of herbage plants is now expressed as the D value, which is the percentage of digestible organic matter in the dry matter.

High digestibility values are desirable as it means that the animal is able to obtain the greatest amount of nutrients from the herbage being fed. It is now accepted that the digestibility of herbage plants is a major factor affecting intake. Dry matter yield increases as the plant grows and thus the ratio of leaf (the most nutritious and digestible part of the plant) to stem widens. This causes a decline

in D value of the whole plant. The rate of decline depends on the grass species and the variety, but it is generally (but not always) determined by the earliness or lateness of flowering (Timothy and some varieties of Italian ryegrass are exceptions). Season and locality can also affect the rate of decline.

At 0.4–0.5% units per day, the rate of decline of D value following the initial high digestibility is fairly constant, although Timothy and some Italian ryegrass varieties decline at less than 0.4%.

With grasses, for conditions approximating to grazing, the D value for perennial ryegrass is about 70%, i.e. 70 D indicates 70 kg of digestible energy producing organic matter per 100 kg of dry feed. Cocksfoot has a D value of approximately 66%. Short-duration ryegrass and meadow fescue have D values just below 70% with Timothy at about 68%. For silage, higher yields will result when the crop is cut at a later stage of growth. With greater emphasis being placed on quality silage, advice is now for a D value of 67% as far as is possible (page 200). For hay, a 60% D value is a realistic figure (page 213).

The D value and rate of decline of regrowth herbage will depend to a great extent on when defoliation first takes place in the spring, e.g. an early graze in the spring will remove a lot of leaf whilst leaving the stem intact. This will result in a lower yielding and stemmier regrowth (with a faster rate of decline of D value). Leafier regrowths (with a slow rate of decline of D value) result when the primary growth in the spring is taken later.

In early May, the D value of white clover can be as high as 78%, with a decline of only about 0.15% unit each day. Red clover at the same time is at about 70% D value, whilst sainfoin and lucerne reach a maximum of 65%. All these three legumes show an approximate 0.36% unit daily rate of decline.

White clover should always be considered seriously in grazing leys of three or more years' duration, even with high nitrogen use.

Lucerne, over the season, will show a higher

yield of digestible dry matter than the other legumes, especially if, when made into hay, the loss of leaf is kept to a minimum.

GRASS VARIETIES

Over 150 different varieties of grasses can be found growing in this country, but only a few are of any importance to the farmer:

(1) Short-duration ryegrasses.
(2) Perennial ryegrass.
(3) Timothy.
(4) Cocksfoot.
(5) Meadow fescue.

The ryegrasses

Tetraploid ryegrasses

The chromosome numbers in the cell nuclei have been doubled. The seeds are larger, and this will normally mean a slightly higher seed rate. They generally produce a bigger plant than the diploid (ordinary grass), but in yield of dry matter there is no great difference.

Most of the recommended tetraploids are from the Continent (although British breeders have produced some short duration ryegrass tetraploids), and they do appear to have better resistance to frost than the British diploids, but this is probably due to the country of origin.

Tetraploids are characteristically deep green, and are slightly more palatable and digestible than diploids. But because of their higher moisture content and larger and thicker-walled cell structure, tetraploids are perhaps not quite so suitable for conservation in the wetter parts of the country, particularly with high nitrogen use.

Sown alone they are not so persistent, as they tend to produce rather too open a sward. Not more than 30% should be included in leys intended for intensive grazing.

Short duration ryegrasses (Fig. 58)

This class consists of varieties which fall into three main groups: Westerwolds ryegrass,

FIG. 58. Short-duration ryegrass.

Italian ryegrass and Hybrid ryegrass. It is usually referred to simply as Italian ryegrass.

Westerwolds. This is an annual and the quickest growing of all grasses. A good crop can often be obtained within 12–14 weeks of sowing. It should not be undersown. It is best direct sown in the spring and summer as it is not at all winter-hardy, except in very mild districts. Dutch bred varieties are used.

Italian ryegrass. This is short-lived (most varieties persist for 18–24 months) and very quick to establish. Sown in the spring, Italian ryegrass can produce good growth in its seeding year and an early graze the following year, but for optimum production in its harvest year (particularly the spring) it is best sown in summer or early autumn. It does well under most conditions, but responds best to fertile soils and plenty of nitrogen. Like all ryegrasses, its winter-hardiness is improved when surplus growth is removed in the autumn. Although stemmy, it is palatable with a high digestibility.

Tribune and *Roberta* (tetraploid) are examples of varieties.

Hybrid ryegrass. Some varieties in this group are similar to the less persistent Italian ryegrass varieties and others are similar to perennial ryegrass. A feature of hybrid ryegrass is the good resistance to disease. *Polly* and *Barvisto* are both tetraploids.

Perennial ryegrass (Fig. 59)

Although it forms the basis of the majority of long leys and is the most important grass found in good permanent pasture, perennial ryegrass is, depending on the variety, also used in three to four year leys. It is quick to establish and yields well in the spring, early summer and autumn. It does best under fertile conditions and responds well to nitrogen. The D value is high.

Timothy (Fig. 60)

This is fairly slow to establish. It is not particularly early in the spring. It is less productive than the other commonly used grasses, but is very palatable, although its digestibility is not as good as the ryegrasses. It is often included in grazing leys with perennial ryegrass, but because of its palatability it tends to get eaten out at the expense of the ryegrass, although the more persistent varieties compete reasonably well. It is winter-hardy and does well under a wide range of conditions, except on very light dry soils.

Early varieties: *Kampe 11, Goliath.*
Intermediate varieties: *Motim, Pecora.*
Late varieties: *S48, Intenso, Olympia.*

FIG. 59. Perennial ryegrass.

FIG. 60. Timothy.

Crown rust disease has been a problem in the past, but now most recommended varieties show good resistance (Table 72).
Examples:
Early varieties: *Frances, Bastion* (tetraploid).
Intermediate varieties: *Corbiere, Merlinda* (tetraploid).
Late varieties: *Hercules, Condesa* (tetraploid).

Cocksfoot (Fig. 61)

This is quick to establish and is fairly early in the spring. It is a high-yielding grass, but unless it is heavily stocked, most of the existing varieties will soon become coarse and unpalatable. Cocksfoot is a deep rooting grass and therefore an excellent drought resister.

FIG. 61. Cocksfoot.

FIG. 62. Meadow fescue.

It does not need really fertile conditions. It produces well in late summer/autumn and most varieties show good winter-hardiness.

It should be considered as a special-purpose grass on the drier lighter soils in areas of low rainfall. Cocksfoot varieties are not as digestible as the other important grasses.

Varieties: *Jesper, Sparta*.

Meadow fescue (Fig. 62)

Meadow fescue is not used very much now. This is because the once famous Timothy/meadow fescue mixture is no longer considered to be so valuable as it is not so productive compared with the ryegrass sward.

Meadow fescue is slow to establish, but once established it is fairly early in the spring, and it has a high digestibility.

Varieties: *Salfat, S.215*.

Other Grasses

Forage brome (e.g. Grassland Matua). These grasses are perennial, but with persistence limited to three to four years. They have been developed mainly for cutting. They are not as digestible as the ryegrasses and there could be some qualification about their palatability. Forage bromes generally have an upright growth habit (except *Deborah*) which inhibits competition against weeds.

Tall fescue is a perennial. Once it is well established, varieties such as *S170* and *Festal* are very useful for early grass in the spring, but production after this is not very high. It is very hardy and it can be grazed in the winter.

The *small fescue* grasses, such as creeping red fescue and sheep's fescue, are useful under hill and marginal land conditions, and in some situations they will produce more than perennial ryegrass. They have no practical value under farm lowland conditions. They are typical "bottom" grasses, and they produce a close and well-knit sward. For this reason they are often included in seed mixtures for lawns and playing fields.

Rough-stalked meadow grass is indigenous to the majority of soils, but it prefers more moist conditions. In the later years of a long ley, it can make a very useful contribution to the total production of the sward after the white clover has been crowded out by heavy nitrogenous manuring. It is never included in seed mixtures now.

Smooth-stalked meadow grass (Kentucky Blue grass) spreads by rhizomes, and can withstand quite dry conditions.

Crested dog's tail is not very palatable because of its wiry inflorescence. Now it is only used in seed mixtures for lawns and playing fields.

Weed Grasses

The majority of grasses naturally growing in this country are of little value. But some of them are extremely persistent and are able to grow under very poor conditions where the more valuable grasses would not thrive. However, their production is always low, they are usually unpalatable and, under most conditions, they can be considered as weeds.

Well-known examples of weed grasses are:

The bents. They are very unproductive and unpalatable and, under most conditions, they can be considered as weeds.

Yorkshire fog. This is extremely unpalatable except when very young. It is prevalent under acid conditions, although it can be found growing anywhere. It is sometimes used in reclaiming hill pastures.

The weed grasses—couch, slender foxtail (blackgrass), wild oat and barren brome—are all associated with intensive cereal growing.

LEGUMES

The clovers of agricultural importance are the red and white clovers.

Red clovers are needed in the short ley for conservation, and white clovers are used for the longer duration grazing ley where they act as "bottom plants". With their creeping habit of growth they knit the sward together and help to keep out weeds. Although the majority of clover is palatable with a high feeding value, they are not so productive as the grasses and they should not be allowed to dominate the sward (aim for 30% clover).

The digestibility of clovers (except white clover), like other legumes, is not as high as the grasses, but the voluntary intake at equal digestibility is higher with the legume.

Because of the ability of legumes to fix atmospheric nitrogen, a balanced mixture of grasses and clover will leave a useful legacy of nitrogen for the following crop.

Red clovers (Fig. 63)

These are short-lived. They are included in short leys and sometimes in long leys to improve bulk in the early years. Although used more for conservation, some of the more persistent

FIG. 63. Red clover.

improved varieties are useful for aftermath grazing.

There are two main groups:

Early red clover (broad red clover: double cut cowgrass). The tetraploid varieties such as *Redhead* and *Deben* are generally more persistent than the diploids. Early red clovers usually show good resistance to sclerotinia (clover rot, Table 72), but they are susceptible to stem eelworm (Table 71).

Late red clover (late flowering red clover). These varieties are later to start growth in the spring than the early red clovers. They are also more persistent and, under good conditions, they can last up to four years. *Grasslands Pawera* is a tetraploid from New Zealand. It is a high yielder in the first harvest year and it also shows good resistance to stem eelworm.

White clovers (Fig. 64)

These should be regarded as the foundation of the grazing ley. They are generally not so productive as the red clovers, but are more persistent. There are four types, classified according to leaf size.

Very large-leaved white clovers. A variety in this group (such as *Aran*) may become more widely used because of its ability to compete better with the large amounts of nitrogen used on grazing swards. Aran is not very persistent and shows poor resistance to sclerotinia (Table 72).

Large-leaved white clovers. Varieties in this group are used in short and medium-term leys. *Blanca* and *Olwen* are typical examples.

Medium-leaved white clovers. These are extremely useful in leys of up to four years and are very quick to establish. The Aberystwyth variety *Donna* is a good all-round example.

Small-leaved white clovers. These are used in the long ley. They are rather slow to establish but can become dominant. *Kent wild white* clover is a good local variety and *S184* is another very productive and creeping type.

OTHER LEGUMES

Lucerne (Fig. 65)

This is a very deep-rooting legume and it is therefore useful on dry soils, although it can be grown successfully under a wide range of soil conditions provided drainage is good. For better establishment lucerne should be grown with a companion grass such as perennial ryegrass, Timothy or meadow fescue. Although it is very productive, lucerne is not very palatable and it is the least digestible of all the important legumes. It is probably best utilized by drying through the grass drier, although excellent barn-dried lucerne hay can be made without much loss of leaf. There are many varieties of lucerne, and in this country they can be grouped into early and mid-season types. Where possible the early type should be grown. At present, *Euver* is the heaviest yielding variety; it does show some resistance to verticillium wilt (Table 72), although it is not as tolerant to the fungus as is *Vertus*. Both varieties show some resistance to stem eelworm.

FIG. 64. White clover.

FIG. 65. Lucerne.

Sainfoin (Fig. 66)

As regards its economic use, sainfoin is very similar to lucerne although it is not so productive. It has not been very popular in the past due to difficulties in establishment (it does not compete very well with weeds) but, sown with grasses such as cocksfoot, this problem has been overcome. For forage purposes especially it can be considered for calcareous soils. Sainfoin hay is highly valued for horses, but it is difficult to make without losing leaf. There are two main types of sainfoin:

English Giant. This is heavy yielding, but short-lived—usually one harvest year.

Common sainfoin. This lasts for several years, but is not so heavy yielding.

HERBS

These are deep-rooting plants which are generally beneficial to pastures. But to be of any value they should be palatable, in no way

FIG. 66. Sainfoin.

FIG. 67. Yarrow.

FIG. 68. Chicory.

FIG. 70. Burnet.

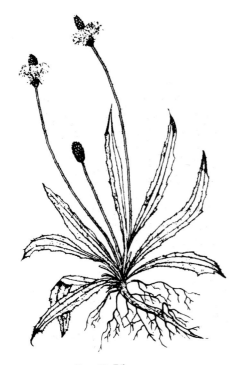

FIG. 69. Rib grass.

harmful to stock, and they should not compete with other species in the sward. They have a high mineral content which may benefit the grazing animal.

Yarrow, chicory, rib grass and burnet (Figs. 67–70) are the most useful of the many herbs which exist. They can be included in a seed mixture for a grazing type of long ley. They are not cheap, however, and, as one or more of these herbs will usually get into the sward on its own accord, there does not seem much point in including them in the first place.

NEW DEVELOPMENTS WITH GRASSES AND CLOVERS

In spite of the development of new grasses for commercial use such as the *Bromus* species, plant breeders will continue to concentrate most of their work on improving the performance of the ryegrasses. The next generation of these grasses could well see more emphasis on persistency, winter-hardiness and drought resistance. New varieties are needed to stand up to the more intensive grazing systems common now, compared with 40 years ago. This intensification of grassland farming, brought about by heavier nitrogen use and increased stocking, has meant that varieties which were persistent for eight to ten years are now dying out much sooner. It is now recognized that long leys need plants better able to withstand this hard management as well as leafier varieties with a more prostrate habit of growth to form a good bottom to the sward. High nitrogen applications are also altering the balance of protein and carbohydrate in the plant. A plant with a higher protein content is resulting in greater susceptibility to winter kill. These new varieties will generally have a higher carbohydrate content. Furthermore, it seems that these carbohydrates could be more effectively used by selecting ryegrass material with a lower rate of respiration. This should lead to an increased dry matter yield, particularly during the summer when temperatures are higher.

More persistent medium-leaved white clovers, which will grow at lower temperatures, have been developed. The new varieties (not yet commercially available) have a larger stolon length. These, with the grass plant to protect them, can produce more leaf the following growing season.

Susceptibility to winter kill has always been a problem with ryegrass varieties. Not only can the plant actually die, but in less severe conditions the winter temperatures can slow down subsequent growth in the spring and summer, resulting in significantly lower production. New varieties bred from selected parents, particularly from Switzerland, are showing much greater cold tolerance and the ability to produce a high rate of earlier spring growth.

Drought-resistant varieties are also being sought, both for short duration and perennial ryegrass. At the Welsh Plant Breeding Station selections have been made from plants with a slower evaporation loss, a more expansive root system (not necessarily a larger one), and where the turgidity of the cells can be maintained that much longer in a drought. Progress has also been made in improving mid-summer production in perennial ryegrass by crossing very early and very late flowering varieties. However, apart from this and the Italian/perennial ryegrass hybrids which have been developed at Aberystwyth, hybrids between other grasses are also showing promise. An Italian ryegrass/meadow fescue cross should mean the combination of the winter-hardiness of the fescue and the rapid growth and earliness of the ryegrass. Perennial ryegrass is also being used with meadow fescue which would obviously show more persistence than the Italian ryegrass cross.

The possibility of a hybrid between Italian ryegrass and tall fescue with the potential of a persistent conservation plant is again being examined. Poor seed production has been a problem in the past. New plant breeding techniques could alter the picture.

Cocksfoot varieties which are easier to manage under grazing conditions and which, at the same time, have higher digestibilities, are still awaited. Material from the Iberian Peninsula is showing promise in both these respects.

As well as the nutritive value of forage legumes being higher than that of grass, the intake is up to 30% more when compared with grass at the same D value. All this is in addition to the obvious higher protein content of the legume. With emphasis now on using less nitrogen for the grass crop, the management of the clover plant in the sward is being re-examined (page 180). Research is also involved with seeing how legumes can be more widely used in the diet for cattle. But there is always the problem of bloat (not with sheep). However, this may be overcome fairly shortly with the breeding (by genetic manipulation) of bloat-free varieties of legumes.

Only recently has it been appreciated that grass and clover diseases and pest attack on swards can have a considerable adverse effect on herbage production. Probably up to 20% production is lost every year.

Resistance to rhynchosporium, ryegrass mosaic virus and mildew of the short duration ryegrasses and crown rust of perennial ryegrass has now developed with an increasing number of varieties. Some cocksfoot varieties are showing reasonable resistance to mastigosporium (leaf fleck). Varietal resistance to these and other diseases such as sclerotinia of clovers will offer the best and cheapest form of control (Table 72).

Growth regulators. The use of growth regulators for grass has, until recently, largely been confined to suppressing grass growth in such places as urban parks and golf courses. Only recently has attention been turned to their possible potential in grassland farming. With grass, the emphasis is on regulating tillering, overcoming apical dominance to produce a closer-knit, more prostrate sward for grazing. This, in turn, should help to maintain white clover in the mixed sward.

By suppressing flowering it should also be possible to maintain the D value at an optimum stage for cutting for silage and/or hay that much longer. The outlook for growth regulators is promising, but continuing research is necessary before they can be used properly as a part of grassland husbandry.

SEED MIXTURES

Many farmers depend on reliable seed firms to supply them with standard seed mixtures, whilst others prefer to plan their own mixtures which the merchant will then make up for them.

The following points must be considered when deciding upon a seed mixture:

The purpose of the ley

Varieties of herbage plants have different growth characteristics, and because of this the type of stock using the ley will influence the choice of seed. This is shown in Table 48.

Soil and climatic conditions

The majority of herbage plants will grow under a wide range of conditions. However, on heavy, wet soils there is no point in growing

TABLE 48

Purpose	Varieties which should be used
For early grazing, i.e. early bite	Mainly Italian ryegrass and hybrid ryegrass
For optimum production throughout the season	Ryegrass (supported by either non-ryegrass or permanent pasture)
For winter-grazing cattle – foggage (provided conditions allow)	Non-ryegrass
For the grazing block	Herbage plants which produce a closely knit sward
For the cutting block	Herbage plants which produce a relatively tall habit of growth

Note: The heading dates will generally coincide if the above points are borne in mind when deciding on a seeds mixture. As far as possible, varieties should correspond because, with only a few exceptions, the digestibility of a plant starts to decline when the ear emerges. A grass crop cannot be so valuable if its various plant components head at different times. If, for example, the crop is cut or grazed at the average date for the plants making up the sward, the earliest heading varieties will have declined in quality, whereas the latest will not have produced any sort of yield. Compatibility is important.

early grasses or planning for foggage (winter) grazing. Like the majority of crops, grasses and particularly legumes will not thrive where there is a lack of lime in the soil.

The length of the ley

Perennial ryegrass, Timothy, cocksfoot and meadow fescue, being persistent, are suitable for long leys, whilst the less persistent and quick-growing early types, of which Italian ryegrass is an outstanding example, will make up the short ley.

Cost of the mixture

This will depend on the variety used and the seed rate. It is false economy to buy unsuitable varieties just because they are cheap. It is equally unwise to be persuaded to buy expensive and often unproved seed. Generally speaking, it will be found that varieties required for the short-term ley are cheaper than those for the longer ley.

Tables 49–51 – give examples of different seed mixtures.

MAKING A NEW LEY

A case can be made for sowing the seeds either in the spring or late summer/early autumn period.

Spring sowing

If direct sowing without a cover crop, the maiden seeds can give valuable production in

TABLE 49   *Short-term leys (1–2 years) (amounts per ha)*

25–30 kg Westerwolds ryegrass	Its only year is its seeding year; not suitable for undersowing; very quick to establish; will produce two good cuts or grazings, especially if the first one is taken before the seeds set.
28 kg *RVP* or similar 30 kg hybrid ryegrass, e.g. *Augusta*	If undersown, reduce the seed rate by about 25%. Although chiefly for grazing, it can be used for conservation.
15 kg hybrid ryegrass, e.g. *Augusta* 9 kg early red clover, e.g. *Merviot*	For general-purpose use; hybrid ryegrass is a good companion plant as it is less aggressive than Italian ryegrass. *Merviot* shows reasonable resistance to both sclerotinia and stem eelworm (see Tables 71, 72).

In the examples, ryegrass is the only variety of grass used. The emphasis must be on quick establishment with good production from an early stage, and with an ability to respond well to nitrogen. As persistency is not important, the short duration ryegrasses will amply fulfil these requirements, provided that management is correct.

TABLE 50   *Medium-term leys (up to 5 years) (amounts per ha)*

6 kg tetraploid Italian ryegrass, e.g. *Multimo* 8 kg hybrid ryegrass, e.g. *Augusta* 15 kg intermediate perennial ryegrass, e.g. *Talbot* 3.5 kg large-leaved white clover, e.g. *Blanca*	A highly productive ley suited to intensive grazing. White clover will improve the quality of the sward. However, with controlled grazing, nitrogen fertilizer much above 250 kg will depress the clover (page 180).
15 kg intermediate perennial ryegrass, e.g. *Talbot* 15 kg late perennial ryegrass, e.g. *Melle*	An easy ley to manage; it can be extremely productive with liberal use of nitrogen. Well suited to set stocking.

Properly managed, these leys should be able to produce almost as much as the short-term leys, although they will not be so early to start growth in the spring. Depending on management and conditions, the intermediate and late perennial ryegrasses should keep the sward going up to 5 years.

TABLE 51 *Long-term leys (amounts per ha)*

8 kg intermediate perennial ryegrass, e.g. *Talbot* 14 kg late perennial ryegrass, e.g. *Melle* 5 kg late Timothy, e.g. *S48* 2.5 kg medium-leaved white clover, e.g. *Donna* 1 kg small-leaved white clover, e.g. *S184*	For general-purpose use. A very productive and hard-wearing ley. The timothy is included to add palatability although it may have difficulty competing with the aggressive ryegrass.
22 kg late perennial ryegrass, e.g. *Melle* 2.5 kg medium-leaved white clover, e.g. *Donna* 1 kg small-leaved white clover, e.g. *S184*	A very simple mixture well suited to intensive grazing and plenty of nitrogen.
8 kg late Timothy, e.g. *S48* 10 kg late meadow fescue, e.g. *S215* 2.5 kg medium-leaved white clover, e.g. *Donna* 1 kg small-leaved white clover, e.g. *S184*	For general-purpose use. This ley should tie in with ryegrass leys producing well when ryegrass has fallen back in production in July. It is not so productive as the ryegrass ley.
14 kg intermediate perennial ryegrass, e.g. *Talbot* 7 kg cocksfoot, e.g. *Prairial* 4 kg late Timothy, e.g. *S48* 3 kg late red clover, e.g. *Britta* 2 kg medium-leaved white clover, e.g. *Donna* 1 kg small-leaved white clover, e.g. *S184*	This is the well known Cockle Park type general-purpose mixture. It is meant to give even production throughout the whole season, but eventually, depending upon soil type and management, it will tend to become either ryegrass or cocksfoot dominant.

It will be noted that small-leaved white clover (the so-called wild white clover) is included in these examples of mixtures. After about 3 years, it should have established sufficiently well to fill the bottom of the ley and give a well-knit sward. Only a relatively small amount is needed to avoid it dominating the whole sward at the expense of the more productive grasses.

the summer. Establishment can be enhanced by stock being able to graze the developing sward within a few weeks of sowing. This is not possible to the same extent with autumn sowing. A limiting factor with spring sowing may be moisture, and in the drier districts the seeds should be sown in March if possible. The plant should thus establish itself sufficiently well to withstand a probable dry period in early summer. Of course, undersowing the cereal crop (which is only possible in the spring) does mean that the fullest possible use is being made of the field, although with the slower-growing grasses, establishment is usually not so good, especially in a dry year.

Late summer/early autumn sowing

With ryegrass as the main constituent of the mixture, the crop is now more normally sown at this time of year. It will generally lead to a heavier crop the following spring than when sowing in the spring.

There is usually some rain at this time; heavy dews have started again, and the soil is warm. But with clovers in the seed mixture earlier sowing may have to be carried out so that the plants have developed a good tap root system before the onset of frosts. Whether earlier sowing is possible depends on when the preceding crop in the field is harvested. Winter barley is the ideal crop to follow.

Direct sowing or undersowing?

There are points for and against the practice, but direct sowing is now widespread. It is certainly preferable for early bite following sowing the previous late summer/early autumn period. With spring sowing it is also a better option where conditions for establishment are not so good and when the extra grass is required in the summer.

Green crops as temporary cover crops are excellent, e.g.

(1) Rape sown at 4–7 kg/ha with the seed mixture, and grazed off in 6–10 weeks.
(2) Oats or barley sown at 60 kg/ha with the seed mixture.

Because of competition from the cereal crop, establishment is generally not so satisfactory when undersowing herbage seeds. Occasionally, in a wet summer, the herbage plants (particularly red clover) can grow up too fast, not only to compete with the cereal crop but also to impede combining. Weed control (especially with clover in the seed mixture) is also more difficult in the cereal crop.

An advantage of undersowing is that, apart from perhaps an extra harrowing, only one seedbed preparation is necessary for the two crops, i.e. the cereal cover crop and the herbage seeds.

SOWING THE CROP

Direct sowing

Reference has already been made to the cultivations necessary for preparing the right sort of seedbed for the seeds. But it must be re-emphasized that grass and clover seeds are small, and therefore they must be sown shallow, and that therefore a fine and firm seedbed is necessary.

With ample moisture the seed can be broadcast. This should be on a ribbed-rolled surface, so that the seeds tend to fall into the small furrows made by the roller. Most fertilizer distributors can be used for broadcasting.

In the drier areas, and on lighter soils, drilling is safer. The seed is then in much closer contact with the soil. The 10-cm coulter spacing of the ordinary grass drill should give a satisfactory cover of seeds, but if using the corn drill with 18-cm spacing the seeds should be cross-drilled. After broadcasting or drilling, except on the wetter seedbeds (when the seeds harrow will be used), rolling will complete the whole operation. Where necessary, 30 kg/ha nitrogen and 50 kg each of phosphate and potash can be broadcast

and worked in during the final seedbed preparation.

Direct drilling

A certain amount of grassland is now reseeded by direct drilling. Although it is no cheaper than conventional reseeding, the practice does allow a much quicker turn-round from the old grass to the new sward. It also reduces poaching and, most important, it enables reseeding in situations where it would not be possible to plough or even do minimum cultivations, e.g. watermeadows.

Glyphosate is increasingly associated with the technique, although paraquat can be used.

Undersowing

Any of the autumn-sown cereals may be used as a cover crop, but they compete more with the seeds than spring-sown cereals. Harrowing of the ground will be necessary and then the seed should preferably be drilled across the corn drills followed by the light harrows. Alternatively, the seed can be broadcast and harrowed in, but this is not so satisfactory.

With a spring-sown cover crop the cereal is sown first to be followed immediately by the seed mixture drilled or broadcast. This is desirable with slow establishing mixtures, but with vigorous species, e.g. Italian ryegrass and red clover, it may be preferable to broadcast the seed after the cereal is established.

FERTILIZERS AND MANURES FOR GRASSLAND

Grass, like all crops, needs plant food for its establishment, maintenance and production.

Nitrogen is vital for the grass crop. Provided other essential plant foods are available, nitrogen and site (the situation in which the crop grows) are the most important factors influencing grass production.

Grass growing sites differ in their potential. Described as *site class*, they are classified in terms of soil texture and summer rainfall (April to September). There are five grassland production site classes as in Table 52.

The target or potential dry matter yield from all-grass swards on the five site classes is shown in Table 55 and Fig. 71.

On all sites and with frequent cutting and grazing, there is generally a linear response up to 300 kg nitrogen/ha. This is the most economic phase. Depending on the amount of extra nitrogen used and the site class, the linear response phase will be followed by a diminishing response (which may still be economic). In practical terms, the better the growth potential of the site, the better the response to nitrogen.

There are other sources of this plant food and, although on their own they are unable to supply all the nitrogen requirements of the crop needed to achieve an economic return, they should not be ignored, particularly in view of increasing concern about nitrogen pollution of waterways and aquifers.

Clover nitrogen. Without fertilizer nitrogen, white clover will make a useful contribution to the requirements of the grass crop. The rhizobium bacteria in the root nodules fix atmospheric nitrogen and, as they die and are replaced, nitrogen is made available for the companion grass plant. The amount of

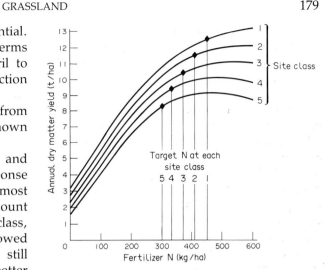

FIG. 71. From Grassland Research Institute (now Institute Grass and Environment Reseach—IGER).

nitrogen produced will be proportional to the amount of clover in the sward, as well as to soil temperature conditions and seasonal weather (rainfall especially). With a 30% clover population and a long growing season and with no fertilizer nitrogen, up to 200 kg nitrogen/ha could be made available to the crop in the year.

White clover will reduce the response of grass to low rates of nitrogen. However, as the amount of fertilizer nitrogen is increased, the clover plant is crowded out by the responding

TABLE 52

Site class.
Classification of grass production site potential

	Average rainfall – April–September		
	> 400 mm	300–400 mm	< 300 mm
Soil texture			
Clay loams & heavy soil	1	2	3
Loams, medium soils & deep soils over chalk	2	3	4
Shallow soils over chalk or rock, gravels & coarse sands	3	4	5

TABLE 53
*Target (potential) dry matter yield from
nitrogen application*

Site class	All grass swards
1	12.7 t/dm/ha using 450 kg N/ha
2	11.6 t/dm/ha using 410 kg N/ha
3	10.5 t/dm/ha using 370 kg N/ha
4	9.5 t/dm/ha using 330 kg N/ha
5	8.4 t/dm/ha using 300 kg N/ha

i.e. POTENTIAL response of dry matter (for the linear response phase) per kg nitrogen is approximately the same on each site class – 28 kg dry matter.
ACTUAL response (linear up to 300 kg N) under all sites with frequent cutting and grazing from a recent study on 21 sites managed uniformly showed a mean response of 23 kg dry matter (14–29).

grass plants (the nitrogen may also hold back the development of the rhizobium), resulting in a steady decrease in clover nitrogen to an eventual nil contribution with fertilizer rates of 200–250 kg/ha. (Very approximately, 2 kg fertilizer nitrogen decreases clover nitrogen by 1 kg/ha, although there can be considerable variations.) Irrigation (page 195) and set stocking (page 190) allow more nitrogen with less prejudice for the clover plant. A large-leaved long-duration white clover plant would compete better with the nitrogen fertilized grass.

Nitrogen applied in early spring and well before clover growth has started should help to maintain the plant in the sward. This should be the objective. Although annual dry matter yield of clover is not as much as that from grass, it has a higher nutritive value and a higher D value maintained over a longer period. In a ryegrass/white clover sward, the clover does help to sustain production in July—the "bad summer gap" period with ryegrass. The clover contribution could be at least 50% of the dry matter yield at this time.

Clover nitrogen on its own cannot sustain an intensive grass management system (a 25–30% population should be the target; a higher clover plant population could be considered for more extensive systems). It will have to be augmented by fertilizer nitrogen—the amount depending on the production sought (nitrogen and stocking rate, page 187).

Recirculated nitrogen. Up to 90% of the nitrogen content of the sward consumed by the animal can be returned to the soil. However, it is not evenly distributed and even at a stocking rate of 2.5 livestock units/ha its effectiveness will probably not exceed 30%. Recirculated nitrogen in slurry and farmyard manure is certainly more effective than that from grazing animals, provided that these organic substances have been properly stored and applied (page 59).

Soil nitrogen. Depending on the organic matter content and the activity of the micro-organisms, nitrogen in the range 40 kg (light arable soil) to 180 kg (very good heavily-stocked permanent pasture) will be released per hectare from the soil each year. This can increase the longer the field is in grass.

Phosphate and potash will both help in the establishment of the sward, and on most soils they are important in helping to maintain the general vigour and well-being of the crop. There is no doubt that the response to nitrogen by the grass crop is far greater if there is an adequate supply of phosphate and potash present.

Phosphate can be applied at any time of the year (although it is not generally advised during the winter non-growing period). However, recent work from the Scottish Agricultural Colleges suggests that it might be preferable to apply phosphate (in combination with nitrogen) annually in the spring to low phosphate soils. Soil phosphate is largely immobile with low early spring soil temperatures, and when root extension begins in early spring the plant needs a supply of this plant food close at hand. How much to apply will depend to some extent on the soil phosphorus index, but it can be the same, irrespective of whether the crop is cut or grazed (Tables 54 and 55).

Potash. Like phosphate, the need for potash increases when higher nitrogen rates are used for the grass crop. With potash it is also important to treat a grazed and a cut crop as two entirely different crops, but in either case the potash should not be applied in the spring. The uptake of magnesium by the plant is always slow at that time of the year and, as potash tends to act as a buffer against magnesium, a spring application will accentuate a shortage of magnesium in the plant. This could well lead to hypomagnesaemia (grass staggers) in the animal.

For grazed swards, little potash is required, although this depends on the potash status of the soil. The grazing animal itself recirculates potash back to the soil, and although distribution may be uneven, at normal stocking rates this can amount to more than 125 kg/ha in the year. If the soil's annual natural contribution (60–250 kg/ha, depending on soil type) is added, it will be seen that it is not always necessary to apply potash. Luxury uptake of potash (the plant containing potash surplus to its needs) should be avoided. It is wasteful, and it does increase the risk of hypomagnesaemia.

The cut crop needs more potash than the grazed sward. A vigorous growing crop, continually cut, will take up about 360 kg/ha in the year. Depending upon soil type, up to 300 kg/ha might be needed to replace that removed. For efficient utilization, it should be applied at intervals through the season **after** the crop has been taken (Tables 54 and 55).

Sulphur. There is now some evidence about the need for sulphur on second and third cuts in the year.

The use of modern fertilizers and a less polluted atmosphere has reduced the "natural" build-up of sulphur in crops and grass (Table 11). However, there is normally no necessity for extra sulphur for the first cut (it will have built up in the grass over the preceding months). A deficiency is only likely in parts of the country well away from heavy industry. A foliar spray may have to be used, but advice should be sought if a sulphur deficiency is suspected, i.e. a lack of vigorous growth in spite of nitrogen application.

Lime. The grass, or grass/clover sward, like any other crop cannot thrive in acid conditions. The legumes, particularly red clover and lucerne, are very sensitive to a low pH. Regular maintenance dressings of calcium carbonate will usually be necessary, although how much and how often will depend on rainfall, soil type and type of management. The pH should be kept at about 6.0. This should be checked every four to five years with a "surface sample" taken, i.e. down to 7–8 cm. For lucerne a minimum pH of 6.5 is necessary through a considerable depth of soil (page 30).

Farmyard manure (FYM). The composition of FYM has already been discussed. If it cannot be used for an arable crop, then it is best applied to those swards which are going to be taken for hay or silage. Cutting will help to even out the uneven effects of growth following FYM application, due to the variable plant food content of the manure.

It should be applied at 25–30 tonnes/ha. With a normal autumn/winter application some of the nitrogen will be lost, but it should be possible to reduce particularly the phosphate and potash requirements of the grass crop the following season.

Slurry. The plant food content of slurry, and the problems of applying it to the grazing sward, have been discussed on page 61. On the all-grass farm unseparated slurry must obviously be used on the cutting block. Used in this way (rather than on a "sacrifice" field), it is best applied at rates of up to 50,000 litres/ha at intervals throughout the summer. This is seldom carried out in practice and, as a means of emptying the store at the end of the winter, the slurry goes on, sometimes in very large amounts, for the following silage crop. The possible adverse effects on soil structure should be noted. Also, if it is applied too near to cutting (allow at least six weeks) there is a risk that some of it will be carried into the silo, producing the wrong fermentation (page 203).

The Code of Good Agricultural Practice concerning the collection, storage and use of organic manures should be followed. All organic manures and waste waters must be spread so that they do not enter waters by run-off or rapid percolation through soil.

MANAGEMENT OF A YOUNG LEY

A direct-sown long ley in its seeding year sown in the spring or summer

The sward should be grazed six to ten weeks after sowing, depending upon weather conditions which naturally affect growth. This early grazing helps to consolidate the developing sward, and encourages the plant to tiller out. It should not be too hard, but equally it is important not to undergraze. The sward should then be rested and followed by periodic grazings becoming more intensive throughout the season.

It is unwise to cut the ley in its seeding year. Plants which are allowed to grow too tall before being cropped never develop very strongly. The essential tillering is not encouraged to the same extent, and the sward is left "very open" into which weeds may soon gain a foothold.

If grazing is not possible, the plants should be topped before they grow too tall.

A direct-sown ley sown in the late summer

This should either be grazed or topped in the autumn.

Frit fly can be a problem (Table 71).

An undersown ley

This should either be grazed or topped in the autumn. If the seeds look "thin" after the corn has been harvested, 40 kg/ha of nitrogen will help to stimulate growth.

TABLE 54   *Management of long leys. Perennial ryegrass – mainly for grazing*

MARCH	Apply 80–100 kg nitrogen for first graze following early bite on a short-term ley. If have to use perennial ryegrass for early bite, nitrogen applied earlier (pages 184, 185).
APRIL	First graze taken.
MAY JUNE JULY AUGUST	Apply nitrogen at a maximum rate of 2.5 kg/ha for each day's interval between grazing, e.g. a 28-day paddock system requires 70 kg nitrogen. When set stocking, the whole field can either be top-dressed with nitrogen every 4 weeks or each quarter of the field receives nitrogen in successive weeks. The August application of nitrogen can be reduced by about 20% to help maintain or improve the sward's clover content. Late nitrogen also increases the protein content of the grass which makes it more vulnerable to winter kill.
SEPTEMBER	If there is too much clover in the sward the grass should be allowed to grow up a little more before taking the final graze.
OCTOBER	It is important to see that eventually the sward is grazed down tight (sheep are ideal, and in any case autumn-calving cows should be off grass by this time) to avoid winter kill. Apply 0–40 kg phosphate and 0–40 kg potash annually; 1–1.25 tonnes/ha of K slag (or its equivalent as ground mineral phosphate) can be used every 4–5 years.
NOVEMBER DECEMBER JANUARY FEBRUARY	Lime if required.

Notes
  (i) Under dry conditions such frequent nitrogen applications are unnecessary in the March–August period, but at least 250 kg should have been applied by early June. This will help to overcome the effects of a soil moisture deficit, as a soil with a high nitrogen content increases the efficiency with which the soil water is used. (See also p. 187, Stocking rate for the dairy herd.
 (ii) Urea, used up to the end of May, has, on most soils, given comparable results to that of ammonium nitrate fertilizer. But if warm, dry weather follows, urea used then as a top-dressing may not perform quite as well.

MANAGEMENT OF THE ESTABLISHED GRASS CROP

The long ley

Ryegrass now makes up more than 90% of the grass species used in leys.

Table 56 outlines the management of the long ley. Obviously this will vary according to the season, growth and stock used. The management chiefly refers to the dairy cow, although it can apply to an intensive beef grazing system.

Swards containing some late varieties of perennial ryegrass with white clover, together with the sensible use of nitrogen, have to some extent lessened the problem of a reduction in grass growth in July and August before it comes away again in early autumn. There is always a dip in the growth curve and it can be quite marked in a dry summer. Direct-sown ryegrass leys seeded down in the spring, and the short ley, will also help to avoid a fall in production later on in the summer, as also will good permanent pasture if properly managed. This all contrasts with previous years when, in order to obtain a fairly uniform supply of grass throughout the season, non-ryegrass swards were managed alongside ryegrass mixtures. But, as perennial ryegrass under most management conditions is the most productive grass, it is obviously sensible to use it wherever possible, although the value of good permanent pasture should not be forgotten. Because it will inevitably carry a proportion of less productive non-ryegrass plants, it will not show the summer production dip characteristic of the ryegrass sward. Of course, there are still some farmers who will use a Timothy/meadow fescue mixture as an insurance against poor production of the ryegrass. And non-ryegrass swards are useful for winter grazing or foggage.

MANAGEMENT OF PERMANENT PASTURE AND NON-RYEGRASS LEYS

On permanent pasture and non-ryegrass leys, management is generally not as intensive as the ryegrass sward. Total nitrogen use will be below 300 kg/ha, although obviously there is no hard and fast rule. Early bite would not be expected but, as far as possible, these swards should be managed so that they can produce a worth-while crop when the ryegrass may have fallen back in production. The swards are also very useful for autumn grazing, and nitrogen applied late August–early September would help for winter grass, provided ground conditions were favourable.

As with the ryegrass swards, attention should be given to maintaining a reasonably correct balance between the clovers (25–30% plants making up the sward) and the grasses. Too much clover in the sward should be avoided, although too little will lead to an unproductive ley.

The short ley

The emphasis with the short ley should be on maximum production. Persistency is not important. The varieties used should respond to high nitrogen application for cutting and/or grazing.

Table 55 outlines the management of the short ley.

Grazing early in the season

The early graze in the spring is usually (but not always) associated with the short ley. As far as possible an "early" field should be chosen for first graze. A field which is sheltered, and with a south-facing slope (but not susceptible to frost) and which has a light, free-draining soil to warm up early in the spring is ideal. This, of course, is not always possible but, whatever field is chosen, it is really only possible to get early grazing from a ryegrass sward. 80–90 kg nitrogen is usually applied for early bite and the timing of this first nitrogen is now considered important. Obviously, if it goes on too early some losses can occur in a wet spring through leaching, as the herbage plants are not in a

TABLE 55   *Management of short leys – short duration ryegrass*

	Seeding year	First year	Second year
1-year ley	After the cover crop has been removed, 30–40 kg/ha N can be applied to give useful grazing in September and October. Apply 50 kg/ha of phosphate and potash after this grazing.	If early bite is required, 80–90 kg/ha N, applied in February. Following early bite, 75 kg N can be applied either for conservation or grazing followed by a further 75 kg/N for a second cut or graze.	Lime should not be necessary. If possible, the arable break in the rotation should receive the lime. Rolling will probably be necessary in early spring as the leys are usually cut, and it is wise to press the stones down firmly.
2-year ley	As for the 1-year ley	As for the 1-year ley, with the addition of 40 kg/ha phosphate and up to 120 kg/ha potash in the autumn (depending on soil K index).	As for the previous year, except that the autumn application of phosphate and potash will not be needed.

condition to take up the nitrogen, although this leaching is not as much as has hitherto been supposed. On the other hand, too late an application will mean that potential growing time has been lost.

Some attention has recently been given to the T Sum method—the use of accumulated mean daily temperature (minus temperatures are disregarded) for the timing of the first nitrogen application for early grazing. It originated in Holland where farmers are advised to apply the first lot of nitrogen when T 200 is reached. However, it should be understood that air temperatures in early spring are lower in Holland than they are in the United Kingdom, and so strict adherence to the T Sum accumulation would generally mean nitrogen being applied earlier than has been normal practice.

Trials in England and Wales now seem to indicate that the date of T 200 is not critical, and that nitrogen application during a three-week period around T 200 usually gives maximum response. Other factors should also be taken into account to help decide the best time to apply the nitrogen. Ground conditions are

particularly important, perhaps more so than T Sum. Irreparable damage can be caused by going on to land which is too wet in the spring.

There will also be regional differences in the time that T 200, for example, is reached. ADAS work shows that in the extreme south-west it is from early February, over much of lowland England—mid-February and on upland sites in northern England—early March. It does not seem to vary much from year to year. T Sums relate to sunshine hours and altitude. The accumulated temperatures normally relate to level sites, and obviously a moderate south-facing slope will record the accumulated figure sooner than a north-facing slope. There could be a difference of up to three weeks.

An alternative method is based on the start of the growing season which, for many crops, is still accepted at 6°C soil temperature, usually at a 15-cm depth. So that the nitrogen is readily available for the plant roots when 6°C is reached, the advice is that it should go on at 5°C which, under average conditions, is about 10 days before the 6°C temperature. Meteorological records in south-west England, excluding Devon and Cornwall, show that the first nitrogen application could be on about 7 March, whilst in northern England it would be 15–20 days later. Adjustments have to be made for altitude, with the start of the growing season delayed by five days for each 100 m above the average height for the area.

Forage rye

Forage rye as a catch crop can also be used for early spring grazing. It is very winter-hardy and is one of the earliest plants to start growth in the spring, being two to three weeks earlier than short duration ryegrass. It is useful for both cattle and sheep, but it can be expensive if only one crop is taken. Varieties for forage purposes only should be grown and these should be sown from August to mid-September. Early August sowing is preferable for the north of the country.

Forage rye does not tiller readily and so a relatively high seed rate of 220–250 kg/ha is used. To improve the palatability, 10 kg/ha of short-duration ryegrass (not Westerwolds) can be included if sowing before September.

Fertilizer requirements for rye are not exacting and will depend on soil indices. Very often, 50 kg phosphate and potash are applied, although they may be unnecessary.

It is claimed that some of the new varieties can be cut or grazed off in November/December without prejudicing growth the following spring.

80 kg nitrogen are applied in February or early March for grazing three to four weeks later. The sooner this first crop is grazed, the better the yield of the second crop. Sometimes there is no grass ready to follow the rye, which means a delayed start and a subsequent very poor second crop.

The nitrogen rate is repeated for the second crop of rye which is often taken for silage. Cattle, having been on grass, find the second graze of rye rather unpalatable. Normally, the field will then be prepared for kale or forage maize.

Triticale

Triticale, sold as a blend of varieties, can also be used for early grazing. The growing and utilization is similar to forage rye, although, as it tillers more than rye, the seed rate can be reduced to 180 kg/ha.

It is very slightly later in the spring than rye. Its advantage is that it is slower to mature and after the initial spring grazing there is a more digestible regrowth, although this second crop is usually taken for silage.

GRAZING BY STOCK

All stock do not graze in the same way. Some are much better grazers than others. The most efficient are store cattle, followed by dairy cows and fattening cattle, sheep and young cattle. Horses are notoriously bad grazers!

Although mixed stocking is ideal, in practice it is seldom possible. But what should be done is to see that the most profitable stock get the best. This usually means the dairy cow or fattening beast and, at certain times of the year, the sheep flock.

But the objective, if it is possible, should be a balance between the requirements of the stock and the need for maximum production of the grass crop. As an aid to management, the sward stick (Fig. 72) can be used to measure the height of the grass needed for different classes of stock.

Observations recorded for the British Grassland Society show that with set stocking (page 90) the sward surface height should be kept between 3 and 10 cm to maximize the net rate of herbage production. This is the difference between the rate of herbage growth and the rate of loss to senescence and death. If the height falls below 3 cm, there is insufficient leaf area to bring about a reasonably rapid regrowth, and an increase above 10 cm will mean poorer utilization of the sward as well as a risk of subsequent deterioration.

TABLE 56   *Suggested target range of sward heights for set stock animals*

SHEEP

In spring and summer

Ewes and lambs (medium growth rates)	4–5 cm
Ewes and lambs (high growth rate)	5–6 cm
Dry ewes	3–4 cm

In autumn

Flushing ewes	6–8 cm
Finishing lambs	6–8 cm
Store lambs	4–6 cm

CATTLE

Dairy cows	7–10 cm
Finishing cattle	7–9 cm
Cows and calves	7–9 cm
Store cattle	6–8 cm
Dairy replacements	6–8 cm
Dry cows	6–8 cm

From the British Grassland Society.

Table 56 indicates the suggested target range of sward heights for set stock systems. As far as possible, these heights should be adhered to by, if necessary, adjusting stock numbers and/or the grazing areas. In the autumn, with ryegrass, the

TABLE 57   *Suggested target grazed sward heights for rotational grazing systems*

	Pre-grazing sward height 15–30 cm 21–28 day rotational grazing
Sheep	grazed down to 4–6 cm
Finishing beef cattle	grazed down to 7–10 cm
Lactating dairy cows	grazed down to 7–10 cm
Store cattle/dry cows	grazed down to 6–8 cm

From the British Grassland Society.

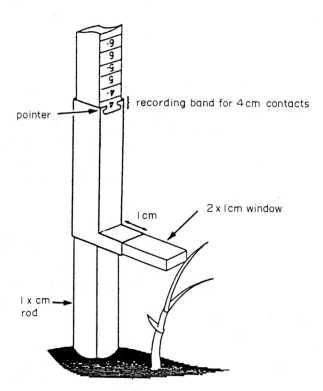

FIG. 72. The sward stick designed by the Hill Farming Research Organization.

sward height should be reduced to about 3 cm to minimize the risk of winter kill.

With rotational grazing (page 188) the residual sward surface height **after** grazing is measured as is shown in Table 57.

It will be noted that the stubble heights are taller than those for set stocking. It is recommended that pre-grazed sward heights should be in the range 15–30 cm.

The sward stick (Fig. 72) is used to take weekly representative measurements throughout the grazing season.

DAIRY CATTLE

Stocking rate for the dairy herd

This will depend on the productivity of the sward, which in turn will depend on site class and the use of nitrogen (Table 53). It will also depend on the requirements of the stock which, in theory, can be calculated in terms of grazing, and silage for winter feed.

Grazing

Note—one cow is equivalent to one livestock unit.

Assume the cow (a Friesian) requires 12 kg DM from grass each grazing day (total daily requirement of dry matter is 18 kg).

100-cow herd = 1200 kg DM required each day from grass.

200-day grazing season = 240,000 kg DM required.

Silage

12 kg DM daily requirement, i.e. 1200 kg DM for 100-cow herd.

165-day winter = 198,000 kg DM required.

Total dry matter required, i.e. grazing and silage, is 438,000 kg DM—say, 440 tonnes.

Assume Site Class 3—370 kg nitrogen for 10.5 tonnes/ha DM.

440 tonnes required.

Therefore 440 (÷ 10.5) = 41.9 ha—say, 42 ha needed.

i.e. 0.42 ha/cow—2.38 cows/ha.

If Site Class 1—450 kg nitrogen for 12.7 tonnes/ha DM.

440 tonnes required.

Therefore 440 (÷ 12.7) = 34.6 ha—say, 35 ha needed.

i.e. 0.35 ha/cow—2.86 cows/ha.

The amount of nitrogen fertilizer suggested in the two examples cannot now be considered realistic and so the target stocking rates would not, in practice, be met. Furthermore, the calculations involving site and use of nitrogen and dry matter production does not allow for wastage. However, any extra production could, in theory, be provided by soil and recirculated nitrogen.

It is accepted that gross margin per hectare increases as the stocking rate increases, but it is unwise to aim for, say, 2.5 livestock units/ha when growing conditions are naturally not so good.

Allocation of grass area

This will vary according to the period in the grazing year. At the beginning of the year grass will be more productive and so the density of stocking should be tighter to allow as much grass as possible to be conserved as silage.

Flexibility is important and until the performance of the growing conditions is known it would be wrong to predetermine the area required for both grazing and conservation.

The use of the buffer system—a feed (usually silage or hay) can be used in addition to grazing when grass may be in short supply.

Livestock unit values

In the examples, the dairy cow has been used as the livestock unit. The correlation of other classes of stock to the dairy cow is as follows:

Friesian dairy cow   1.00 livestock unit (L.U.)
  (Ayrshire 0.9,
  Guernsey 0.8,
  Jersey 0.75)

Suckler cow (dry)	1.00
Spring calving cow with calf	1.10
Autumn calving cow with calf	1.25
Grazing cattle:	
0–11 months	0.25
12–20 months	0.50
21–30 months	0.75
Weaned calf:	
spring born	0.20
autumn born	0.40
Ewe (depending on weight)	0.10–0.20
Weaned lambs	0.08–0.10

Breeding ewes (lactating) add 0.01 per ewe for each 25% lambing percentage.

(From the British Grassland Society)

In the previous example the suggested stocking rate was 2.38 L.U. Therefore at, say, 0.75 (21–30 months) the stocking rate should be 3.17 L.U.

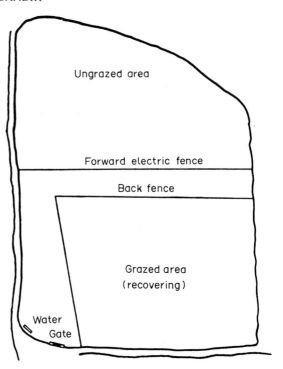

FIG. 73. Strip grazing.

Grazing systems for the dairy herd

Basically, there is a choice of three grazing systems which can be used for the dairy herd.

Strip grazing (Fig. 73)

Although now not so popular, strip grazing is particularly suitable for the smaller dairy herd—suggested less than sixty cows. The system allows the animal access to a limited area of fresh crop either twice daily, daily, or for longer intervals. Wherever possible, the back fence should be used, whereby the area once grazed is almost immediately fenced off. This is to protect the recovering sward from constantly being nibbled over. Without the back fence the recovery rate is slower, although few strip grazing systems use it these days. Properly managed, it is the most efficient of any grazing system for the dairy herd. However, it is labour-consuming, and it also requires a daily decision as to how much grass is needed for the cows. Until experienced (and even then it is not easy), there is a tendency to allow too little to satisfy the herd's requirements. Production could then suffer, although buffer feeding can help to remedy any shortfall of grass. In addition, wet soil conditions can lead to serious poaching along the line of the fence.

Paddock grazing (Fig. 74)

The principle of the paddock system is rotational grazing alternating with rest periods. The grazing area is divided into equal sized paddocks which will normally occupy the fields near the milking unit where access is easy.

The 21-day system was the first of the modern paddock grazing systems which have been evolved in the last 40 years. Twenty–one one-day paddocks are used, because it is calculated that, by allowing each cow the target 0.22

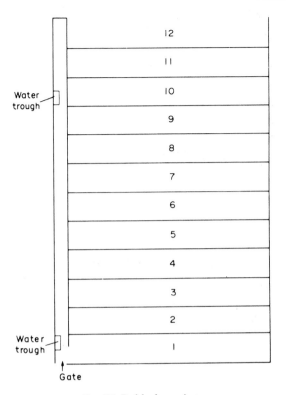

Water
trough

12

11

10

9

8

7

6

5

4

3

2

Water
trough

1

Gate

FIG. 74. Paddock grazing.

in drier areas where a 28-day cycle is considered to be more appropriate.

The Morrey one-day, one-sward system has 36 paddocks with a cycle of 32 days, and the Wye College 28-day system of grazing comprises four plots, each of which is divided into seven sub-plots, a fresh one being grazed daily, but the previously grazed sub-plots are not back-fenced off.

Similar length grazing cycles are also being used with larger sized paddocks (up to 12) which are grazed for two to three days. In the early part of the season it may be necessary to strip graze individual paddocks, but because fewer paddocks are required, the main fencing can, with advantage, be more expensive and reliable. It is also more flexible than using one-day paddocks.

At the peak of spring growth it may be possible to use fewer paddocks, in which case any surplus grass should, if possible, be cut for silage or even hay. (Extra potash—average 70 kg—should be applied after the crop has been taken.) Naturally, adjustments will have to be made when the crop is not so productive, with other grass being brought into the system. Usually this will come from the "cutting block" following the silage or hay cut.

As far as possible, any uneaten herbage should be removed by topping after two or three cycles of the paddocks. Tighter grazing (above the rates already suggested) is not the answer with dairy cows, unless on the Leader/Follower system. However, it is important to prevent an accumulation of uneaten herbage as otherwise it will cause a rapid deterioration of the sward, as well as leading to aerial tillering. This is when the uneaten herbage prevents the light getting to the tiller buds (which thus remain dormant) at the base of the plant. The uppermost buds on the stems develop into new tillers which often only develop weak adventitious roots or no root system at all. Quite quickly a field can be littered with dead pieces of grass.

The *Leader/Follower* system can be superimposed on the one-day system.

It is best used for calves and heifers (page 191),

hectare per season for grazing and applying 70 kg nitrogen after each grazing, average growing conditions will bring the grass back to the right stage for grazing in three weeks.

The paddock size will obviously depend on the number of cows in the herd, but as a rule of thumb and on the basis of daily dry matter intake, 100 cows will need about 0.8 hectare (50 cows/acre) per day of fresh grass. Thus, on this basis, as an example, 20 hectares divided into 21 paddocks will be needed for a 100-cow herd. This is tight stocking and it does not allow for wastage and some flexibility will be needed; buffer feeding will help.

There can well be a loss of nitrogen efficiency if the grass is cropped before the full response from the plant food has been allowed to take place, and this has prompted a definite move away from the 21-day grazing cycle, particularly but with dairy cows it consists basically of dividing the herd up into four groups of cows (and so it is really only applicable for the larger herds) which will use four paddocks at any one time. The freshly calved cows are placed in the first paddock, followed next day (when the first group of cows has moved on) by the mid-lactation group, and then the stale milkers, with the dry cows grazing on the first paddock on the fourth day. The stocking rate is as before—0.22 hectare per cow. With this system, the third and fourth groups are being used as scavengers, and good utilization of the sward is being achieved, but obviously management is more complicated.

Set stocking (continuous stocking)

Set stocking, as understood from the past, was associated with extensive grazing, a corresponding low rate of stocking and little or no nitrogen.

Intensive set stocking uses the same area of grass as for paddock grazing. The only difference is that there are no internal fences.

To help prevent the grass getting out of control, it has been found necessary to start grazing in the spring, a little earlier than would be the case with a controlled system. A stocking rate of 0.22 ha is advised at the start, but as the season proceeds and aftermath grass from the cutting area becomes available, so the grazing can become less intensive. In any case, compared with controlled grazing, there tends to be more of a decline in grass production after mid-summer. It is also preferable to have persistent and prostrate perennial ryegrass varieties as the main constituents of the sward.

Nitrogen fertilizer is applied every four weeks over the whole grazing area or to a quarter of the area each week. Because the grass is kept fairly short all the time, even with 300 kg nitrogen in the season, the clover seems to stay in the sward that much better compared with controlled grazing.

Provided there is sufficient grass available for the cows, any grazing system is as efficient as any other from the point of view of animal performance. But whatever method of grazing, the yardstick must be the production and condition of the stock, coupled as far as possible with maintaining the productive capacity of the sward. Given a reasonable rest period, a good sward will soon recover although, as already mentioned, it may be necessary to top over the uneven effects of grazing, but this is not often the case with set stocking. Trials have shown that set stocking under intensive conditions does not compare so favourably with controlled grazing systems, particularly from mid-summer onwards. Under less intensive conditions of two or less cows/ha, there is little to choose between controlled grazing and set stocking. Once the correct allocation of grass has been made, management should be easier with a set stocked system.

Buffer feeding

Because of the vagaries of the season due to weather conditions, it is never possible to predict with any certainty the annual output from the grass sward. A buffer feed—usually silage (but it can be hay or other feed), to which the cows have daily access, enables the farmer to overcome the problem of fluctuating production. It further means that, whatever the grazing system, a fairly tight stocking density can be maintained, certainly in the first half of the grazing season.

According to the work done at Crichton Royal Farms, West of Scotland College of Agriculture, the cows consume about 14 kg silage a day, i.e. about two tonnes during the grazing year. For the extra silage, about one-third of the grazing area is set aside for conservation which, if silage, will be cut at the start of the silage-making period.

After the buffer area has recovered with the help of nitrogen, it is added to the grazing area unless there is plenty of grass already available when a second cut could be taken.

Irrespective of the grazing system, it is sometimes considered preferable if the grazing block is kept separate from the conservation area (also about 0.22 ha per cow). This is known as the two-sward system. Management is claimed to be easier and different types of swards can be grown to suit the different methods of utilization. However, as the year goes on, much of the cutting block is brought-in for grazing. Grass is obviously not so productive when compared with the earlier part of the season, and an increased area may be very necessary for the grazing stock. Sometimes, a hitherto grazed area is put up for a second or third silage cut. It can provide an excellent opportunity to tidy up the uneven effects of some earlier poor grazing management. It may not be quite such good silage!

Zero grazing

The feeding of fresh cut grass to stock is an old practice known as zero grazing. It used to be called green soiling, then being associated with a much wider range of crops than is the case these days. Where it is practised, grass or grass/clover mixtures are mainly used, and if zero grazing were to become more popular again other green crops could, in addition, be grown. It is not easy to see a revival of interest in the system. However, when maximum stocking rate has been achieved by orthodox grazing, zero grazing, in most situations, can increase stocking intensity by 5–10%. But it is a twice-a-day all-the-season operation involving expensive machinery which of necessity must be extremely reliable. In addition, there is the problem of dealing with the slurry or farmyard manure from stock housed all the year round. Overall extra costs incurred could amount to well over £40 per cow per year. Very few herds are zero grazed. There might be a case for it with a large herd on a farm where access to the majority of fields is not possible.

Storage feeding

Recently some interest has been shown in the feeding of silage all-the-year round to the housed dairy herd. About 12 tonnes silage/cow would have to be made each year and, as with zero grazing, the problem of disposing of large amounts of slurry or farmyard manure could be very expensive.

Dairy followers

It is seldom easy to fit the followers in behind the dairy cow. In fact, whatever method of grazing, management is easier if the followers are kept separate from the herd grazing area. They are usually set stocked, although paddocks can be used. The stocking density with followers must obviously depend on the class of stock being grazed (lower with in-calf heifers compared with bulling heifers) and the time of the year of grazing (more in the spring). It is in the range of 12 to 5 head/ha.

The *Leader/Follower* system has now also been developed for calves and heifers on grass (although a similar principle was used in New Zealand many years ago). In this system the calves rotationally graze paddocks in front of older heifers which are in their second grazing season. Compared with conventional grazing, the calves have been shown to produce higher growth rates. This is because there is less of a build-up of stomach worms in the sward, and the intake of grass is greater. The growth rate of the heifers may be slightly less. Stocking density can be up to four calves and four heifers/ha.

Because of the greater control of stomach worms, the system is suitable for permanent pasture. Clean land is not required; the same field can be used for a number of years.

BEEF ENTERPRISES

A major problem with the grass crop and its utilization by beef animals is that the seasonal pattern of grass growth does not match the increasing appetite and nutritional requirements of growing and finishing cattle over the grazing season. Also, it does not fit in with the chang-

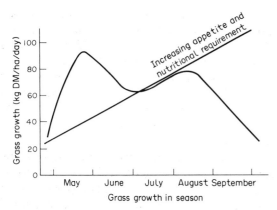

FIG. 75. The incompatibility of grass growth with the increasing appetite and nutritional requirements of growing and finishing cattle.

ing requirements of suckler cows and calves (Fig. 75).

An adjustment of stocking density as the season proceeds will help to overcome the difficulty, and it may be necessary to introduce supplementary feed for the stock.

Grass swards, primarily for beef, should be based on high tillering varieties of perennial ryegrass with small-leaved white clover.

Finishing cattle at grass

This system involves beef cattle (very often dairy cross beef or even dairy bred) in their second summer at grass. The stocking density at turn-out would be approximately:

5.5/ha autumn born
7.0/ha winter born
10.0/ha spring born

Nitrogen is normally applied at four-week intervals up to a total of 200 kg, although this will depend on the season.

As the stock are finished from mid-season onwards, and with the use of aftermaths, so the rate of stocking is reduced to fit in with the increased appetite of the cattle.

The traditional system

The traditional system of finishing off stores on first grade fattening pastures which used to be found in the Midlands is now very seldom seen. Where it is practised, the cattle are turned into the field (which may have received up to 100 kg/ha nitrogen) at the beginning of the season and are kept there all the time. For finishing, a stocking density rate of 2.5 head/ha is used, and they should be able to put on a liveweight increase of 100 kg in the season May until July. The field, with more nitrogen (up to 100 kg), then takes a second group of cattle, although they can seldom be finished off without supplementary feeding.

The suckler herd

The cows and calves are set stocked with a suggested average stocking (autumn and spring calving herds) of 3.5 + calves per hectare. Nitrogen, up to a total of 200 kg, is normally applied at four-week intervals.

Grazing pressure will increase as the calves grow and the cows put on condition and grass production begins to fall. However, this is relieved as the calves are weaned on to the ungrazed aftermaths of the previously conserved areas. Sward height should be kept at 7–9 cm throughout (Table 56).

The semi-intensive system (the one-third, two-thirds system)

This comprises 18–20 months birth to slaughter with six months as calves, six months on grass and six months finished on silage. The six-month period on grass was at one time associated with paddock grazing, but this has now been replaced by set stocking.

Usually the field is, to start with, divided into one-third for grazing and two-thirds for silage. The grazing area is normally grazed continuously until July and then the stock are moved on to the silage area to graze the aftermath. The hitherto grazed area is then put up either for silage or hay and, when this has been cut, the stock, at what will then be at the end of the season, graze the whole field. Nitrogen should be used at 50–60 kg on the grazing area every four weeks. For silage, 120–140 kg nitrogen should be sufficient, but for hay not more than 50 kg should be used, particularly in the wetter parts of the country. Depending upon the soil indices, phosphate and potash will normally be necessary, the potash particularly, after the conservation cuts at about 70 kg respectively.

Stocking at about 11 to the hectare will be higher at the start of the season, but the increase in weight and appetite of the stock coincides with a fall in production from the grass as the season continues. This should mean a reduction in stocking to about 6/ha and then 4/ha which will fit in well with the new grazing area made available in July and September. Liveweight gains of about 0.7 kg per day should be obtained.

SHEEP

For some part of the year (after the lambs have been weaned and after tupping) the grass sheep flock should be regarded as a scavenger flock. There is no reason to treat it otherwise.

It is impossible to give any hard and fast rules on the rates of stocking. The grass will obviously not be at its best, as other stock could have already grazed it; the sheep are simply clearing up. Therefore the condition of the stock must be watched and stocking rates adjusted accordingly. But the pasture should not be overgrazed—a sward height of 3 cm is recommended.

Better grass is needed for ewes being flushed and for ewes and lambs for fat lamb production.

For all practical purposes *set stocking* is now the only method used for the ewes and their lambs during the fattening period. Rotational and forward creep grazing are no longer considered necessary now that anthelmintics are so widely used. Worm infestation is not now such a serious risk, but the pasture should, as far as possible, be "clean", i.e. where worm infestation

has died out, e.g. a new ley following arable or a field which has not carried sheep in the previous 12 months. If not clean, it should be at least a "safe" pasture—where possible infection levels are low enough to be considered safe for ewes and lambs, e.g. pastures grazed by dosed ewes the previous autumn. However, there could still be a nematodirus problem on these "safe" pastures.

Set stocking is the least complicated of any form of fat lamb production off grass, and under good growing conditions a stocking density of at least 20 ewes and their lambs per hectare should be possible.

75–125 kg nitrogen can be applied at the start of the season, but subsequent growth will dictate if any more should be applied later on in the year. Far less nitrogen fertilizer is used on sheep pastures compared with cow pastures. Closer grazing at this time of the year by sheep and lambs (sward height at 5–6 cm) should mean a higher proportion of white clover in the sward. This will go some way to replacing fertilizer nitrogen, but will not help the sward in the spring, when clover growth is slow, nor when clover can be checked in a dry late summer.

The *"Follow N" System* is, in a sense, a compromise between rotational grazing and set stocking. Nitrogen is applied to each quarter of the field in successive weeks, but the flock will tend to keep off the most recently top-dressed quarter and concentrate on that part of the field which had its nitrogen three weeks earlier. In this way the field is grazed in successive quarters. It has the benefits of a rotational grazing system, but with less stress and no internal fencing.

GRASSLAND RENEWAL AND IMPROVEMENT

A poor permanent pasture can be recognized in a number of ways, one of the most obvious features being a mat of old and decaying vegetation at the base of the sward. Although a certain amount is inevitable in a permanent pasture, a heavy mat is not required. This will choke out the better grasses and clovers, leaving the sward mainly consisting of poorer-performing indigenous plants.

Before any programme of improvement is attempted, it is important to understand why the grass has degenerated into its present poor state. Apart from natural limitations such as a difficult soil, poor drainage and weather stress, i.e. drought, waterlogging and frost kill, there are other possible causes. These include weak initial establishment, indifferent management such as understocking, weed infestation, acidity and a deficiency of phosphate and potash.

Reseeding

Ploughing and reseeding is still the main method of renewing a sward. It does mean a "clean" reseed, albeit a rather slow-establishing one. Although it can be justified when the sward obviously fails to respond to fertilizer—particularly nitrogen—reseeding should not now be regarded as routine. (In certain situations, it can be a difficult if not impossible operation.)

The cost of conventional reseeding is approximately £200/ha (1993 figures), but in addition there is the loss of production between destroying the old grass and production from the new sward. Recent trial work shows that there can be a **gross** loss of dry matter of up to 5 tonnes/ha. This can be reduced to approximately 3 tonnes herbage dry matter loss by relating a saving of fertilizer nitrogen (say, 140 kg) in the non-productive period. If a silage cut follows an application of glyphosate (as an aid to better plant kill prior to ploughing) to the grass before it dies, a **net** loss of 2 tonnes dry matter is in theory possible. This can be equated to 6 tonnes silage—worth about £100. The total cost could amount to approximately £300/ha. This can only be met in subsequent years if superior production can be maintained.

Minimum cultivations as an alternative to ploughing can be carried out with the rotary cultivator, for example, ideally in conjunction with either paraquat or glyphosate. Both have the advantage of leaving no toxic residue in the soil,

although sufficient time (14 days) should be allowed for the chemical to move through the plant to kill the grass. Paraquat does not control broadleaved weeds very satisfactorily, and so it is recommended that, if necessary and where it is being used, MCPA or 2,4–D is applied six weeks before the paraquat.

Paraquat or glyphosate should be sprayed on to 10–12 cm of actively growing herbage with application rates per hectare, either paraquat up to 7 litres in 200 litres water (but, if necessary, with a thick old sward, 4 litres in the autumn and 3 litres the following spring may be used), or glyphosate at 4–6 litres in 200 litres water.

After about two weeks, a shallow rotavation (normally two passes) will complete the destruction of the old sward. The field is then rolled, after which drilling can take place.

This method of grassland renewal is usually an early summer operation to be followed by rotavation and seeding.

Direct drilling (page 41) is another method of renewing an old sward. The technique is now usually associated with either paraquat or glyphosate, both of which have the very real advantage of rapid inactivation in the soil which means that there can be a relatively quick turn-round from the old pasture to the new sward. It also produces a firmer sward compared with the plough or rotary cultivator, and there is, of course, less loss of moisture compared with other methods. However, direct-drilling will mean more regeneration of grasses and the cost is approximately the same as using the plough.

It is usually carried out in the summer with an application of glyphosate (4–6 litres/ha) or paraquat (up to 7 litres) on, as far as possible, an even regrowth of grass. After 14 days, nitrogen fertilizer at 50–60 kg can be applied prior to drilling the seeds. It will be necessary to protect any summer/autumn reseed from frit fly and leatherjacket. Slug pellets are also advised (Table 71). Thorough rolling should complete the operation. It ought to be possible to graze the new herbage within six to eight weeks of sowing.

Improvement

There are many situations when the performance of a sward can be improved significantly by methods other than reseeding. However, before undertaking any programme of grassland improvement it is important to check the nutrient status of the soil—pH 6.0, phosphate and potash at index 2.

Ethofumesate, which was originally marketed for weed control in the sugar beet crop, can be used to improve a poor grass sward with little adverse effect on the better grasses such as ryegrass. It is particularly useful against blackgrass, barleygrass, brome, annual meadowgrass and rough-stalked meadowgrass, but not couch or watergrass. It also controls some broadleaved weeds, notably chickweed and cleavers, but it will kill out any clover present.

Ethofumesate is best applied at 10 litres/ha in the late autumn to swards where there is at least one perennial ryegrass plant to 30 cm^2. It is slow acting and it can be some eight weeks before any visible effect is seen. The cost is also high at about £70/ha.

Partial sward replacement using the Weed Research Organization's *one-pass* technique at about £70/ha is considerably cheaper than conventional ploughing and reseeding. In one operation it involves spraying a 10-cm wide band of herbicide—usually glyphosate, turning aside a strip of turf about 2.5 cm wide and deep in the middle of the sprayed band and sowing into the trench—seed, fertilizer and slug pellets. Row widths (distance between the band spraying) can, of course, vary. The Gibbs slot-seeder and the Hunter rotary strip-seeder are examples of machines which can be used. It is normally a contractor operation.

The sward should be closely defoliated before sowing at any time from April to September. Grazing can start again a few days after sowing.

Slit-seeding (strip-seeding, slot-seeding and "stitching-in") is primarily concerned with the introduction of white clover into the sward, although the technique can be used to increase the ryegrass population.

It is a progression from the one-pass technique and, with a suitable multi-purpose drill, it can be carried out by the farmer.

Ideally the new seed should be drilled in June when the existing grasses are less competitive but, in areas of low summer rainfall or on drought-prone light soils, earlier drilling should be considered.

When sowing in June the sward should be sprayed with 1.5 litres paraquat to check the grass, and to kill out any poa species and bottom plants. White clover at 4 kg with 3 kg slug pellets can then be drilled.

Grazing should be carried out 4–6 weeks later with a 4–5-cm sward height. Overgrazing should be avoided.

Oversowing—one of the problems of this technique, whereby the seed is either drilled or broadcast following surface cultivations, is in obtaining a suitable depth of tilth. It is probably best considered for patching up around feed troughs, etc., in a field which may have been used for out-wintering stock.

IRRIGATION

Although for many parts of England the benefits of irrigation for grass are very obvious, it is not easily justified for that crop alone. Rather it should be considered where an irrigation system is already in use for arable cash crops on the farm, and even then only for the more intensive dairy enterprise. And normally, although there can be exceptions, it is only after the end of May that any need for extra water is likely.

If a full irrigation programme is to be implemented, then water should be applied to the grazing area at a soil moisture deficit (SMD) of about 30 mm, but up to 40 mm is acceptable for the less responsive conservation area. Because the deficit can increase very quickly in dry conditions in June and July, irrigation may have to be started before the SMDs are reached to avoid a severe deficit.

It is important to avoid poaching, and with paddocks the water should be applied after grazing. Irrigation could be difficult on a set stock system, but this will depend on soil type.

With irrigation and nitrogen fertilizer the farmer has almost 100% control over grass growth. Above average amounts of nitrogen can be used in the year—up to a total of 400 kg/ha, and it is normally possible to improve stocking rates.

GRASSLAND RECORDING

Although grass may be the most important crop grown in the United Kingdom, it is, nevertheless, a very neglected one. One of the main reasons for this is that, unlike other crops, it is difficult to measure its production. The results of intensifying grass management are usually seen as an increase in stocking rate with a consequent improvement in gross margin. However, whilst that on its own may be sufficient, it is not easy to calculate exactly what extra grass has been obtained when, for example, an additional amount of nitrogen has been applied to the sward. No method of recording grass can do this, but it can indicate possible weaknesses in management which might otherwise go unobserved.

The Utilized Metabolizable Energy (UME) system is the most widely used of a number of grassland recording methods, all of which measure the utilization of the grass by stock, rather than the actual production of grass.

In the UME system all inputs and outputs are converted into one factor, metabolizable energy. This is the feed energy which is actually used by the animal. It is measured in *megajoules* (MJ), or *gigajoules* (GJ), where 1 GJ=1000 MJ.

The calculations assume that if the total annual ME requirements for the livestock on the farm are known, as well as the annual amount of ME fed to the stock from bought feed and concentrates, then the remaining ME necessary must come from homegrown forage, i.e. grass for grazing and conservation. Any hay and/or silage and other forage grown on the farm

does not have to be taken into account unless, of course, it has been bought in.

Example

To record UME production from grass on a dairy farm, the following calculations have to be made:

(1) The annual ME requirement of one dairy cow.
    A Friesian cow needs 25,000 MJ per year (a Jersey cow—22,000 MJ) for maintenance plus 5.3 MJ per litre of milk produced.
(2) The annual purchased feed ME for one cow is:
    (a) Concentrates — assumed ME value 11.5 MJ/kg.
    (b) For other purchased feeds it is necessary to ascertain
        (i) the percentage dry matter,
        (ii) the ME value in the dry matter.

$$\frac{ME\ in\ dry\ matter}{100} \times dry\ matter$$

   = ME value of other purchased feeds.
   Therefore, total ME for purchased feed is (a) + (b).
(3) The annual ME requirement per cow minus the ME value for purchased feed = UME obtained from grass per cow.
(4) The annual UME from grass per hectare = UME per hectare × stocking rate.

By taking into account the site class of the recorded grass, and assessing growing conditions, the probable ME production at different nitrogen levels can be assessed (ADAS/IGAP Experiment).

If, in Table 58, the growing conditions are classed as fair (site class 4, page 179) and, say, 250 kg/ha nitrogen have been used, ME production should, in theory, total 87 GJ/ha.

In this case the efficiency level is 89%.

An example with hay and straw, as well as concentrates having been bought in, shows:

TABLE 58

Maintenance		Milk production		ME requirements	
25,000	+	(6471 × 5.3)	=	59,296	
Concentrates		Hay		Straw	ME purchased/cow
1665 × 11.5	+	$\dfrac{75 \times (9 \times 15)}{100}$	+	$\dfrac{324 \times (8 \times 15)}{100}$	=	19,637
ME requirements		ME purchased/cow		UME/cow	
59,296	−	19,637	=	39,659 MJ	
UME/cow		Stocking rate		UME/hectare	
39,659	×	1.95	=	77,335	
					= 77.3 GJ/ha

It is generally, but not entirely, accepted that a comparison of efficiency of utilization is only possible if grass-growing conditions are similar, i.e. soil type, amount of rainfall and amount of nitrogen used.

Under very good grass-growing conditions (see page 179) a stocking rate of 2.5 livestock units could be achieved which, in the above example, would yield 99 GJ/ha. Stocking rate, as ever, has a most emphatic role in determining output of grass.

GRASS AND CLOVER SEED PRODUCTION

Grass seed production can be very profitable in favourable seasons. It requires considerable skill and perseverance, and good yields are very dependent on good growing conditions, and dry weather during the critical harvesting period. Grasses and clovers for seed are grown most successfully in low rainfall areas such as in the south and east of England. The Italian and hybrid ryegrasses are harvested for one season only, but most of the perennial grasses will produce two or more seed crops. Grass for seed production will, in addition, provide hay and some grazing (the latter is improved if white clover is sown with the grass). It is very important that there are no grass weeds in the field—especially blackgrass and other cultivated grass species with similar sized seed. High weed infestations can lead to a failure of the crop to pass field inspection.

Where there is a danger of cross-pollination, i.e. when grass crops contain the same species, their flowering period overlaps and they are of the same ploidy (e.g. both diploids). Then a gap of 50 m must be cut between the seed crop and its neighbour **before** flowering begins. If this is not possible then, when flowering is over, a strip of 30 metres must be cut out of the seed crop next to the neighbouring field, and this must be discarded.

It is important to remember that grass seed crops can be laid down for three or four years and careful planning is required to avoid cross-contamination. Areas of waste ground, motorway verges, etc., can also be sources of pollen. If there is a danger of cross-pollination, a 2-metre gap or a physical barrier is sufficient.

The ryegrass and fescues are usually undersown in spring cereals or oilseed rape, either in narrow rows or broadcast, whereas cocksfoot and Timothy are usually sown in wide rows (about 50 cm) and are not undersown. A good ryegrass crop will usually "lodge" about a fortnight before harvest and this will reduce losses by wind.

Harvesting is in July or early August (Timothy in late August or September), and the crop may be combined direct or from windrows—timing is critical. The seed must be carefully dried and cleaned (NIAB Seed Growers' Leaflet No. 8).

Red clover. This is a good arable break crop which grows best on well-drained, high pH (6.5) soils with high phosphate and potash levels, but moderately low nitrogen (high nitrogen favours foliage growth). The crop must be isolated from other red clover by at least 50 metres. The seed yield is very dependent on pollination by bees. The seed is best undersown in 15–18-cm rows at 8–10 kg/ha (11–13 kg for tetraploids). At indices below 2, 30 kg/ha each of phosphate and potash should be applied after the cereal has been harvested.

Red clover should be ready for harvesting in late August or early September, when the heads turn brown and the seed (then purple) is easily rubbed out. It usually ripens unevenly, so it may be necessary to desiccate the crop and combine direct two to three days later. Alternatively, it can be cut with a mower and the swath combined 7–10 days later. The combine must be set carefully to avoid damaging the seed and, if necessary, the seed should be dried carefully to a 12% moisture content.

White clover (only 10% is now homegrown). This is usually sown and harvested for seed with a late ryegrass variety.

It is susceptible to drought and so a moisture-retentive soil or irrigation is necessary. It should be cut or grazed up to the beginning of May when the flower stalks start to elongate. It is in full flower by mid-June and it is worth while having one hive of honey bees per hectare for pollination, brought in at the end of May. White clover seed is ready for harvest when 80–90% of the heads are brown, usually in August. The crop is mown very close to the ground, windrowed and picked up with the combine about a week later.

GRASS CONSERVATION

Grass is by far the most commonly used crop for conserving either as hay, silage or dried

grass. In terms of dry matter, the approximate percentage of the total amount conserved is:

*Silage   (approx. 10.5 m tonnes DM) 78.0%
Hay   (approx. 3.0 m tonnes DM) 21.5%
Dried grass (approx. 70,000 tonnes DM) 0.5%

*Additionally, approximately 400,000 tonnes silage DM is made annually from arable crops.

Any form of green crop conservation will lead to a loss of digestible nutrients and Table 59 shows the percentage loss in terms of metabolizable energy and digestible crude protein.

TABLE 59

| | Average percentage loss | |
	ME	DCP
Hay	45	41
Silage	23	24
Dried crop	5	16

Note: ME—metabolizable energy
       DCP—digestible crude protein

Green crop drying is not generally applicable to the average farm system; it should be considered separately (page 217).

Both silage and hay-making techniques have improved in the last 15 years. This is particularly the case with silage and, in the same period, there has been a marked swing towards this method of conservation at the expense of hay. Silage is now playing a much more important role in the diet of ruminants, not only because of the cost of concentrates but, since milk quotas, as a means of helping to maintain financial margins with a lower milk output.

SILAGE

Silage is produced by conserving green crops in a succulent state. The actual conservation process is known as ensilage and the container in which the material is placed is a silo.

Crops for silage

Grass and/or *grass/clover* mixtures are ideal crops for silage provided they are cut at the right stage (Digestibility, page 200) and the fermentation is satisfactory (Table 61).

Lucerne. For ensilage it is preferable to grow lucerne with a companion grass crop, but even then, in spite of wilting and using acid additives, it is not easy to make really well-fermented silage. The reason is that the alkaline minerals contained in the plant:

(1) tend to neutralize the desirable acidity required which will increase the risk of a clostridial butyric fermentation (page 203) which, in turn, will normally mean a poorer intake by stock;

(2) will make the silage unpalatable, which again will lead to a reduced intake by the animal.

Lucerne for silage is normally cut at the pre-flowering stage. The total annual yield of dry matter can be up to 15,000 kg/ha.

Red clover as a crop is not usually taken on its own for silage. However, early red clover can be used with the seed taken from the second growth, usually in September or October.

Compared with lucerne, a more palatable silage should be made from red clover but, again, wilting is necessary as well as an additive. The total annual yield of dry matter from two cuts is about 13,000 kg/ha.

The D values of both lucerne and red clover are lower than grass, but intake is higher from legumes than from grass at the same D value.

Arable crops. A typical mixture is 140 kg oats and 50 kg vetches. It is more often sown in the spring using the appropriate varieties, although an autumn-sown crop should yield up to 30 tonnes/ha (8000 kg dry matter/ha), 10 tonnes more than when spring-sown.

These crops are harvested at the end of June to mid-July (depending on time of sowing) in the milky stage of the oat with the straw still green.

Arable silage is preferably fed to beef cattle, rather than the dairy cow. It is not very "milky".

Whole-crops. Whole-crop silage should be distinguished from arable silage because, apart from its higher dry matter content, much of the nutrient value is within the grain. It was, in the past, associated with high dry matter in the tower silo, but satisfactory clamp silage can be made with barley. Oats have been used and although they may be slightly higher yielding they contain a higher proportion of straw which will reduce the overall feed value.

Barley whole-crop is ensiled in the tower at 34–40% dry matter—the mealy-ripe stage of the grain, but at 25–30% in the clamp—the milk-ripe stage.

Field wilting is not necessary and with the high sugar content fermentation is quite satisfactory. However, digestibility and protein levels are lower than good grass silage at an average D value 60 and 5% digestible crude protein in the dry matter. Yield (winter barley) should average 9000 kg DM/ha.

Urea-treated wheat is now being ensiled as a whole crop (triticale can be used). As with barley, the crop is grown conventionally using normal winter varieties. It is harvested at the cheesey-doughy stage—50–60% DM. The crop at this stage is predominantly yellow with traces of green and this will be some two to three weeks before the grain is ripe for normal combining. Urea (the safest alkali for farm use) is added at harvest at the rate of 4% of the wheat dry matter—about 20 kg per tonne of fresh crop.

Analysis shows an average ME 10.5, crude protein 16% and D value 65. Yield at 8500 kg DM. The pH is about 8 and as such it is usually fed with an acid grass silage.

Fodder beet silage. Following its use on the Continent, whole-crop fodder beet silage is now being grown in this country. Any variety used should have a low dirt tare, as soil contamination, leading to possible butyric fermentation, can be a problem.

The crop is harvested with a combined fodder beet harvester chopper. The whole plant (beet and tops) is chopped with the production of a considerable amount of effluent. Absorbents such as ammonia-treated straw, chopped hay and beet pellets can be used. 12 kg for 100 kg fresh material should minimize effluent loss.

Present recommendations are that the beet and absorbents should be ensiled in layers, i.e. chopped straw at the base followed by chopped beet and tops, beet pellets, chopped straw and with chopped beet and tops as the top layer. Yield is at 20/25,000 kg DM/ha.

Forage maize (also page 157) makes a very palatable and highly digestible silage, high in energy, but low in protein.

For clamp silage, maize should be ensiled when the dry matter content is between 30% and 35% and, at this stage with the D value about 70, the grain is hard and is beginning to dent (dimple) at the top. Earlier ripening varieties, whilst lower yielding than those varieties maturing later, have brought the harvest forward to late September/early October. This can be a considerable advantage on heavier soil which usually gets progressively wetter as autumn proceeds. It will also help for the sowing of winter cereals as the following crop.

A moderate frost before the crop is harvested will not seriously affect the yield or nutritive value. In fact, it can help to increase the dry matter of the crop simply by damaging the plant cell and increasing the rate of evaporation of cell moisture.

Precision-chop harvesters must be used when the crop is ensiled. In order to get really good consolidation the aim should be a chop between 10–15 mm. This should also rupture the grain to ensure maximum digestibility. Additives are generally considered to be unnecessary for fermentation, but they can be used (i.e. based on propionic acid) to slow down aerobic deterioration when the clamp is opened for feeding. A narrow feeding face will also help to reduce aerobic deterioration.

For tower silos, 35–45% dry matter contents are preferred and this will mean later harvesting. In these cases, the chop should be less than 10 mm.

The yield of maize silage varies between 25 and 60 tonnes/ha (average 13,500 kg DM/ha).

By-product silages, e.g. that produced from sugar beet tops and pea haulms, can be of high quality. The difficulty of ensiling them clean should not be minimized.

Silage digestibility

There are a number of factors to be considered when deciding on the quality of a sample of silage (Silage analysis page 211), but its digestibility and thus the metabolizable energy value can be said to be very important.

It is estimated that the majority of silage made on farms in the United Kingdom averages no higher than 60 D, whereas the figure should be nearer 67 D. However, as the grass crop begins to mature, so its digestibility starts to decline by between 1.5 and 2.5 units each week, depending upon variety. The actual silage-making process will also bring about a fall in D value and this does continue, albeit slowly, whilst in store. In theory, therefore, it may be necessary to start cutting when the D value is at least 70. But the higher the D value, the lower the yield, and there has to be a mean between yield and quality. 67 D is now accepted as a suitable compromise for grass silage.

Most ADAS regions have a "Herbage Quality Monitoring Service" which will give the farmer up-to-date information on D values of herbage plants in the area as well as sugar levels in the crop. Some estimation of D values can also be obtained from the NIAB Herbage Variety Leaflet No. 4.

Depending on the amount of silage to be made, there should be at least two different cutting swards on the farm which reach the optimum D value in sequence.

Silos

The sizes of silos vary; the density of settled silage is in the range of 600–1000 kg/m³ according to the degree of compaction.

Permanent roofed and walled clamp silo—the Dutch barn silo (Fig. 76)

This is generally accepted as the most satisfactory type of silo. The roofing allows protection from the rain whilst filling, during storage and when the silage is being used, but plastic sheeting is still necessary directly covering the silage. The roof should be not less than 5.5 m from floor to eaves to allow adequate tractor movement when filling. The walls can be constructed of various materials—reinforced concrete probably being the most popular now that railway sleepers are so expensive. It also helps to make more of an airtight seal at the sides of the silo.

Unroofed, unwalled silos (Figs. 77–78)

Unwalled clamp silos are not commonly made these days. Generally the losses are high and, within reason, the larger the silo the better, as this should mean a smaller proportion of wastage. Heavy gauge plastic sheeting drawn right over the top and down the sides will help to reduce the losses by preventing convection currents going through the clamp, causing excessive oxidation.

All clamps must now have a leak-proof base and channels to prevent effluent run-off and a tank to store the effluent before disposal (page 202).

FIG. 77. Run-over clamp.

FIG. 76. Dutch barn silo.

FIG. 78. Wedge clamp.

On a gross capital cost per tonne of silage stored, these silos are expensive but, in addition, they can be used for storage of hay and straw as well as possibly being part of the building housing the stock.

Unroofed walled clamp silo.

This is the cheapest of the commonly used silos. The silage should be carefully covered on the top and shoulders to keep the rain out during storage. With careful making, the side wastage can be kept to a minimum.

Safety rails above the walls are necessary.

Trench or pit silo (Fig.79)

This is not seen so often now. It was, in the past, associated with the buckrake system of harvesting.

Pit silos are cheaply constructed; they may perhaps be considered for out-wintered stock away from the buildings, in low rainfall areas and on well-drained soils.

FIG. 79. Pit silo.

Tower silo (Fig. 80)

This is normally made of galvanized steel and is glass-lined or treated with a protective paint to make it airtight and acid resistant. A domed metal top completes the seal.

Tower silos are either top or bottom unloaded. The latter, whilst being more expensive, does mean that the silo is more airtight when it is in use (when being emptied) and so there is less risk of aerobic deterioration. It has a plastic breathing bag and pressure-relief valve inside at the top to compensate for the difference in pressure between the outside and inside of the silo.

The top unloader has the advantage in that the feed auger is less prone to blockages and, if there is a breakdown, it should be possible to unload it by hand. Additionally, aerobic deterioration of the silage is virtually minimal if no less than 5 cm thickness is removed each day.

FIG. 80. Airtight tower silo.

Conventional towers of wood, concrete or galvanized steel (but without protective covering to make them more airtight) are rarely used now. The silage is removed from the top.

Drainage of silos

Unless the silo is well drained, the bottom layers of silage will soon putrify. Any effluent must be collected. A simple drain is all that is necessary to get it away. The easiest method of collection is for the silo floor to have a slope with a cross-drain at the lower end leading to a collection tank. This last should have at least a 3 m³ capacity for each 100 tonnes ensiled. There are more expensive modifications.

It is essential that no effluent is allowed to run into any watercourse and all silos must be sited at least 10 m from a watercourse.

Harvesting the crop for silage

Cutting the crop direct into a trailer without wilting and using a double chop forage harvester is the simplest and cheapest system, but these days the crop is usually wilted in the field prior to its being ensiled and, in this respect, the mower conditioner is ideal for initial cutting.

The double chop harvester is still used. Chop length is variable (up to 150 mm) and there can be some soil contamination. It is less expensive than the more widely used precision or meter chop (20–60 mm theoretical length) or the fine chop (20–150 mm theoretical, but less accurate chop length) forage harvester.

The precision chop flywheel mechanism is becoming more popular. The advantages claimed are that, compared with the cylinder-cutting mechanism, more output is obtained from the same tractor power and there is a better output in an uneven windrow.

A shorter chop should mean better silage because, compared with a longer chop, a quicker fermentation takes place as the fermentable juices are more quickly released. However, the intake and performance of precision chopped silage (as opposed to a longer chop) can be affected when fed with a large amount of concentrates to cattle. There is apparently insufficient fibre relative to concentrate. But this is unlikely to be a problem because now most dairy cow feeding practice is based on low concentrate use.

The forage harvester is used with high-sided trailers which collect the grass as it is blown from the "spout" of the harvester. Normally the trailer will dump its load by the silo which, in the case of the clamp, is then filled or built with the small buckrake. In the tower system, the crop is thrown into a dump box and then blown into the silo.

SILAGE MAKING (ENSILAGE)

The fermentation process

In silage making this is essentially a matter of the breakdown of the carbohydrate and protein.

The plant cells in the crop are not immediately killed as soon as it is cut. Respiration will continue for some time by the cells taking in oxygen from the air in the ensiled crop, and carbon dioxide is given off, and then as the cells lose their rigidity the carbohydrate starts to oxidize and the protein begins to break down. This respiration and breakdown of the cells bring about a rise in temperature. The more air there is present, the more the respiration and breakdown.

At the same time, the bacteria which are always present on the green crop act on the carbohydrate to produce organic acids. Two main acids can be produced by their respective bacteria—lactic acid, which is highly desirable, and butyric acid, which is very undesirable.

Thus the stage is reached when the green crop is "pickled" in the acid; this is silage and what type of acid dominates depends upon the type of fermentation. This can be controlled to a large extent by the farmer.

As a result of the different fermentations, three main types of silage can be produced as in Table 61.

The development of heat brought about by respiration should, as far as is possible, be kept to a minimum. Provided the crop is wilted and/or dry, and/or additives are used to create the right conditions for the lactic bacteria, good silage can be made at temperatures below 16°C with little risk of butyric bacteria predominating. At this temperature inevitably there will be some loss of the sugars, but a low temperature and rapid acidification should mean that a reasonable silage is produced.

Silage effluent

Silage effluent is a very strong pollutant. Its BOD (biological oxygen demand) is 10–15 times that of cattle slurry and 250 times more potent than domestic sewerage. It is very necessary to collect such effluent. Because of the Control of Pollution (silage, slurry and agricultural fuel oil) Regulations 1992, prosecution is liable if it is allowed to run into a watercourse. Shortly after ensiling commences, effluent will start to run out from the crop in the silo. The amount will depend on the degree of wilting, as in Table 60. With advice now for less field wilting or, in fact, direct cutting, effluent control is more of a problem.

TABLE 60

Percentage dry matter of crop at ensiling	Amount of effluent per tonne silage
10–15	360–450 litres
16–20	90–230 litres
Over 25	Virtually nil

It is a valuable commodity if properly used. In many situations the most practical way of dealing with effluent is to allow it to run into the slurry store. It should **not** be added if the store is under cover or in a confined space. This is because of the danger of releasing hydrogen sulphide which can cause human and animal fatalities. If it is possible, effluent, diluted with an equal volume of water (to avoid scorching the crop), should be put back on to the land at

TABLE 61 *Types of silage*

Silage	Sample	Feeding value	Reasons	Prevention
Overheated	Colour: brown to black Smell: burnt sugar Texture: dryish	Although palatable, nutritionally is poor. Carbohydrates have been burnt up and protein digestibility considerably impaired by the high temperature	Temperature remains at 49°C or more, due to an appreciable amount of air present in the silo. This happens with stemmy and/or over-wilted material	Do not let the crop get too mature before ensiling. When necessary fill the silo quickly and keep the air out
Butyric acid	Colour: drab, olive-green Smell: unpleasant and rancid Texture: Slimy, soft tissues easily rubbed from fibres Taste: not sharp, pH 5.0 or over	Reasonably palatable and nutritionally quite good, but this depends on the stage of butyric acid fermentation. With very butyric silage, palatability will be poor and much of the protein will have been broken down by the spoiling bacteria. In extreme cases the silage may become toxic, especially to younger stock	The butyric acid bacteria are allowed to dominate, conditions being unfavourable for the growth of the desirable bacteria, i.e. when young, leafy and unwilted crops with a high moisture content are put into the silo, and also when soil-contaminated crops (butyric acid bacteria most commonly occur in the soil) are ensiled. The growth of the lactic acid bacteria is slow under these conditions, and therefore they do not produce sufficient acid to prevent the butyric acid bacteria from maintaining and increasing their presence	Create unfavourable conditions for the butyric acid bacteria, i.e. encourage the lactic acid bacteria. Ensile dry, wilted crops, and, if necessary (when less than 3% sugar on a fresh weight basis), use an additive
Lactic acid	Colour: bright light green to yellow-green Smell: sharp to odourless Texture: firm soft tissue not easily rubbed from fibres Taste, sharply acid to neutral Physical properties of lactic silage now less clear cut with new additives used.	Good, and palatability should be excellent	The lactic acid bacteria have dominated the ensiling process. They have grown rapidly to produce sufficient acid to keep out the spoiling bacteria. Dry conditions have favoured the lactic acid bacteria and if the ensiled crop has been young and leafy, an additive has been added to help the desirable bacteria	Do not prevent, encourage!

a rate not exceeding, in one application, 25,000 litres/ha (25 m³). This will contain about 75 kg nitrogen, 25 kg phosphate and 100 kg potash. A three-week interval should be allowed between applications, which must always be away from watercourses and not on land with cracked soil or land drains.

Silage effluent is also a valuable livestock feed. Depending on the quality of the crop being ensiled, and with an average 7% dry matter content—approximately 20 litres have a feed value equivalent to 1 kg barley. It is normally preserved with 0.3% formalin and fed to cattle at amounts varying between 15 and 50 litres per day.

An alternative method of dealing with silage effluent is to keep it in the silo by using an absorbent. Chopped straw and molassed sugar beet shreds appear to be the most satisfactory absorbents at present in use (synthetic polymers—up to 50% more absorbent than beet pulp—could be used in the future). Rather than as a single layer on the floor of the silo, the absorbent is best applied mixed in the crop in the silo; this can slow down the actual ensiling. Molassed sugar beet pulp at 70 kg per tonne grass ensiled (necessary for grass at 18% dry matter) will take up 20% more space in the silo, i.e. less grass can be ensiled. The cost of the absorbent is up to £12 per tonne silage. However, no other additive is necessary and the feed value of the silage should be improved.

Field wilting will:

(1) Create more favourable conditions for the lactobacilli. By wilting, water is removed which results in a higher concentration of sugars in the crop when it is ensiled. Ideally, it should contain at least 3% sugar on a fresh weight basis to produce a desirable lactic fermentation (below 3% an additive is suggested to help get the right fermentation). There is no simple field test for determining sugar levels, but ADAS regions offer a telephone service to farmers which will give a daily indication of the general pattern of changes of sugar levels in the grass crop, and the need or otherwise for an additive.

(2) Reduce silage effluent. Table 64 indicates that the drier the crop the less the effluent. Grass at 10–15% dry matter ensiled in a 300-tonne clamp will produce about 135,000 litres (30,000 gallons) of effluent, and most of this will be discharged during the first 10–14 days after ensiling.

(3) Enable, on average, up to one-third more crop to be carried in from the field in the trailer compared with the heavier, unwilted crop.

The disadvantages of wilting are:

(1) When wilted to more than 28% dry matter for subsequent ensiling in a clamp, feed trials indicate that there can be a reduction in dry matter intake and subsequent animal performance.

(2) It will bring about a greater loss of digestible nutrients as in Table 62, because respiration has been allowed to continue that much longer.

TABLE 62

Moisture content of the crop when ensiled (%)	Loss of digestible nutrients in the field (%)
70–75	10
60–70	12–14
Under 60	15–16

(3) It is more difficult to consolidate a crop in the clamp silo, although this is largely overcome by using a precision-chop harvester.

The advantages of wilting generally outweigh the disadvantages. The best advice when making clamp silage is to carry out wilting (up to 24 hours) so long as it does not hold up actual silage making.

The rate of wilting will depend on the prevailing weather conditions, as well as the treatment that the crop receives, which will depend on the type of crop. The moisture content can be reduced by up to 10% (from an initial 80–85%) in the first 24 hours under **good** conditions where the crop receives no treatment after cutting. But if a mower conditioner is used, there can be a 15% reduction in six hours.

Table 63 gives an indication as to the dry matter content of a grass crop for ensiling.

Additives

With the exception of absorbents and palatability enhancers, silage additives are used to improve fermentation. They are, however, becoming more expensive. Excluding absorbents, 1993 figures show a range from £0.50 to £5.50 per tonne crop ensiled. At an average £3.00, this amounts to £3000 for, say, 800 tonnes of silage (allowing for 20% wastage). It is not easy to quantify any net gain in production from this outlay.

Properly used and **when necessary**, an additive should ensure a satisfactorily fermented silage. And since milk quotas and consequent less use of concentrates, the dairy farmer has perforce to depend on better quality silage. Additionally, with renewed interest in direct cutting the crop, i.e. no field wilting, an additive will be necessary in most cases. However, it is not always necessary. It will depend on the condition of the crop (Table 63 for assessing dry matter). Refractometers can also be used as an aid to estimating sugar levels but they should not be considered in isolation. It is preferable to use the ADAS monitoring service (page 205). The Star System initiated by Liscombe EHF (Table 64) will also help to decide on the need or otherwise for an additive.

There are three main groups of additives:

(1) **Inhibitors**

(i) Strong acid, e.g. formic acid 85%; sulphuric acid 45%. These give rapid acidification for the desirable lactobacilli at the expense of the clostridial bacteria, i.e. from pH 6.0 as grass to pH 4.5 and then with increased lactobacilli activity pH 4.2

TABLE 63 *Hand assessment of crop dry matter content*

Treatment of crop	Assessment
Crop in swath following Disc or drum mowing	Grass dullish green, limp when laid over palm of hand: dry matter 23–26%.
Conditioned or flail cut	Juice squeezed out when grass is wrung in hand; dry matter below 25%.
Chopped material (crop is rolled into a ball and tightly squeezed for 30 seconds)	Ball retains shape and some juice released; dry matter below 25%.
	Ball retains shape – no juice released; dry matter 25–30%.
	Ball unfolds slowly; dry matter over 30%.
	Ball unfolds quickly, grass breaks; dry matter over 45%.

to 4.0 as silage. Very effective but corrosive and unpleasant to handle. Application rates from 2 to 5 litres per tonne fresh crop depending on its condition.

(ii) Strong acid/formalin, e.g. (a) sulphuric acid 20%; formalin 50%. (b) formic acid 27%; formalin 62%. Gives rapid acidification and partial sterilization to the crop. The clostridial bacteria and, to an extent, the lactobacilli are inhibited, but there is still sufficient lactic acid produced for a stable silage — pH 4.5 to 4.8. The formalin partly prevents protein breakdown both in the silo and in the rumen. Effective, but corrosive and unpleasant to handle. Application rates vary (according to the condition of the crop) at 2 to 5 litres/tonne of fresh crop.

(iii) Acid mixtures:
(a) Chiefly of sodium nitrite and calcium formate, e.g. 12% and 16% by weight respectively. The acid, to inhibit the clostridia and allow the

TABLE 64 *The Star System*

The Star System will help to decide on the need or otherwise of an additive. Low levels of grass sugars provide insufficient material at too low a concentration for rapid and effective acid production. The need for additives is lessened when the sugars in the plant are either concentrated by wilting or their levels are improved by sunshine. The Star System simply assesses the likelihood of an acceptable fermentation by taking into account the variety of grass; the level of nitrogen fertilizing; the weather prior to, and at, cutting; the amount of wilting; and the length of the crop to be ensiled. Each factor is given a star rating and if five stars are obtained an additive should not be necessary. But for every star short of five, a recommended rate of additive should be applied.

			Examples	
			1	2
Grass variety (sugar content)	Timothy/meadow fescue	*		
	perennial ryegrasses	**	**	–
	Italian ryegrasses	***		***
Growth stage	leafy silage	0	0	0
	stemmy mature	*		
Fertilizer	heavy (125 kg/ha +)	–*	–*	–*
nitrogen	average (40–125 kg/ha)	0		
	light (below 40 kg/ha)	*		
Weather conditions	dull, wet	–*		
	dry, clear	0	0	0
	brilliant, sunny	*		
Wilting	none (15% DM)	–*		
	light (20% DM)	0	0	
	good (25% DM)	*		*
	heavy (30% DM)	**		
Chopping and/or bruising	disc/drum cutting	0		
	flail cutting	*	**	
	double chop	**		
	meter/twin chop	***		***

For example: (1) A perennial ryegrass sward (**), in leafy silage growth stage (0) and heavily manured (–*), being ensiled in dry weather (0) and only lightly wilted (0) with pick-up by double chop (**), gives a total score of *** and will show a benefit from 2.25 litres formic acid (**). Thus the final score is brought to *****.

(2) An Italian ryegrass sward (***), in a leafy stage (0), which is heavily manured (–*), in average weather (0) but well wilted, 25% DM (*) and meter chopped (***) will give a total score of ***** and will not require additive.

development of the lactobacilli, is released on contact with the green crop. Non-corrosive. It is available as a solid or liquid with application rates at 1.8 kg and 2.25 litres respectively per tonne of fresh crop.

(b) Salts of formic acid and carboxylic acids (80% of a partially neutralized blend of formic, propionic and octanoic acid). This is a new development where restriction, by the use of organic acids rather than encourage-ment of fermentation, takes place. This should result in a lower level of lactic and other organic acids, but it is still sufficient to give a stable fermentation and a higher level of sugars in the silage compared with normal fermentation. Because of the reduced fermentation there is very little smell to the silage. The pH is about 4.0–4.5. It is non-corrosive. Application rate at 6 litres per tonne of fresh crop.

(c) For big bale silage—mixtures containing propionic acid, e.g. at 35% and 3% acrylic acid as a mould inhibitor. Safe to handle.

Application rate 1 to 2 kg per tonne fresh crop.

(2) **Stimulants** (Enhancers)

(i) Molasses (active ingredient—sucrose 43–48%) has been used for many years. It supplements the fermentable sugars in the crop to ensure that there is sufficient sugar (3% by weight) for the lactobacilli to produce more acid to bring the pH more rapidly down to 4.0. It is safe to use and improved handling equipment is now available for applying the molasses (10–15 litres /tonne fresh crop) either in the field or at the silo. It is not as effective as an acid but it can be used for an organic system.

(ii) Biological additives are used to inoculate the crop with suitable bacteria to help convert the sugars in the crop to lactic acid to achieve a stable pH. However, a suitable substrate (water-soluble carbohydrate) has to be present in sufficient quantity at the start for the conversion to take place. If sufficient bacteria is added (*Lactobacilli plantarum* plus, with some additives, *Pediococcus acidiloctici*—a faster growing bacteria at least 10 times as many as on the crop—10,000 billion per tonne fresh crop) this will bring about a more rapid fermentation than with an untreated crop.

This, in turn, should increase the amount of lactic acid, at the expense of acetic and butyric acid (less protein breakdown), from a relatively small amount of the water-soluble carbohydrate.

Inoculants are safe and easy to apply; there are obviously no corrosive effects and, depending on the crop, the application rate is in the range: liquid 0.5–3 litres, or solid 0.15–0.5 kg per tonne fresh crop.

Inoculants are mainly freeze-dried, but a "live (reactivated) system" has been developed for DIY on the farm. It is quicker acting, which could mean that a lower sugar concentration in the crop is acceptable and, as less culture is needed, it is cheaper.

(iii) Enzymes are natural proteins produced by all living cells. Additives containing cellulose and hemicellulose help to break down the plant cell wall material, thus increasing the soluble sugars for the acidifying bacteria, notably the lactobacilli already present in the crop. More rapid fermentation to a pH 4.0–4.2 should be achieved. It is formulated either as a liquid or solid with application rates in the range 0.2–2 litres and 0.25–0.5 kg respectively per tonne of fresh crop.

(3) **Absorbents**

These are used principally to reduce or even eliminate the silage effluent problem (page 203).

(4) **Palatability enhancers**

There is a group of additives designed to improve the palatability of silage. They are not widely used.

COSH—Control of Substances Hazardous to Health

Regulations state that measures to be taken for the safer use of additives involve three priorities:

(1) Use a safer additive.

(2) Use methods (e.g. a hand pump to transfer the additive) to control exposure.

(3) Reduce exposure.

Suitable protective clothing must be worn if these measures do not give adequate control.

Sealed silage (Waltham or Dorset wedge system)

To produce silage with less waste it is essential to kill the respiring plant cell as quickly as possible. Thus air should be prevented from entering the crop in the silo. The oxygen originally present is used up and, with no further air and with adequate acid production, the plant cell will die.

The walls of the silo should be made as airtight as possible to prevent air getting into the ensiled crop. With a sleeper wall, a plastic sheet attached to the inside is quite satisfactory.

The silo is filled in the form of a wedge as in Fig. 81. The first trailer loads are tipped up against the end wall and subsequent loads are built up with a buckrake to form a wedge. At the end of the day the wedge is covered with plastic sheeting (500 gauge) attached to the back wall of the silo. The next day, after the sheet has been rolled back, more crop is added to the wedge, extending its length until the silo is filled. Care is taken to keep the completed part of the wedge covered all the time and the uncompleted section covered after each day's loading. After the silo has been filled the cover should be well weighed down; old car tyres are

often used. To reduce side wastage particular attention must be paid to pressing the crop down at the sides. The tractor and buckrake will normally give sufficient consolidation, although extra may be necessary at the start of each day's work.

The wedge-shaped principle can also be used for a clamp silo without walls. In this case a plastic sheet is used to cover the wedge completely after each day's building and at the completion of the silo.

Big bale silage

There are situations where big bale silage can be seriously considered:

(1) Where previously silage has not been made on the farm. Capital cost and storage requirements can be quite low. No expensive building is required. The bales can in theory be stored almost anywhere, although some consideration should be given to choice of site.

(2) Where the purchase of a baler is unnecessary, it already being used for other crops on the farm.

(3) Where existing silo capacity on the farm needs to be supplemented.

More than 25% of silage made in the UK is now baled. A comparison on an annual cost per tonne of silage made will generally indicate that 1000 tonnes conventional silage made is cheaper than the big bale system. At 500 tonnes there is little to choose between the two systems whilst, at 250 tonnes, big bale silage is less expensive.

However, the conventional silage operation is at least 15% faster than the big bale system. This could be critical at silage-making time.

Round bales are mainly used for silage. The baled grass is either stored in a separate polythene bag (500-gauge) or is individually wrapped in plastic film.

When big bale silage was first practised, unbagged bales—stacked ideally in a clamp—were an alternative to bagging. This is seldom seen

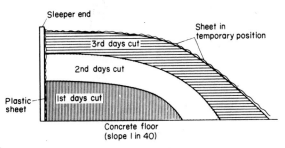

FIG. 81. Sealed silage.

now (the silage stores better in bags) except possibly for large square bales which can be stored unwrapped in "feed lots" enveloped with plastic sheeting. However, wrapping of the big square bale has been introduced with higher work rates than with the round baler. Handling of this bale might present some problems.

The cut grass should be wilted to between 25% and 45% dry matter. It is necessary to reduce the effluent, particularly with wrapped bales.

It is generally recommended that the width of the swath for baling should be the same as the baler pick-up. It is very important (especially when bagging) that an even rectangular swath is prepared to produce a neat compact drum-shaped bale of even density. Cone or barrel-shaped bales with lower densities at one or both ends can mean aerobic losses with over-heating. Subsequent stacking is also more difficult.

Evidence as to the benefits of using an additive is conflicting. Because of the higher dry matter involved, there does not appear to be the same necessity for an additive as an aid to fermentation. A propionic-based mould inhibitor could help to prevent aerobic deterioration when the bales are fed, but this type of additive is not widely used.

A large number of bale wrapping machines is now in use — both trailed and tractor-mounted. Wrapped bales can stay out in the field without deterioration, but not for too long if a quick recovery of the aftermath is required.

In spite of the wrapped bale costing more (up to £1 per bag), its use is becoming more widespread at the expense of the bag. A big advantage of wrapping is that the wrapping machine can be operated by one person from the tractor seat. Bagging is a more labour-intensive, onerous operation. Ideally, three persons are needed.

Wastage (air getting into the bale) should be less with wrapping, and bale size and shape are not so critical compared with bagging. In favour of the bag, it is possible, although not ideal, to bale at 25% or less dry matter. An airtight seal cannot be guaranteed when the wrapped bale settles; any effluent produced by a wetter ensiled crop is held in the bag, but could escape through the overlap of a wrapped bale.

Although holes can be sealed, bales—whether bagged or wrapped—should not be spiked. Bale handlers—a gripper or cradle loader—can be used. For storing the bales, an existing clamp silo is ideal but, in any case, the site should be carefully chosen and free from stones or other objects likely to pierce the bales. To minimize vermin, it is preferable to site away from places which may harbour rats and mice. However, bait around the stack may be necessary.

Bales should not be stored within 10 metres of a watercourse.

The bags are best stacked in pyramid style as in Fig. 82. So that the air is properly removed the bales in the first layer should not be tied until a layer of bales has been placed on them; then tie, using soft polypropylene twine.

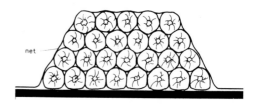

net

FIG. 82. Bagged silage bales.

The height of the stack, both for bagged and wrapped bales, should depend on their dry matter content. Below 25% (too wet for wrapping) they will contain effluent and, to avoid this being squeezed out, the bales should be stored in single layers; between 25% and 35% DM stacked two layers high; 35% and 45% DM three layers, and over 45% DM four layers.

Unless stored in a silo, bagged bales are best double-bagged on the outside layer of the stack. Polypropylene netting, weighted down, should be used as a cover to minimize wind damage. Wrapped bales should not require any further protection.

High dry matter silage

High dry-matter silage of between 35% and 60% dry matter (depending on the crop) is normally made in a tower as illustrated in Fig. 00, although on a large scale it is possible to make it in a sealed clamp.

Silage made in a tower follows the same principle as moist grain stored in a sealed silo. The air which gets into the silo with the crop is rapidly used up by the respiring plant cells; very quickly there is an 80% or more carbon dioxide atmosphere with little oxygen and the plant cell dies. There has been some bacterial activity before the death of the plant and so there is lactic acid production with a pH of about 5.0.

To justify the high capital cost involved of tower, field and feeding equipment, the silage produced should have a high nutritive value capable not only of maintaining the animal but also of making a significant contribution to its production ration. Compared with clamp silage, the in-silo losses are less, but more wilting is necessary with a consequent higher dry matter loss in the field (see Table 62), particularly when wet weather holds up field operations. However, "whole cereal" crops can be ensiled in towers, i.e. forage maize, barley and urea-treated wheat. They are, in a sense, "pre-wilted" before cutting (page 205).

Grass silage from towers, whilst very palatable and showing little visible waste, is sometimes disappointing, especially as a milk-producing feed. Although feeding is usually a very simple operation, because of the expense of the tower system, it is difficult to see it having any great impact in the future.

The "Ag-Bag system"

This system of ensiling green crops has been developed from the Eberhardt silo-press which was used, albeit on a limited scale, in this country in the mid-1970s.

The contractor-operated system involves feeding the cut crop on to a reception table from which a rota forces it into a 1000-gauge plastic sack (2.7, 3.0 and 3.3 m diameter) containing up to 375 tonnes made silage.

As with the tower system, energy losses should be lower than conventionally made silage as air is extruded from the crop as it is forced into the bag. Fermentation and subsequent losses are therefore reduced. One other advantage claimed for the system is that it is possible to make different "quality" packs of silage (e.g. grass ensiled at different times of the season) which could be fed as separate lots.

Contractors' charges average £8.50 (1993) per tonne of silage.

Silage analysis

Physical evaluation of the silage as indicated in Table 61 will help to determine its value. In addition, the following will give a guide as to dry matter content when a sample is squeezed in the hand.

Wet silage—squeezed out easily; dry matter content less than 20%.

Medium dry silage—a little moisture expressed when squeezed firmly; dry matter content 20–25%.

Dry silage—moisture not squeezed out; dry matter content more than 25%

Chemical analysis. An example of an analysis of a silage sample (made from a grass/clover sward) from a clamp is as follows:

DRY MATTER BASIS

Dry matter %	24.0
pH	4.1
Total crude protein %	15.2
Ammonia nitrogen as % total N	8.0
Ammonia as crude protein %	0.6
Neutral detergent fibre (NDF) %	47.7
Total ash %	8.1
Enzymic DOMD % (NCGD)	63.1

Estimated feeding value

Digestibility (D value)   67

Metabolizable energy

(ME)   10.8 MJ/kg DM

Digestible crude protein

(DCP)   102   g/kg DM

Interpretation of analysis

Dry matter %—this is corrected to take into account the loss of some organic acids during drying (particularly the case with wet silage). ADAS add 1.9 to dry matter percentage as a correction.

A dry matter of 24–26% is very satisfactory for grass clover silage made in a clamp. There tends to be a poorer animal performance above this level. Exceptions are whole crop silage (including maize) in the range 25–60% DM (page 199).

The pH is perhaps not quite such a useful indicator of silage fermentation as previously. The acidity figure will depend on the level of additive used, the type of additive and the crop ensiled. It can be in the range pH 3.8 to 4.2 (grass clover silage—acid additive), 5.0 (less fermentation) to 8.0 (urea-treated whole crop). Intake is adversely affected below 3.6 and at 8.0 the silage should be fed with an acid grass silage.

Total crude protein—range 10–17% grass clover silage. It is a measure of the total nitrogen in the silage. It includes the crude protein lost in the drying process added to that in the dried sample. 15.2% is highly satisfactory. Maize silage would be about half this amount.

Ammonia N as % total nitrogen. It should be less than 10%. It gives a guide as to the amount of protein breakdown that has taken place during fermentation. It is an important indicator of silage fermentation quality. If there is too much butyric fermentation, the protein is converted into less valuable non-protein nitrogen as ammonia.

Ammonia as crude protein %. This is poorly utilized by the animal.

Neutral detergent fibre %. This should be in the range 35–65%. It indicates the amount of plant cell wall and cell contents. There should be a higher intake of silage dry matter with a lower NDF.

Total ash %. This shows the mineral matter. Above 9.5% usually indicates soil contamination which causes poor fermentation and a resultant lower silage intake. It can also cause Listeriosis, particularly if fed to sheep.

Enzymic DOMD: NCGD (neutral cellulose gamalase digestibility). This determines the carbohydrate and fibre fraction of the silage.

Estimated feeding value

Digestibility — D value. Ideally, this should be 67 or more. Until recently it was estimated from the modified acid detergent (MAD) fibre. It is a convenient way to predict energy content. However, it is not a wholly accurate method, even when an allowance is made for loss of nutrients during the drying of the sample for analysis.

The introduction of near infra red reflectant spectroscopy—NIR—to predict fibre and protein value is a major improvement on the MAD fibre measure. Calibrated from results of data from feeding trials, the NIR computer predicts the organic matter digestibility of the silage which is then used to calculate the energy value. Early cut well-made silage has been shown to have an ME value about 0.6 MJ (e.g. 12 MJ/kg DM) higher than when compared with the MAD fibre prediction. Poor silage, on the other hand, tends to have ME values about 0.3 MJ lower than previously predicted.

Digestible crude protein, is still the standard measure of the protein value of silage. It is calculated from the crude protein content. It may in the future be changed to show the protein as rumen degradable protein and undegradable protein. This would help in formulating the animal's ration.

Any silage should be properly analysed in the laboratory. A chemical analysis is very reliable, but one of the problems is to make sure that the sample analysed is truly representative of the silage to be fed.

Because of possible secondary fermentation (it develops slowly), silage should not be analysed until at least 10 weeks from making.

Core samples, in several places, should be taken throughout the depth of the silage or from a number of bales—two cores for each sample.

Analysis should, if possible, commence within 12 hours of taking the sample; otherwise refrigerate.

HAY

An analysis of good quality hay should ideally have a minimum value as follows:

Digestibility %—D value	60
Metabolizable energy MJ/kg	9.0
Modified acid detergent (MAD) fibre %	34
Crude protein g/kg DM	100

Unless it is barn dried (page 215), little hay of this quality is made in the United Kingdom. Most hay is of low quality. Apart from indifferent weather, poor-quality herbage and inefficient methods of making are mainly responsible. Although crops such as lucerne, sainfoin and cereal/pulse mixtures can be used for hay, the cheapest and generally the most satisfactory crop is the grass/clover mixture. Whilst a Timothy/white clover sward has long been recognized as producing a very palatable hay, its D value is unlikely to be as high as a ryegrass-based sward which will, at the same time, be higher yielding.

Digestibility of hay

In theory, optimum digestibility is suggested at 65 D which, with the majority of grasses, is when the flowering heads have just emerged (the regional ADAS office will give up-to-date D values for different grasses). This allows for the fact that unsuitable weather may delay curing in the field and that, at this growth stage, D value falls by about three points in a week. Good

hay should have a D value of at least 60, but the digestibility of most hay is well below this figure.

The critical period for hay occurs when the crop is partly dried in the field and therefore there is a *golden rule* for haymaking. No more hay should be ready for picking up in one day than can be dealt with by the equipment and staff. If the cutting outstrips the drying and collection, the hazards of weather damage are greatly increased. If the baler can only deal with, say, 8 hectares in the day, then cutting should be in near 8-hectare lots.

The objective in haymaking should be to dry the crop as rapidly as possible without too much exposure to sun and the least possible movement after the crop is partially dry. Consequently, there are only two methods of making hay worth considering:

(1) The quick haymaking method whereby the crop is baled in the quickest possible time consistent with its safety.
(2) Barn drying of hay.

Quick haymaking

With this method there can be two or three stages in the making of hay.

(1) **Fresh crop** 75–80% moisture content — **curing in the field**.
(2) Approximately 25% moisture content baled followed by drying in the bale in the field, or
(3) Approximately 20–23% moisture content baled followed by drying in the stack.
(4) **Hay** 18% moisture content—**safe for storage**.

Curing in the field. The crop should be cut when it is dry and when the weather appears to be set fine. The local meteorological office may be able to give a weather forecast for two to three days ahead. If possible the headlands of the field should be cut earlier for silage. Grass on the headland usually takes longer to dry out than the rest of the field. So that the crop should be cut quickly when the forecast is right, high-speed uninterrupted mowing is imperative.

The flail cutter has been developed from the forage harvester. It will certainly increase the rate of drying in the field, but unless special care is taken, dry matter loss through shattered leaf can be very high.

The drum or disc mower has a high performance under all conditions. It works fast and has the additional important advantage of causing little loss of leaf and, although there is little curing effect, the swath is left in good condition for subsequent treatment.

Mower conditioners can set the pattern for quick curing without a heavy loss of leaf in the field. The principle is that, after the crop has been cut, it is immediately treated with a conditioner in tandem. There are many types now available such as a disc drum cutter with a rotor fitted with swinging flails or tedder-like tines, or a drum rotary mower with spirally-ribbed steel crushing rollers at the rear.

The quick haymaking method should work on the principle that as soon as the crop is cut it is moved and thereafter, as soon as the broken swath is drier on top than underneath, it is moved again. Under most conditions this could involve at least one pass through the crop in the day (excluding the cutting). The tedder is the ideal implement to use on the first day before the leaf dries to any extent. The next morning, when the dew is off the ground, it should again

be moved as on the previous day, but as it gets drier (and/or depending on the amount of leaf present) more gentle handling is necessary, using only the turner. On the third day, when the dew has gone, the crop should be turned, perhaps twice, and then it may possibly be ready for baling. But this does depend on the weather, the size and type of crop, and it may be **at least** another day before the hay is fit to bale.

It is important not to bale too tightly as further drying has to take place. Excluding the high-density baler, the density of the bale should be about 200 kg/m³. It should be possible, with difficulty, to get the palm of the hand into the bale.

Only general principles have been discussed for field curing. The importance of preserving the leaf cannot be over-emphasized. In the past there has perhaps been a tendency to stress the green colour as the only index of well-cured hay. Sometimes this has been achieved at the expense of the leaf.

Although it is convenient to talk in terms of moisture content in connection with stages of drying, there is no reliable moisture meter to test the moisture content of the crop in the field. It is a question of experience in deciding when the hay is fit to bale. As a guide, Table 65 from the Silsoe Research Institute shows the condition of the hay crop at varying stages of moisture content.

Another way of assessing moisture content is to select and weigh at intervals random lengths of the swath in the field. When the swath weighs

TABLE 65 *Assessment of moisture content of hay*

Moisture content (%)	Conditions
50–50	Little surface moisture – leaves flaccid, juices easily extruded from stems or from leaves if pressed hard
40–50	No surface moisture – parts of leaves becoming brittle. Juice easily extracted from stems if twisted in a small bundle
30–40	Leaves begin to rustle and do not give up moisture unless rubbed hard. Moisture easily extruded from stems using thumb-nail or pen-knife, or with more difficulty by twisting in the hands
25–30	Hay rustles – a bundle twisted in the hands will snap with difficulty, but should extrude no surface moisture. Thick stems extrude moisture if scraped with thumb-nail
20–25	Hay rustles readily – a bundle will snap easily if twisted – leaves may shatter – a few juicy stems
15–20	Swath-made hay fractures easily – snaps easily when twisted – juice difficult to extrude

25% of its fresh weight it is fit for baling in the field. When it weighs 30% it can be treated with an additive (page 216). When it weighs 35–40% it can be baled for barn drying.

Drying in the bale. It is a matter of opinion as to whether the bales should be left in the field after baling to continue drying. The moisture is only slowly lost when in the bale, yet the bale will pick up a lot of moisture if it rains. Certainly mechanical handling of bales has resulted in fewer of them being left out in the field as a recognized stage of the drying process. However, it should be understood that the immediate carrying of the bales must mean baling at a lower moisture content—say 23%.

Round bales are virtually weatherproof and they can be left to dry slowly in the field for some weeks, although obviously aftermath recovery of the grass will be very much retarded under the bales.

Drying in the stack. There are various ways in which the bales can be picked up from the field prior to stacking. They include a number of systems ranging from the tractor front-end loader to the more sophisticated mounted bale transporter.

The stack is usually built under a Dutch barn. With any type of stack a good level bottom is necessary; one formed of substantial rough timbers is ideal to keep the first layer of bales well clear of the ground. The stack walls must be built carefully and firmly. Within the walls, the bales should be stacked leaving air spaces between them; they should not be squeezed in. There are two reasons for this:

(1) The moisture content at stacking will generally be about 22%. It will eventually drop to about 16%. This moisture must be allowed to escape.
(2) As the bales are not very dense they will, in the lower layers of the stack, tend to be squeezed out by the weight of the bales above. Therefore, spaces between the bales will reduce the tendency of the walls to "belly out".

A layer of straw should be put on top of the stack to soak up the escaping moisture. This will help to prevent the top layers of bales going mouldy.

The big bale

The most obvious advantage of the big bale is its facility for speeding up baling and carting from the field. It means that more grass can be cut and cured at any one time, i.e. more hay can be made in fewer days.

There are two types of big bale—the round and the rectangular bale—and, with both, present experience suggests that even more care is needed to see that the crop is uniformly cured in the field, and then formed into a swath of rectangular cross-section to match the width of the baler intake.

For field-cured, rather than barn-dried, hay it should be dried down to 20–25% moisture content and then baled to about 34.5 bars (500 lb/in^2) pressure; the higher the moisture content the more compact should be the bale. There is some evidence that, on baling, the round bale requires a lower moisture content. Actual baling is also slower with the round bale, but it is more weatherproof and it can safely be left out in the field, although this should never be for too long. Because big bales weigh anything from 350 to 700 kg their handling must obviously be mechanized. "Grippers" fitted to the front end loader on the tractor are suitable for the rectangular bale. The round bale can be handled in a similar way and it can also be lifted with a single spike driven into one end of the bale.

It is not easy to make really good quality hay in the big bale, nor is it easy to feed, and the practice is not widespread.

The barn drying of hay

This is a process whereby partially-dried herbage is dried sufficiently for storage by blowing air through it. When making hay in the field, it is in the final stage of curing—the reduction of moisture from about 35 to 25%—that demands all the skill and attention of the farmer. It is at this critical period that the major losses of dry matter take place through too severe handling and damage by the weather. If the hay can be carried at an earlier stage for curing to be completed in the barn, it should be much leafier and more nutritious.

Basically, storage drying, using unheated air, is now the only system of drying. The principle involved is that air is forced through semi-dried bales which are stacked to ensure that it passes through and not between them.

The bales can be stacked in a walled (airtight sides) barn, placed on a flat evenly-ventilated floor or in an open barn with the bales placed round a central air duct so that air flows out from the centre of the stack. This is similar to drying in an open barn with an above or below ground main duct running the length of the barn. The bales are stacked around and over the duct.

There are various modifications of these barn-drying systems, with drying normally taking up to 21 days.

Whatever method is used for drying, the importance of even wilting in the field cannot be overemphasized. This is to ensure that subsequent drying of the bales will be as uniform as possible. Under normal conditions, 1–1½ days' curing in the field prior to baling will be necessary for the hay to be reduced to a moisture content of between 35–45%. The crop should be baled at a density of about 100 kg/m³.

Blowing should be continuous with any form of barn drying. With the mobile moisture extractor unit especially, there may be a temptation to dry for a few days and then to move on to another stack for a time before coming back to the first one. In the intervening period the partially-dried hay will have started to heat up and valuable digestible nutrients will have been lost.

The drying of big bales

The drying of big bales (as in tunnel drying) was feasible with the square bale. However, it is not easy with the round bale and if any drying is done it is with the soft-centred round bale placed on end over a plenum chamber. The top end of the bale is capped with a plastic cover to slow down the escape of air being pushed through the bale. It can be described as batch drying because when the bales have been dried (up to three weeks is needed) they are removed for a further batch.

Hay additives (preservatives)

Hay additives are a possible compromise between quick haymaking and barn drying in that their use permits the hay to be baled at a higher moisture content than would be possible with quick haymaking, although lower than if for barn drying. Compared with quick haymaking, leafier, better-quality hay should result simply because it is brought in more quickly from the hazards of the weather.

The chemical used as the additive is based either on propionic acid or ammonium bispropionate. It is normally applied as a spray on to the hay entering the baler at an application rate, depending on formulation and the moisture content of the hay, varying between 1 and 15 litres/tonne hay baled. Properly applied, the additive will inhibit bacterial fermentation and will also suppress mould growth during storage. These micro-organisms will consume valuable nutrients and so reduce the feeding value of the hay. At moisture contents over 25% and temperature in excess of 40°C they can also cause farmer's lung disease and mycotic abortion in cattle. One of the problems of hay additives has been the difficulty of distributing a small quantity evenly throughout a large mass of material. And, as the rate applied depends on the moisture content, there is the additional problem of a variation not only in the throughput of the hay going into the baler, but also its moisture content which can easily vary throughout the day. The importance of the material being evenly treated with the correct amount of additive cannot be over-emphasized. Application of the additive to the swath before being picked up by the baler is being looked at again.

Once the difficulties have been overcome, the use of additives could be a valuable aid to haymaking. There is, however, no substitute for sound haymaking techniques.

Tripoding of hay

Although good-quality hay can be made by curing the crop on wooden frameworks in the field, it is a very slow and laborious method. It is generally only seen now on small stock farms in the wetter western parts of the British Isles.

GREEN CROP DRYING

This is more commonly referred to as grass drying. It is the most efficient method of conservation because, compared with silage and haymaking, there is far less loss of digestible materials (Table 59).

With very few exceptions, green crop drying these days can no longer be considered as a farming enterprise. It is, in fact, an industrial operation which uses an agricultural crop.

Grass and lucerne are the raw materials most commonly used for drying, although there is no reason why other crops such as whole cereals, forage rye, maize and beans should not be used.

However, many crop drying enterprises have now diversified into compounding or blending the dried material, or it is sold for the same purpose to animal feed compounders. Relatively little dried crop is used on the farm as a feed.

0.5% (approximately 70,000 tonnes—1992) of the total amount of green crop conserved as dry matter in the United Kingdom is processed through a drier (page 198).

Most of the large driers operate in the east of the country. Grass and lucerne are treated as cash and/or break crops. To obtain optimum feeding value in terms of energy and digestible crude protein, the crop is cut on a five to six week cycle to give a dried product with a typical analysis of ME 11.5 MJ/kg DM and 16% crude protein.

High temperature (up to 1100°C) triple pass rotary drum driers are used. The principle with this type of drier is that the wet crop (75–85% moisture content) chopped into short lengths is introduced into a stream of hot air. As the drum rotates, the material, losing moisture and getting lighter all the time, moves along the drum until in about three minutes it reaches the other end, where the temperature is down to about 120°C and the moisture content has been reduced to about 12%. Moisture content is further reduced to 10% when the nearly dried material is first milled, cubed and then cooled prior to storage. Depending on the capacity of the drum, the output of dried material can be up to 10 tonnes per hour.

The very steep increase in the price of oil over the last 20 years has posed a problem for the dried crop industry. The equivalent of about 200 litres oil per tonne of dried material is needed. Some driers have now been converted to coal firing; savings of up to one-third have been achieved.

Wilting the crop prior to picking up from the field is still frowned upon because of the adverse effect it has on carotene. However, there can be more than a 20% saving of fuel through reducing the moisture content by, say, 80% when cut, down to 70% when picked up for drying.

For the grass crop, nitrogen usage up to 600 kg/ha per year (with the first 100 kg/ha going on in February) will normally be required. This in turn will necessitate potash application of up to 250 kg/ha depending on the soil type. Lucerne should not need any nitrogen, but it may require a total of 375 kg/ha of potash throughout the year, depending on soil type.

In addition to the capital requirements (up to £2 million), the high inputs involved (harvesting, transport, drying and cubing) to produce a tonne of dried material will, to a great extent, preclude any large-scale expansion of the industry.

FURTHER READING

Crop Conservation and Storage, N. J. Nash, Pergamon Press.

Forage Conservation and Feeding, Raymond Shepperson and Waltham, Farming Press.

Grass. Its Production and Utilization 2nd Edition, Ed. Holmes, Blackwell.

Grass Farming, Mcg. Cooper and Morris, Farming Press.

Grasses and Legumes in British Agriculture, Spedding and Diekmahns, Publ. Commonwealth Agric. Bureau.

Silage UK 6th Edition, M. Wilkinson, Chalcomb Publications.

11

WEEDS

WEEDS are plants which are growing where they are not wanted. Several crop plants can become serious weeds in other crops, e.g. volunteer potatoes, weed beet.

HARMFUL EFFECTS OF WEEDS

(1) Weeds reduce yields by shading and smothering crops.
(2) Weeds compete with crops for plant nutrients and water.
(3) Weeds can spoil the quality of a crop and so lower its value, e.g. wild oats in seed wheat; black nightshade berries in vining peas.
(4) Weeds can act as host plants for various pests and diseases of crop plants, e.g. charlock is a host for flea beetles and club-root which attack brassica crops; fat hen and knotgrass are hosts for virus yellows and root nematodes of sugar beet; couch grasses are hosts for take-all and eyespot of cereals.
(5) Weeds such as bindweed, cleavers and thistles can hinder cereal harvesting and increase the cost of drying the grain.
(6) Weeds such as thistles, buttercups, docks and ragwort can reduce the grazing area and feeding value of pastures. Some grassland weeds may taint milk when eaten by cows, e.g. buttercups and wild onion.
(7) Weeds such as ragwort, horsetails, nightshade, foxgloves and hemlock are poisonous and if eaten by stock are likely to cause unthriftiness or death. Fortunately, stock normally do not eat poisonous weeds.

Spread of weeds

Weeds become established in various ways such as:

(1) *From seeds*:
   (a) sown with crop seeds—this is most likely where a farmer uses his own seed which is not properly cleaned;
   (b) shed in previous years—some weed seeds can remain dormant in the soil for up to 30 years, e.g. wild oats;
   (c) in farmyard manure, e.g. docks and fat hen;
   (d) carried on to the field by birds, animals, the wind or on machines such as combine harvesters.
(2) *Vegetatively from pieces of*:
   (a) rhizomes (underground stems), e.g. couch, black bent, corn mint, ground elder and coltsfoot;
   (b) stolons (mainly surface runners), e.g. creeping bent (watergrass) and creeping buttercup;
   (c) deep creeping roots, e.g. creeping thistle and field bindweed;
   (d) tap roots, e.g. docks;
   (e) bulbs and bulbils, e.g. wild onion;
   (f) bulbous shoot bases, e.g. onion couch (false oat grass).

Assessing weed problems in the field

The seriousness of weed problems can be judged by:

219

(1) The possible weeds which are likely to appear, for example in a root crop. A record of the weeds present in the field in recent years can be a very good guide when selecting a suitable soil-acting residual herbicide. Weather and soil conditions affect weed populations emerging each year.

(2) The correct identification of weed seedlings growing up in a crop; young grass seedlings are particularly difficult to identify properly.

(3) The numbers of weeds present, e.g. per square metre or hectare.

(4) Proper field inspection. Weed populations tend to be patchy over the field.

The effects on yield will depend on the type of weed and its aggressiveness and on the density and vigour of the crop. Other harmful effects which must be considered include:

(1) Problems at harvest with climbing weeds such as cleavers and bindweed, and increased drying costs.

(2) Crop quality; weeds can affect the amount of lodging in combinable crops. Increased lodging can lead to poorer grain quality.

(3) The future. If allowed to grow and seed, a small population of weeds can create a large problem in a very short time, e.g. wild oats, blackgrass and barren (sterile) brome.

It is not very practical or easy to apply threshold levels for weeds compared with pests and diseases.

Where there are more than 100 weed seedlings per square metre, the damaging effects on cereal crops can be very serious, and one or more carefully timed herbicidal treatments may be necessary. Between 10 and 100 weed seedlings per square metre, spraying costs are usually justified by the possible yield loss and contamination of grain problems. Lower populations, e.g. 1–10 weed seedlings per square metre, are much less damaging, especially in good vigorous crops. However, spraying can be worthwhile if the crop is thin and backward or aggressive weeds

such as cleavers are present. At less than one weed seedling per square metre, there is very little competition and spraying can usually only be justified to prevent seeding and future problems, and for appearance sake, e.g. wild oats.

Weeds in crops vary considerably in their competitiveness. On an individual plant basis, cleavers, for example, are regarded as being up to three times as competitive as wild oats and barren brome, twelve times as competitive as poppies, fourteen times worse than blackgrass or chickweed and fifty times worse than annual meadow grass or pansies. Some grass weeds can be very aggressive; for example, when very large numbers of barren brome or blackgrass appear in parts of fields they can smother a cereal crop.

Weeds such as chickweed and annual meadow grass will germinate throughout the year, but most have main germinating periods, e.g. knotgrass (in the spring), parsley-piert (early autumn), poppy (spring and autumn), fat hen and black nightshade (spring and summer). Crane's-bill (late summer) is encouraged by early sown oilseed rape.

In grassland, it is often difficult to decide on which plants should be regarded as weeds. Several weeds can produce a reasonable yield but not necessarily of high quality. Thistles, rushes, bracken, tussock grass and poisonous weeds such as ragwort, horsetails and hemlock should be destroyed.

WEED IDENTIFICATION

For optimum control of weeds, both culturally and chemically, it is very important to be able to identify the weeds correctly. Grass weeds are the most difficult, especially at the seedling stage when many herbicides are applied.

Grass weeds

There are several factors (one of which is whether the weed is an annual or a perennial) which can aid identification. Annuals are spread by seed, whereas perennials are spread either

TABLE 66   *Guide to the recognition of some common arable grass weeds by their vegetative characters*

1. **Shoots flattened**. Leaves with boat-shaped tips and "tramlines" on upper surface.

    A.   Young leaves with non-glossy undersurface;
non-stoloniferous; ligule tall and rounded.   **ANNUAL MEADOW GRASS**

    B.   Young leaves glossy on undersurface;
stoloniferous; ligule pointed;
heading mid-May onwards.   **ROUGH STALKED
MEADOW GRASS**

2. **Shoots cylindrical**

    A.   Leaves and leaf sheaths hairy.

        i)   Auricles and rhizomes present;
heading June.   **COMMON COUCH**

        ii)   Auricles and rhizomes absent.

            Basal leaf sheath with red coloration
restricted to veins; dense short hairs;
heading May.   **YORKSHIRE FOG**

            Basal leaf sheaths with red colour
not restricted to veins; long hairs
and ragged ligule;
heading from mid-May.   **BARREN BROME**

        iii)   Auricles absent, bulbous base present;
slightly hairy leaf blades;
tufted habit.   **ONION COUCH**

        iv)   Auricles absent;
no stem-based coloration;
hairs present on leaf margin;
long blunt ligule.   **WILD OATS**

    B.   Leaves, leaf sheaths and stems non-hairy,
no auricles.

        i)   Prostrate habit; stoloniferous; internodes
with strong colour; ligule long;
heading July.   **CREEPING BENT**

        ii)   Upright habit; rhizomatous; ligule long and blunt;
heading June.   **BLACK BENT**

        iii)   Upper surface of leaves with distinctly
marked veins; no rhizomes or stolons;
leaf sheaths purplish; ligule blunt;
heading from May.   **BLACKGRASS**

        iv)   Annual; ligule long and oblong;
mainly found in the east and south-east
of England.   **LOOSE SILKY BENT**

        v)   Similar to blackgrass; leaf sheath base
pinkish; red sap oozes out of stem base
when squeezed.   **AWNED CANARY GRASS**

by seed or vegetative parts. Some of the stem characteristics help with identification such as presence of stolons or rhizomes (Fig. 17).

Leaves. Some grasses have folded leaves in the leaf sheath which help in recognition. Leaf shape and size are important, as is the presence of auricles. Size and shape of the ligule usually

FIG. 83. Broadleaved weeds — Identification of common weeds.

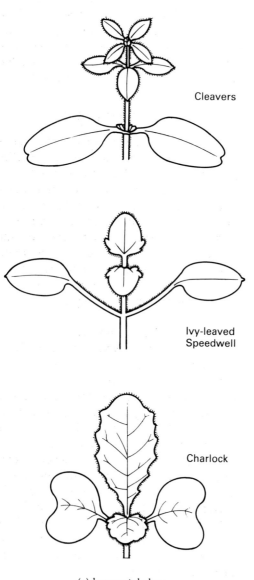

(a) large cotyledons

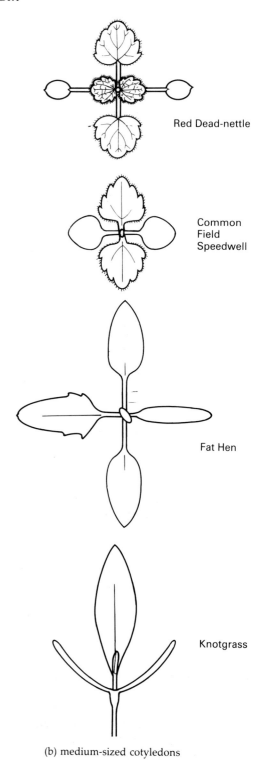

(b) medium-sized cotyledons

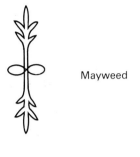

Mayweed

Forget-me-not

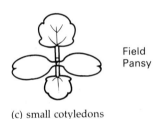

Field Pansy

(c) small cotyledons

Common Chickweed

(d) similar shaped cotyledons and true leaves.

aid identification of species in the same family (page 162).

Presence or absence of *leaf/stem hairs*. A few grasses have very hairy leaves, such as barren brome and Yorkshire fog. Wild oats only have hairs on the leaf margin.

Stem base coloration. A few important grasses have coloured stem bases; ryegrass is a deep red, blackgrass has a purplish blotch and awned canary grass is pink.

Broadleaved weeds

Control of annual broadleaved weeds is usually most effective at the seedling stage. Many broadleaved weeds have very characteristic cotyledons or true leaves which aid identification. Size, shape, colour, if stalked, presence of hairs or prominent leaf wax all help recognition. Figures 83a–d show the characteristics of some common broadleaved weeds.

CONTROL OF WEEDS

In recent years, the introduction of chemical herbicides or weedicides has greatly simplified the problems of controlling many weeds. Most of these chemicals can act in a selective manner by killing weeds growing in arable crops and grassland. The control of weeds with herbicides is now an established practice on most farms. Nevertheless, it is worth remembering that other good husbandry methods can still play an important part in controlling weeds.

Methods used to control weeds are:

(1) Crop hygiene. Ensuring sowing clean seed, hand roguing and avoidance of machinery contamination can help to prevent weed increase, e.g. with wild oats.

(2) Cultivations. Ploughing can be very effective at containing or reducing some annual grass weed problems such as blackgrass and barren brome. In some years, if there is enough soil moisture, stubble cultivations and the "stale" seedbed technique can be a useful aid to weed control. Inter-row cultivations are used in some crops, particularly if there is a difficult weed to control such as weed beet in sugar beet. There is some interest in the use of in-crop weeders as a method of reducing herbicide inputs.

(3) Cutting, e.g. bracken, rushes, ragwort and

thistles. This weakens the plants and prevents seeding. The results are often disappointing if not repeated. Timing is important with thistles.

(4) Drainage. This is a very important method of controlling those weeds which thrive in waterlogged soils. Lowering the water table by good drainage will help to control weeds such as rushes, sedges and creeping buttercup.

(5) Rotations. By growing leys and various arable crops there is an opportunity to tackle weeds in many ways and at various times of the year. This method is still useful if there are difficult weeds such as barren brome, volunteer potatoes and weed beet to control.

(6) Maintenance of good fertility. Arable crops and good grass require a high level of fertility, i.e. the soil must be adequately supplied with lime, nitrogen, phosphates, potash and organic matter. Under these conditions crops can compete strongly with most weeds.

(7) Chemical control. There are now nearly 100 different herbicides and over 1000 products at present on the market. The following is a summary of the main chemicals and methods which are used.

Most of the chemicals used have a **selective** effect, i.e. they are substances which stunt or kill weeds and have little or no harmful effects on the crop in which the weeds are growing. A severe check of weed growth is usually sufficient to prevent seeding and to allow the crop to grow away strongly.

Herbicides are usually sold under a wide range of proprietary names which can be very confusing, but the common name of the active material must now be stated on the container. In the text of this book, the common name of the chemical is used when referring to herbicides and the trade name may also be given where there is only one proprietary product. The annual Crop Chemicals Guide is a good reference to products available.

The selectivity of a herbicide depends on:

(1) The chemical itself and its formulation, e.g. whether it is in water-soluble, emulsion or granular form; also, whether wetters, stickers or spreaders have been added.

(2) The amount of the active ingredient applied and the quantity of carrier (water, oil or solid). Most herbicides are applied in water solution.

(3) The stage of growth of the crop and the weeds. In general, weeds are easier to kill in the young stages of growth. However, treatment may have to be delayed until the crop is far enough advanced to be resistant to damage.

(4) Weather conditions. The action of some chemicals is reduced by cold air temperatures and rain after spraying.

The chemicals now commonly used as herbicides can be grouped as follows:

(1) Contact herbicides. These will kill most plant tissue by a contact action with little or no movement through the plant; shoots of perennials may be killed but regrowth from the underground parts usually occurs. Some examples of contact herbicides are phenmedipham, sulphuric acid, bentazone, ioxynil, bromoxynil, diquat and paraquat. Contact herbicides have very little or no residual action in the soil.

(2) Soil-acting residual herbicides. These chemicals act through the roots or other underground parts of the plant after being applied to the soil surface or worked into the soil (the more volatile types). Some herbicides are also absorbed by foliage, e.g. linuron. These act by interfering with photosynthesis or cell division. Some examples are atrazine, simazine, ametryne, prometryn, chloridazon, monolinuron, trifluralin, propachlor, carbetamide, propyzamide, pendimethalin, triallate, isoproturon and chlorotoluron. These

herbicides vary in their soil persistency from weeks to months.

(3) Translocated herbicides are those which can move through the plant before acting on one or more of the growth processes.

(3a) Growth regulator ("hormone") herbicides. These are a special group of translocated chemicals which are similar to substances produced naturally by plants; they can regulate or control the growth of some plants. Susceptible plants usually produce distorted growth before dying. They are mainly used for controlling weeds in cereals and grassland. The more important ones are MCPA, 2,4-D, mecoprop, dichlorprop, dicamba, triclopyr, benazolin and clopyralid.

(3b) Growth inhibitors. These limit or stop the cell division and elongation of the growing point. Some of the herbicides called "graminicides" work in this way. These herbicides kill grass weeds in broadleaved crops, e.g. fluazifop-P-butyl (Fusilade).

(4) Safeners. Several companies are looking at the use of safeners or crop protectants to increase the range of crops that can be treated or allow the use of higher dose rates. One example is the use of graminicides plus safeners for grass weed control in cereals.

Herbicides may be classified by their chemical groups, or sometimes as a group which can be used to deal with some particular problem, e.g. wild oat herbicides.

Note. Under the Control of Pesticides Regulations 1986 it is illegal to use any pesticide except as officially approved.

It is vital to check the manufacturer's recommendations before buying and applying herbicides.

WEED CONTROL IN CEREALS

Cereals used to be regarded as the dirty crops in the rotation until the introduction of MCPA in 1945. This, and the very similar 2,4-D, easily killed all the troublesome and aggressive broadleaved weeds at that time, especially charlock, poppy and fat hen. However, it was necessary to use these chemicals for many years to destroy seedlings developing each year from the large numbers of dormant seeds in the soil. After a time, these weeds virtually disappeared from many fields, but others, which were resistant to MCPA, were able to grow and set seed. Weeds such as chickweed and cleavers then became troublesome and, in 1957, two herbicides—CMPP (now known as mecoprop) and TBA/MCPA mixture—were introduced. These proved effective against chickweed and cleavers as well as the weeds which were controlled by MCPA.

Then the resistant polygonum group (redshank, pale persicaria, black bindweed and knotgrass) became problems and, to deal with them, 2,4-DP (dichloprop) and dicamba/MCPA were introduced in 1961. Later, other resistant weeds which were previously rare on arable land became numerous, e.g. mayweed, speedwell, field pansy, forget-me-not, groundsel and parsley-piert, and also corn marigold and spurrey on acid land. To control these, other herbicides and many mixtures have been introduced, e.g. ioxynil, bromoxynil, clopyralid, cyanazine and bentazone.

Where clovers are undersown in cereals, the safe butyric chemicals MCPB and 2,4-DB can be used as well as mixtures containing bentazone or benazolin.

With better broadleaved weed control and the increasing areas of early-sown winter cereals, annual grass weeds have increased rapidly, especially blackgrass, meadow grasses, barren brome, loose silky bent and awned canary grass. To control these, more chemicals and mixtures are being used, e.g. triallate, trifluralin, metoxuron, chlorotoluron, isoproturon, fenoxaprop-ethyl, methabenzthiazuron, pendimethalin and diclofop-methyl. The soil-acting effect of some of these is considerably reduced in soils with more than 10% organic matter (adsorption).

Since the introduction of MCPA, it has been

TABLE 67 *Grass and broadleaved weed control in autumn-sown cereals*

Crops (w=wheat, b=barley, o=oats, r=rye, d=durum, t=triticale)	Loose silky bent	Awned canary grass	Seedling ryegrasses	Rough-stalked meadow	Annual meadow grass	Barren brome	Blackgrass	Wild oats	Active ingredients (1. pre-emergence, 2. post-emergence, *control at high rate)	Code	Trade names – see product literature for rates, timing and safe use	Charlock	Chickweed	Cleavers	Crane's-bill	Dead-nettle	Forget-me-not	Field pansy	Fumitory	Mayweed	Poppy	Redshank	Shepherd's purse	Speedwell
w b d t r	–	–	–	s	s	s	s	S	tri-allate liquid granules	1	Avadex BW	r	r	r	r	r	r	r	r	r	r	R	R	r
w b d t r	–	S	S	S	R	–	S	S	diclofop-methyl	*2	Hoegrass	R	R	R	R	R	R	R	R	R	R	R	R	R
w	s	S	–	S	R	–	S	S	fenoxaprop-ethyl	2	Cheetah	R	R	R	R	R	R	R	R	R	R	R	R	R
w b r t	S	–	S	S	S	s	S	s	isoproturon (IPU)	*1,2	various	S	S	R	R	–	–	R	–	S	S	–	S	R
most vars w b t d	s	–	S	S	S	s	S	s	chlorotoluron	*1,2	various	S	S	–	–	r	–	–	–	–	S	R	S	S
most vars w	s	–	S	S	S	s	S	–	chlorotoluron + bifenox	*1,2	Dicurane Duo	S	S	s	s	S	S	s	S	S	S	–	S	S
w b t	–	–	S	S	S	s	s	R	linuron + trifluralin	1	various	S	S	s	–	s	S	s	S	S	S	S	S	S
w b o d t r	–	–	–	S	S	–	S	s	terbutryn	1	Prebane	–	S	R	S	S	S	S	S	S	S	–	S	r
w b d t r	S	S	–	S	S	–	S	s	pendimethalin	*1	Stomp	S	S	S	–	S	S	r	–	–	S	S	S	S
w b t r d	S	–	R	S	S	–	R	R	methabenzthiazuron	1,2	Tribunil	S	S	R	S	S	S	S	S	S	S	–	S	S
w b	S	–	S	S	S	–	S	s	IPU + trifluralin	*1,2	Autumn Kite	S	S	r	s	S	S	s	–	–	S	–	s	S
w b r t	S	–	s	S	S	–	S	–	diflufenican + IPU	*1,2	Javelin Gold	S	S	r	s	s	S	s	r	S	S	s	–	S
w b r t	S	–	s	S	S	–	S	–	isoxaben + IPU	*1,2	Ipso, Fanfare	S	S	r	–	S	S	s	s	S	S	r	S	s

Symbols: S = susceptible; s = moderately susceptible; R = resistant; r = moderately resistant.
Note: Several of the above chemicals can be mixed or are available in mixtures with chemicals in Table.

TABLE 68 *General guide to broadleaved weed control in cereals*

Crops	Herbicides (Active ingredients – all post-emergence sprays)	Trade names – see products leaflets for rates, timing, and safe use	Black bindweed	Charlock	Chickweed	Cleavers	Crane's-bill	Dead-nettle (red)	Fat hen	Forget-me-not	Fumitory	Hemp-nettle	Knotgrass	Marigold, corn	Mayweed	Nettle, small	Pansy, field	Poppy, common	Redshank	Shepherd's purse	Sow-thistle	Speedwell, common	Speedwell, ivy-leaved	Spurrey, corn
w b o r	MCPA	various	r	S	–	R	r	R	S	–	S	S	S	R	S	S	–	–	r	S	S	–	–	r
w b o	mecoprop (CMPP)	various	r	R	S	S	r	r	r	S	R	r	r	R	S	S	R	S	S	S	S	r	–	r
w b o t r d	fluroxypyr	Starane 2	s	S	S	S	–	s	S	R	s	S	s	s	r	r	–	R	s	r	s	r	r	–
w b o t r	ioxynil + bromoxynil	various	S	S	s	r	–	S	R	S	S	r	S	R	s	R	r	R	r	R	R	r	r	–
w b o t r d	ioxynil + bromoxynil + mecoprop	Swipe	S	S	S	S	S	S	S	S	S	S	S	s	r	S	S	S	S	S	S	s	s	S
w b o r	dicamba + MCPA + mecoprop	various	s	S	–	–	–	r	–	–	–	–	–	R	S	–	R	–	–	–	–	–	–	–
w b o	clopyralid	Shield	r	S	–	R	S	S	s	s	S	S	s	S	r	S	s	S	r	S	S	R	R	S
w b o d t	metsulfuron-methyl	Ally	S	S	S	s	S	S	s	S	–	–	S	S	S	S	s	S	S	S	S	–	R	–
w spring b	metsulfuron-methyl + thifensulfuron-methyl	Harmony M	S	S	S	S	S	S	s	S	–	S	S	S	S	S	s	S	S	S	S	S	s	S

Symbols: S = susceptible; s = moderately susceptible; R = resistant; r = moderately resistant.

TABLE 69   General table of weed susceptibility to some herbicides in common use in broadleaved crops

Crops	Herbicides	Time of application	Annual broadleaved weeds																	Grass weeds				
			Black bindweed	Charlock	Chickweed	Cleavers	Fat hen	Hemp-nettle	Knotgrass	Marigold, corn	Mayweed	Nettle, small	Nightshade, black	Pennycress	Poppy, common	Redshank	Shepherd's purse	Speedwell	Thistle, Creeping	Blackgrass	Barren brome	Annual meadow grass	Volunteer Cereals	Wild oats
Potatoes	Paraquat plus Diquat	Contact only	S	S	S	r	S	S	r	s	S	r	r	S	S	S	S	S	r	S	s	S	S	S
	Metribuzin "Sencorex"	Pre- and post-emergence	s	s	S	R	r	S	S	S	S	S	R	S	S	S	S	S	R	S	–	S	–	r
	Monolinuron, "Arresin"	Soil-applied pre-emergence	S	S	S	S	S	S	S	S	S	S	R	S	S	S	S	S	R	R	R	S	R	R
Sugar and Fodder beet	Chloridazon – various	Pre-emergence	S	S	S	r	S	S	S	S	S	S	S	S	S	S	S	S	R	R	–	–	–	–
	Metamitron "Goltix"	Pre- or post-emergence	r	s	S	R	R	s	S	S	S	S	r	S	S	S	S	S	r	r	–	S	–	R
	Phenmedipham, e.g. "Betanal E"	Post-emergence	S	S	S	r	S	S	s	s	r	S	r	s	S	S	s	S	R	S	R	R	R	R
	Ethofumesate "Nortron"	Pre- or post-emergence	s	S	S	S	S	s	S	–	S	S	s	S	S	r	S	s	R	S	–	S	S	s
Oilseed rape	Metazachlor, "Butisan S"	Pre- and post-emergence	s	r	S	s	s	r	R	R	S	s	s	R	S	s	S	S	–	S	S	S	r	r
	Benazolin + Clopyralid "Benazalox"	Post-emergence	s	S	S	s	s	–	s	S	S	–	–	–	s	s	–	R	–	S	–	S	–	r
	Propyzamide, "Kerb"	Post-emergence	S	R	S	s	S	–	S	–	R	R	S	–	R	S	R	s	–	S	R	R	R	S
	Clopyralid, "Shield"	Post-emergence	s	–	–	–	–	–	r	S	S	–	S	–	–	r	–	–	S	–	–	R	R	R
Swedes, turnips, kale	Trifluralin-various	Soil incorporated	S	S	S	R	S	S	S	R	R	s	R	R	–	–	–	S	R	S	R	R	–	s
	Propachlor-various	Pre-emergence	R	R	S	S	S	S	R	S	S	S	s	R	–	R	R	S	–	S	–	S	R	R
Kale	Desmetryn, "Semeron"	Post-emergence	R	s	S	–	S	S	R	–	R	S	R	R	S	S	R	s	R	R	R	R	R	R
Peas	Terbutryn + Terbuthylazine, "Opogard"	Pre-emergence	S	S	S	R	S	–	S	–	S	S	S	–	S	S	S	S	R	S	–	S	R	R
	Pendimethalin, "Stomp"	Pre-emergence	S	–	S	s	S	s	S	–	–	S	S	–	S	S	S	S	–	S	–	S	R	S
	Bentazone + MCPB "Pulsar"	Post-emergence	S	S	S	s	S	s	s	–	S	S	S	S	S	S	S	S	S	R	–	R	R	R
Field beans	Simazine-various	Pre-emergence	S	S	S	R	S	S	S	S	S	S	S	–	S	S	S	S	–	S	–	S	R	r
	Trietazine + Simazine "Remtal" or "Aventox"	Pre-emergence	s	S	S	R	S	s	S	–	S	S	S	S	S	S	S	S	–	R	–	R	R	R

TABLE 69 cont'd

Crops	Herbicides	Time of application	Annual broadleaved weeds																	Grass weeds				
			Black bindweed	Charlock	Chickweed	Cleavers	Fat hen	Hemp-nettle	Knotgrass	Marigold, corn	Mayweeds	Nettle, small	Nightshade, black	Pennycress	Poppy, common	Redshank	Shepherd's purse	Speedwell	Thistle, Creeping	Blackgrass	Barren brome	Annual meadow grass	Volunteer Cereals	Wild oats
All above crops	Dictofop-methyl, "Hoegrass"	Post-emergence	R	R	R	R	R	R	R	R	R	R	R	R	R	R	R	R	R	S	–	R	R	S
Most of above crops	Graminicides. e.g. Cycloxydim "Laser" Sethoxydim "Checkmate"	Post-emergence	R	R	R	R	R	R	R	R	R	R	R	R	R	R	R	R	R	S	S	R	S	S

Symbols: S – susceptible, s – moderately susceptible, R – resistant, r – moderately resistant.
See product leaflets for timing, application rates and safe use.

expected that resistant broadleaved weed strains would appear. However, this has not happened until recently. Some chickweed plants have shown resistance to mecoprop but not to the more effective fluroxypyr. This latter will also deal with large cleaver plants which resist mecoprop. In cereals there is only a small range of chemicals available for the control of blackgrass and there are now a few cases of resistance to some commonly-used blackgrass herbicides. The build-up of resistance is faster with new products such as fenoxaprop-ethyl than the standard herbicide isoproturon. Resistance appears to be due to the repeated use of the same chemicals plus the use of non-ploughing techniques.

Tables 67, 68 and 69 give general guidance to the weeds controlled by currently available herbicides. Dose rate recommendations are often based on the assumption that all farm sprayers and weather conditions may not be perfect. In ideal situations, lower rates may be very satisfactory, but companies will only consider compensation claims for poor control when the recommended rates have been used. Advisers, with wide experience of treatments and results, and BASIS qualified, are best able to give sound advice.

Broadleaved weeds

Perennial broadleaved weeds are more difficult to control than annual weeds. This is especially the case with thistles, field bindweed, wild onion and docks, mainly because the foliage usually appears after the normal time for spraying annuals. Due to the mass of green foliage, field bindweed can be very troublesome, causing lodging and difficulty when combining. However, all perennial broadleaved and grass weeds can be effectively controlled by spraying glyphosate on the nearly mature crop at least one week before harvest. The weeds must be green and actively growing (grain moisture less than 30%) and sprayer booms set to give good weed coverage. Tramlines, high clearance wheels and a sheet under the tractor minimize crop damage. This technique is approved be-

cause of the very low mammalian toxicity of glyphosate, but the straw should not be used as a horticultural growth medium or mulch. Stubble treatment is often less effective.

Annual weeds

These are usually much easier to kill when germinating or as young seedlings. The safest time for spraying a crop is clearly set out in the product literature and must be followed. Some herbicides can be applied at the 3–4-leaf stage, but many of the "hormone" types, e.g. MCPA, 2,4-D and dicamba, should be applied between the 5-leaf stage and jointing (page 86). If applied too soon (before the spikelets of the young ears have been determined), then some ears are likely to be deformed. If sprayed too late (when or after the cells are dividing to form the pollen and ovules), some of the upper spikelets become sterile and give the ear a "rat-tailed" appearance and reduced yield accordingly.

The efficacy of herbicides is often affected by the growth stage of the weed. A weed growth stage key has now been produced (*Annals of Applied Biology* 1987).

Description of weed growth stages.
Pre-emergence.
Early cotyledons.
Expanded cotyledons.
One expanded true leaf.
Two expanded true leaves.
Four expanded true leaves.
Six expanded true leaves.
Plants up to 25 mm across/high.
Plants up to 50 mm across/high.
Plants up to 100 mm across/high.
Plants up to 150 mm across/high.
Plants up to 250 mm across/high.
Flower buds visible.
Plant flowering.
Plant senescing.

Weed growth stages in relation to herbicide efficiency are explained on the product labels. Timing of control of broadleaved weeds in cereals is not as critical as for annual grasses. If dealing with grass weeds in the autumn in the winter cereal crop, then broadleaved weeds are normally also controlled (Table 69).

Some broadleaved weeds, such as field pansy, are more easily controlled by a number of the autumn herbicides, e.g. diflufenican. If an autumn treatment is used, then it may be possible to reduce chemical rates in the spring. Some of the newer broadleaved herbicides can be applied over a long period without crop loss, unlike the "hormone" weed killers. Metsulfuron-methyl (broad spectrum) and fluroxypyr (mainly for cleavers), depending on the crop, can be applied between crop GS 12 and 39.

Grass weeds

These are a major concern for cereal growers. Once established, complete elimination is very expensive and not always possible. The following are the main grass problems and some possible methods of containing them:

Couch (Fig. 84a) has been a problem for a very long time. It can spread rapidly from pieces of rhizomes and can cause serious yield losses as well as perpetuating diseases and making harvesting difficult. Pre-harvest glyphosate is recommended also in peas, beans, linseed and oilseed rape, but not in crops for seed. Although expensive, it does provide an effective method of killing couch. If there are several clones of couch in a field, then viable seed may be produced to start new plants later.

Onion couch (false oat-grass) (Fig. 84b) is a more serious problem to deal with than couch. It has bulbous rhizomes and senesces too early for the pre-harvest glyphosate technique in some crops, e.g. wheat. It may have to be dealt with by spraying regrowth in the stubble. It can produce much viable seed. It is encouraged by continuous winter cereals and minimum

FIG. 84a. Couch grass.

FIG. 84b. Onion couch.

FIG. 84c. Creeping bent grass.

FIG. 84d Black bent grass.

cultivations. Some of the wild oat herbicides can help reduce bulbils and seeding.

Watergrass(creeping bent) (Fig. 84c) is not so difficult to deal with. It spreads by seed and surface rooting stolons which can be controlled by good ploughing or spraying with paraquat when green in the stubble.

Black bent grass (Fig. 84d) is common in some areas. It has rhizomes like ordinary couch and can be dealt with in the same way, but it does produce much more seed.

Meadow grasses (Fig. 84f) can cause problems when present in large numbers. They can be effectively controlled by the blackgrass herbicides, often at reduced rates (Table 67).

Wild oats (Fig. 84e) are found on over 90% of cereal-growing farms in this country. 80% of infested fields (over 60% of all cereal fields) have numbers which are too high to be rogued. The spring germinating type is found all over the country, but the winter (autumn) germinating type is found mainly in southern and eastern areas. Once established in a field, no matter what control measures are taken, it is likely to persist for a very long time because of dormant seed in the soil. This dormancy problem is made worse by stubble cultivations after harvest which bury the seed. Most of the shed seeds, if left on the surface, are destroyed or disappear. The seed can arrive in a field in many ways, e.g. in the crop seed, dropped by birds, in farmyard manure made with infested straw, and from the combine harvester. The first to appear in a field should be removed by roguing. Hand roguing can be justified when numbers are less than 500 plants/ha or possibly 1000/ha if a roguing glove is used. (This applies some glyphosate to the heads of the plants and the seeds will then be sterile—the plants do not have to be removed.) Later, if numbers are allowed to increase, it may be necessary to use herbicides. It is difficult to decide when this can be justified.

FIG. 84e. Common wild oat.

FIG. 84f. Annual meadow grass.

It could be based on yield response only or likely grain contamination, or to prevent a build-up of numbers in the future. (Left uncontrolled, wild oats can increase threefold each year.) Deeply buried seed, which may have fallen down cracks on clay soils, for example, can remain viable for decades. Seed germinates from up to 25 cm depth of soil.

Yield losses may be up to 90% in very bad cases, and 25% is common with moderate infestations. Premiums can be lost for seed, malting, bread-making and EC Intervention purchases. Straw carrying wild oat seeds may be difficult to sell. The presence of wild oats on a farm may incur a high dilapidations claim at the end of a tenancy.

The main herbicides which are giving good results are mostly foliar acting and include:

Fenoxaprop-ethyl, tri-allate, flamprop-*m*-isopropyl, difenzoquat and diclofop-methyl.

There may be a restriction on the use of some of these herbicides. There can, for instance,

be a certain amount of toxicity if used on particular cereal crops/varieties, or at specific growth stages in the cereal plant. Some of the herbicides will not mix with other chemicals. However, most of them will give some control of other weeds.

Blackgrass herbicides which give useful control of wild oats include chlorotoluron and isoproturon alone, and in mixtures. Timing of control of wild oats will depend on the time of germination, either in the autumn or spring or both, and whether other weed problems are present.

Blackgrass (Fig. 84h) is now the most important grass weed on cereal farms in this country (100 plants/m² can reduce yield by one tonne/ha). It used to be associated with heavy, wet soils where winter cereals were grown, but it is now widespread on most types of soil where autumn-sown cereals predominate. Blackgrass produces a very great number of viable seeds (individual plants can produce up to

FIG. 84g. Soft brome grass.

FIG. 84h. Blackgrass.

150 heads and many thousands of seeds). Most seed germinates in the first three years but some can remain dormant for nine years. The seeds germinate mainly in early autumn, but in heavily infested fields spring germination can also be important. A high percentage control must be achieved each year in continuous cereals to contain the problem, and this usually means that cultural practices are necessary to supplement herbicides. Vigorous crop competition is very important. Ploughing, to bury the seed deeply, is preferable to shallow cultivations. Blackgrass only germinates in the top 5 cm soil. Spreading the seeds to clean fields should be avoided by using weed-free seed and thorough cleaning of the combine harvester. Delaying drilling or changing to spring cereals can be helpful, but the resultant lower yields may not be acceptable. A number of herbicides is available (Table 69) and although expensive, the cost can be justified where there are more than two blackgrass plants/m². The most common chemicals used are chlorotoluron (this may damage some varieties and should not be used on the oat crop) and isoproturon (should not be used on the oat crop nor for Durum wheat). Early post-emergence spraying (about mid-October) is preferable. If applied in September, the herbicides may break down in the soil before the later blackgrass germinates and so control is poor. The activity of these residual herbicides will be reduced in cloddy seedbeds and where there is a high trash or organic matter content in the soil. The herbicides also control some broadleaved weeds, and often mixtures are used to widen the weed spectrum and/or improve control.

Barren brome (Fig. 84i). This became a serious problem on many cereal farms after the very dry years in the mid-1970s. It spread from hedgerows and waste corners and was encouraged by increased early sowing, established by minimal cultivations. The seeds are anything but barren and germinate very readily in the autumn. The seed has little dormancy.

FIG. 84i. Barren brome grass.

Ploughing can be very effective in controlling this weed if the seed is buried more than 13 cm. Stubble cultivation, in a damp autumn, can also aid control. Where only headlands are infested they are sometimes left for drilling in spring after the germinated brome seedlings have been destroyed. The best herbicide treatment is tri-allate in the seedbed followed by isoproturon or chlorotoluron—early post-emergence. Several herbicides used in other crops are also effective.

One or two other brome species, such as rye brome and meadow brome, can also be a problem.

Awned canary grass is a relatively new problem. It is not widespread but it can cause serious yield losses and harvesting problems where it does occur. The heads resemble large Timothy heads. Numbers have built up because it is resistant to many of the usual grass herbicides.

Diclofop-methyl and fenoxaprop-ethyl can be used post-emergence.

Loose silky bent is a local problem on sandy or light loam soils in some eastern and southern areas. It is controlled by several herbicides (Table 67).

Grass weed control can prove more difficult in cereals than broadleaved weeds. The cost of grass weed herbicides is usually higher than herbicides for broadleaved weeds.

WEED CONTROL IN OTHER COMBINABLE CROPS

Oilseed rape

Winter oilseed rape is a very competitive crop. A fairly high weed population, e.g. 50–100 weeds m^2, can have little effect on yield. Many farmers grow oilseed rape as a cleaning crop. This is because there is a wide range of herbicides available which will control weeds that are otherwise difficult to control elsewhere in the rotation, e.g. brome. Cleavers and other brassicae are difficult to control and should be dealt with in other crops. Volunteer cereals are one of the main problems, especially if the rape is established using minimum cultivations after winter barley. Several graminicides, e.g. "Fusilade" and "Pilot", will control the cereals.

In well-established crops, the time of weed removal is not as important as in backward thin crops. Most broadleaved weed herbicides are applied in the autumn. A number are very persistent and the manufacturer's recommendations should be followed concerning the cultivations before the next crop. There is only a limited range of products which can be used in spring rape. Trifluralin (incorporated) is commonly applied.

Peas

Both dried and vining peas are uncompetitive crops.

Poor weed control can reduce yields as well as affecting harvesting, drying and crop quality. The main method of weed control is with herbicides. Care must be taken in the choice of chemicals, as there are restrictions on variety and soil types. The majority of crops are treated with a pre-emergence residual herbicide. Those commonly used include pendimethalin, simazine + trietazine, prometryn, terbuthylazine + terbutryn. There are only a limited number of post-emergence broadleaved weed herbicides approved for use in peas. These include bentazone, MCPA + MCPB, and cyanazine. Post-emergence treatments are used if the pre-emergence treatment fails, or the peas are growing on an organic soil. Care must be taken with post-emergence treatment as it can cause crop damage. The amount of leaf wax should be analysed using the crystal violet test.

A number of graminicides are approved for use in peas.

Field beans

Initially, field beans are susceptible to weed competition, although later they are a very competitive crop. Most growers rely on herbicides for weed control, although in-crop harrowing can be used as an aid to keep the weeds under control. Most herbicides used are residual and are applied pre-emergence. Simazine is the standard chemical applied to the winter bean crop as long as the latter is sown deeper than 7.5 cm. In spring beans, several of the pea herbicides are approved pre-emergence. Only bentazone, applied post-emergence, is recommended for broadleaved weed control in field beans.

A number of graminicides are approved for grass weed control.

Linseed

Linseed competes poorly with weeds. Any weed problem not controlled during the growing season can be desiccated at harvest using diquat or glufosinate-ammonium (as in oilseed rape, dried peas and field beans). Glyphosate can be used pre-harvest for the control of per-

ennial weeds. As linseed has been a relatively minor crop until recently in the United Kingdom, there are at present only a small number of chemicals which are approved for use. These include bentazone, bromoxymil + clopyralid and metsulfuron-methyl (where volunteer potatoes are present). Any grass weeds can be controlled with diclofop-methyl or sethoxydim.

WEED CONTROL IN ROOT CROPS

Potatoes

Potatoes are a very competitive crop once they meet across the rows, but early weed emergence which is not controlled can reduce yields. Weeds can also affect potato quality and ease of harvesting.

Weed control by cultivations may be impossible in wet seasons and there can be considerable loss of valuable moisture and damage to crop roots in dry seasons. Cultivations can also produce clods in some soil conditions. Consequently, in most potato crops, weeds are now controlled by herbicides applied before and/or after planting.

Pre-planting. Perennial weeds should, if possible, be controlled by glyphosate applied pre-harvest in the previous cereal, oilseed rape, pea, or bean crop (the weeds must be green and actively growing) or by treatment in the autumn.

Pre-emergence of potato shoots. Diquat + paraquat or glufosinate-ammonium will kill most emerged seedling weeds. These chemicals work by contact action. Other residual herbicides can also be used, e.g. linuron or monolinuron. Metribuzin is more persistent, but there are varietal restrictions. On organic soils it can be incorporated into the ridge to improve the control of weeds.

Post-emergence of potato shoots. There are very few products which can be used post-emergence in potatoes. Metribuzin can be applied to some crops and varieties, as can the foliar-acting herbicide bentazone. Several graminicides are approved for grass weed control.

Haulm destruction. Chemicals are normally applied to destroy the haulm; this encourages skin set and reduces the spread of blight to the tubers. The following may be used: diquat but not in dry conditions; metoxuron plus a tin fungicide; sulphuric acid or glufosinate-ammonium.

Volunteer potatoes from tubers and seed are an increasing problem, particularly in close rotations. Glyphosate, pre-harvest in other arable crops, can aid control.

Sugar beet (fodder beet, mangels)

Couch and other perennial weeds should be controlled in the year prior to planting the crop. Weeds can seriously reduce beet yields if not controlled, particularly in the first eight weeks following the sowing of the crop.

Traditionally, weeds were controlled by inter-row cultivations and hand hoeing combined with singling. Following this, band spraying combined with inter-row cultivations were used.

Now, most annual weeds are controlled by overall herbicide treatments applied pre- and/or post-emergence of the crop, using the low-dose technique. It is important that the correct type and amount of herbicide is applied (according to the weed growth stage) and the soil conditions are suitable, i.e. fine and moist. This technique relies on treatment when the weeds are at the cotyledon stage.

Pre-sowing. Pre-sowing treatments are now rarely used as growers try to reduce the amount of wheeling damage to the seedbed.

Pre-emergence. The majority of crops receive a pre-emergence treatment. Choice of chemical will depend on potential weed problems and soil type. Chloridazon + ethofumesate or metamitron are commonly used.

Post-emergence. It may be satisfactory or necessary to leave all annual weed control to this time. However, bad weather conditions may delay spraying and many weeds could grow past the susceptible stages. Using the low-dose technique, there is less restriction on timing of the post-emergence herbicides in relation to crop growth stage. Mixtures of herbicides with different modes of action are commonly used, e.g. the contact herbicide, phenmedipham with or without metamitron or other residual types such as ethofumesate. A clopyralid spray can be used to give better control of established thistles and mayweeds, as well as suppressing volunteer potatoes. Graminicides, such as sethoxydim, quizalofop-ethyl or fluazifop-*P*-butyl, can be used to control wild oats and some grasses, and severely check couch grass. Glyphosate can be applied with a rope-wick or roller machine to control weed beet and other tall weeds (page 134).

A standard herbicide programme in sugar beet is one pre-emergence product followed by at least two low-dose post-emergence treatments. Soils rich in organic matter are unsuitable for some residual herbicides. Care is also required on very light soils. The product recommendations should be followed very carefully (Table 69).

Brassicae (Brussels sprouts, cabbages, cauliflower, kale, forage rape, swedes and turnips).

Weeds in horticultural brassicae are mainly controlled with herbicides and some cultivations. Weeds in transplanted crops are controlled with contact herbicides, e.g. paraquat, or by cultivations before the crop is planted. Little weed control is required in the quick-growing forage brassicae, whereas swedes and kale are very sensitive to weed competition.

Couch and other perennial weeds can be controlled with glyphosate in the previous crop pre-harvest or on the stubble. Tri-allate applied to the seedbed should be used for the control of blackgrass, wild oats and ryegrass seedlings. An alternative treatment (depending on the crop), used post-emergence, could be a graminicide such as alloxydim-sodium.

Trifluralin incorporated into the seedbed plus propachlor (or propachlor + tebutam) pre-emergence, will give good control of a wide range of annual weeds, but not charlock, in many of these crops.

Post-emergence desmetryn in Brussels

sprouts, cabbages, kale and forage rape will control many broadleaved weeds, especially fat hen, but not mayweeds, knotgrass, black bindweed or shepherd's purse. Clopyralid is useful for controlling mayweeds and thistles in all crops. Where any of these crops, e.g. kale, are direct drilled into a chemically-destroyed grass sward, annual weeds are unlikely to be a problem.

WEED CONTROL IN GRASSLAND

In arable crops most damage is caused by annual weeds but in established grassland biennial and perennial weeds are responsible for most of the trouble. The presence of the weeds can bring about a reduction in the yield, nutrient quality and palatability of the sward. Stock do not like grazing near buttercups, thistles and wild onions. Some weeds are poisonous, e.g. ragwort and horsetails, and some can taint milk if eaten, e.g. buttercups and wild onion.

Weeds in grassland are encouraged by such factors as:

(1) Bad drainage, e.g. rushes, sedges, horsetails and creeping buttercup.
(2) Lime deficiency, e.g. poor grasses (Yorkshire fog and the bent grasses) and sorrels.
(3) Low fertility. Many weeds can live in conditions which are too poor for the better types of grasses and clovers.
(4) Poaching (treading in wet weather). The useful species are killed and weeds grow on the bare spaces.
(5) Over-grazing. This exhausts the productive species and allows poor, unpalatable plants such as Yorkshire fog, thistles and ragwort to become established.
(6) Continuous cutting for hay encourages weeds such as soft brome, yellow rattle, knapweed and meadow barley grass. These weeds seed before the crop is harvested.

Chemicals are a useful aid to controlling grassland weeds, but should not be regarded as an alternative to good management.

Many weeds are checked by inexpensive products such as MCPA, mecoprop, 2,4-D and dicamba, though treatment may need repeating. Newer products such as those with clopyralid or triclopyr can be more effective, although more expensive. These chemicals are useful for spot treatment.

The main chemicals used to control broadleaved weeds in swards where clover is important are MCPB, 2,4-D and benazolin. Asulam is used to control bracken and docks; it also kills some grasses. Glyphosate, when applied with a rope-wick or roller machine, will selectively kill any tall weeds which are growing actively.

Table 72 is a guide to the control of the more troublesome weeds in grassland.

Weed control when sowing grass

When seeding a grass sward without a cover crop, weeds, especially annuals such as charlock, chickweed and fat hen, can be a problem. Additionally, grass sown in an arable rotation can have more problems with arable weeds such as blackgrass, cleavers and speedwell.

Annual weeds with upright stems can be killed by mowing or early grazing. However, chickweed should be sprayed with, for example, mecoprop, or a benazolin mixture where clovers are present and the crop is at a safe growth stage.

Where grassland has to be reseeded, and ploughing is not possible or desirable, the old sward can be killed by a glyphosate spray from June to October when it is 25–60 cm in height and the green leaves are growing actively. The treated sward can be grazed or conserved about 10 days after application with the feeding value retained. Alternatively, two to three weeks after applying glyphosate and when the sward is brown and desiccated, the next crop can be direct-drilled. However, if there is a thick mat of decaying surface vegetation it should be lightly rotavated; otherwise the new seeds may be killed by toxic substances.

TABLE 70

Weed	Control
Bracken	This weed is a very serious problem which is increasing, especially in hill areas. The fronds (leaves) should be cut or crushed twice a year when they are almost fully open. If possible, the field should be ploughed deep and then cropped with potatoes, rape or kale before reseeding. Rotavating about 25 cm deep chops up and destroys the rhizomes. Very good chemical control is possible with asulam applied in the summer when the fronds are just fully expanded. Glyphosate can be used as a non-selective spray before reseeding or, selectively, with a rope-wick machine. Researchers are looking at better chemicals and the use of biological control using mycoherbicides (weed-eating fungi).
Buttercups	Spray with MCPA or MCPB. The bulbous buttercup is the most resistant type. Improved drainage and soil fertility will lessen this problem.
Chickweed	Spray with mecoprop (this kills clover) or benazolin mixture.
Docks	Grazing can be helpful. Asulam gives good control when sprayed on the expanded leaves in the spring or autumn. Triclopyr, fluroxypyr, dicamba, mecoprop, MCPA alone or in mixtures will give good control, although treatment may need repeating. If reseeding, the old sward should be sprayed with glyphosate before ploughing.
Horsetails	If possible, drainage should be improved. Spraying with MCPA or 2,4-D will kill aerial parts only and regrowth will occur. However, if it is done two to three weeks before cutting, any hay crop should be safe for feeding.
Nettles	Spray with triclopyr alone or in mixtures; glyphosate can also be used with a rope-wick.
Ragwort	Cut before the buds develop to prevent seeding. Spray with 2,4-D or MCPA in early May or in the autumn. Grazing with sheep in winter is helpful.
Rushes	If possible, drainage should be improved. Spray with MCPA or 2,4-D for the common or soft rush. The hard and jointed rush should be cut several times a year. Apply glyphosate with a rope-wick when the rushes are growing well. Grasses and clovers should be encouraged by good management.
Sorrel	Lime should be applied if necessary. Spray with MCPA or 2,4-D.
Thistles	Spray with clopyralid, MCPA or MCPB. The more resistant creeping type should be sprayed in the early flower bud stage. Over-grazing should be avoided.
Tussock grass	Drainage should be improved. The "tussocks" should be cut off with a flail harvester or topper. A hay or silage crop could be taken. Glyphosate can be applied with a rope-wick when the tussock grass is growing well.

Problems can also arise when seedling weed grasses, such as meadowgrass or blackgrass, establish with the sown seeds. Few herbicides can be used for grass control in new leys, but an exception is ethofumesate.

As a residual herbicide ethofumesate can also be applied in one or two doses to establish grass between mid-October and February to control many weeds such as chickweed, annual meadowgrasses, blackgrass, sterile and soft brome, wall barley grass and volunteer cereals.

Reseeding is not a cheap operation and so it is important that weed control is not skimped. Weeds can affect crop establishment and the longevity of the sward.

SPRAYING WITH HERBICIDES

This is a skilled operation and should only be carried out by trained operators. A recognized certificate of competence is required (pages 302, 303).

The following precautions should be taken when spraying:

(1) Make a careful survey of the field to determine the weeds to be controlled. Choose the most suitable and safest chemical and the best time for spraying.

(2) Check carefully the amount of chemical to be applied and the volume of water to be used (220–330 l/ha is a common range). Make sure the chemical is thoroughly mixed with the water before starting. Soluble and wettable powders should be mixed with some water before adding to the tank. Use the agitator if necessary.

The rate of application is mainly controlled by the forward speed of the tractor (use a speedometer) and the size of nozzle and, to a lesser extent, by the pressure (follow the maker's instructions). Always use clean, preferably not hard, water. Always use a filter. An accurate dipstick is necessary when refilling the tank if it is not emptied each time.

(3) Wear the correct protective clothing. Read carefully the instructions issued for that product when handling the concentrate and when spraying.

(4) Do not spray on a windy day, especially with the hormone type of herbicides and if the spray is likely to blow into susceptible crops or gardens. It can be helpful to keep the boom as low as possible and to use a plastic spray guard. To avoid spray drift, it is preferable to use a high volume application (larger droplets) than a very low volume mist.

(5) Spinning disc applicators, which produce uniform size droplets, only cause drift problems when very small droplet sizes are used.

(6) Make sure that the boom is level and that the spray cones or fans meet just above the level of the weeds to be controlled.

(7) Spray the headlands first. If using a wide boom, it is advisable to use markers. Slight overlapping avoids "misses". Tramlines are very helpful in this respect.

(8) Wash out the sprayer thoroughly. Follow the guidelines for safe disposal.

(9) Only use approved tank mixes. Follow the manufacturer's recommendations for mixing.

FURTHER READING

British Poisonous Plants, Bulletin No. 161, HMSO.
Crop Chemicals Guide, ACP Publishers Ltd.
Research Reports, AFRC, AAB, BCPC.
Schering Guide to Grass and Broadleaved Weed Identification.
The Arable Weeds of Europe, Martin Hanf, BASF.
The UK Pesticide Guide 1993, Ed. G. W. Ivens, British Crop Protection Council.
UK Pesticides for Farmers and Growers, The Royal Society of Chemistry.
Weed Biology and Control, Gwynne and Murray, Batsford Technical.
Weed Control Handbook, Blackwell.

12

PESTS AND DISEASES OF FARM CROPS

PESTS are responsible for millions of pounds of damage to agricultural crops in this country every year.

Before discussing the various methods used to control pests, it is important to understand something of their structure and general habits.

One of the major groups of pests is the **Insect**. Insects are invertebrates, i.e. they belong to a group of animals which do not possess an internal skeleton. Their bodies are supported by a hard external covering—the exoskeleton. It is composed chiefly of chitin, and is segmented so that the insect is able to move. From the diagram of the external structure of an adult insect (Fig. 85) it can be seen that the segments are grouped into three main parts:

(1) *The head*, on which is found:
    (a) The antennae or feelers carrying sense organs for touching and smelling.
    (b) The eyes; a number of simple and a pair of compound eyes are present in most species.
    (c) The mouth parts (Fig. 86); two main types are found in insects:
      (i) the biting type used for grazing on foliage;
      (ii) the sucking type—insects in this group suck the sap from the plant and do not eat the foliage.

The type of mouth part possessed by the

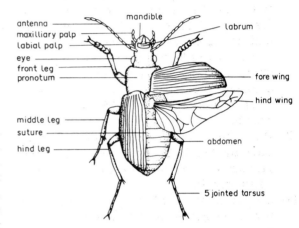

FIG. 85. Structure of an insect.

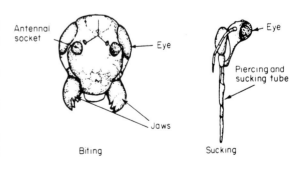

FIG. 86. Insect mouthparts.

insect is of considerable importance in deciding on the method of control.

(2) *The thorax*, which bears:
   (a) The legs; there are always three pairs of jointed legs on adult insects.
   (b) The wings; these are found on most, but not on all, species.

(3) The *abdomen*; this has no structures attached to it except in certain female species where the egg-laying apparatus may protrude from the end.

LIFE-CYCLES

A knowledge of the life-cycles of insects can be of great help in deciding on the best stage at which the insects will be most susceptible to control methods.

Most insects begin life as a result of an egg having been laid by the female. What emerges from the egg, according to the species, may or may not look like the adult insect.

There are two main types of life-cycles:

(1) The "complete" or four-stage life-cycle (Fig. 87).
   (a) The egg.
   (b) The larva (plural larvæ)—entirely different in appearance from the adult. This is the active eating and growing stage. The larvæ usually possess biting mouth parts, and it is at this stage with many insects that they are most destructive to the crops on which they feed (Fig. 86).
   (c) The pupa—the resting state. The larvæ pupate and undergo a complete change from which emerges—
   (d) The adult insect—this may feed on the crop, e.g. flea beetle, but in many cases it does far less damage than the larvæ, e.g. flies.

(2) The "incomplete" or three-stage life-cycle (Fig. 88).
   (a) The egg.
   (b) The nymph—this is very similar in appearance to the adult, although it is

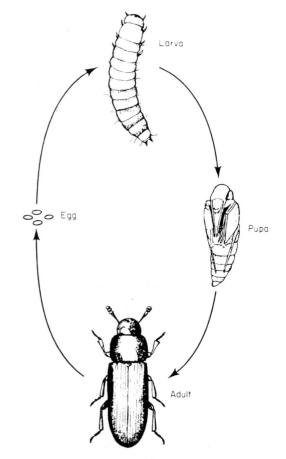

Fig. 87. Four-stage life-cycle.

smaller and may not possess wings. It is the active eating and growing stage.
   (c) The adult insect—invariably this stage will also feed on and damage the crop, e.g. aphids.

Most insects and/or larvæ and nymphs depend on the crops on which they feed for part or all of their existence. The crop is the host plant, whilst the insect is the parasite to the host. Not all insects are harmful to crops; some are beneficial in that they prey on or parasitize crop pests, e.g. the ladybird is particularly useful because, both as the larva and adult, it feeds on aphids which are responsible for transmitting

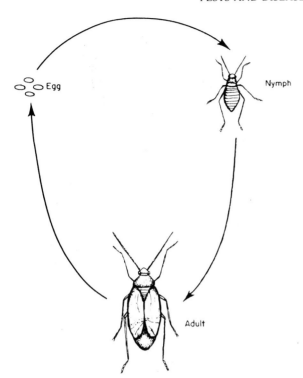

FIG. 88. Three-stage life-cycle.

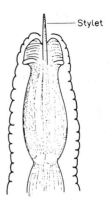

FIG. 89. Mouth-parts of plant-feeding nematode.

certain virus diseases in plants as well as causing physical damage to plants.

A second major group of pests is the **Nematoda**.

Nematodes have unsegmented elongated worm-shaped bodies surrounded by a tough cuticle. They vary from 0.1 mm to 2.0 mm in length with an average size of 1.0 mm. Quite often they are too small to be seen with the naked eye.

As far as is known, most species of nematodes are free-living and beneficial, but there are a number of important species which are either parasitic in crop plants (endoparasitic) or feed on the surface tissues of crop roots (ectoparasitic).

The mouth parts of these plant-feeding nematodes consist essentially of a cavity—the stoma—in which is positioned the mouth spear or stylet (Fig. 89). It is this which pierces the cellular tissue for sucking out the cell contents of the parasitized plant.

The life-cycle is relatively simple, consisting generally of an egg hatching into a larval form. There are several juvenile stages (shedding a cuticle each time) before reaching the sexually mature male and female stage. In many cases, part of the life-cycle takes place in the soil.

With many pest species there is a stage in the life-cycle (e.g. a cyst containing eggs) which may remain dormant, sometimes for several years, only becoming active again when conditions are suitable. Some species, however, have more than one generation in the year, and are only inactive during the winter months. Both the juvenile and adult stages are collectively the cause of damage to the host plant. Colonies develop, creating progressive damage which is gradually expressed in the form of characteristic symptoms. These symptoms, supported by plant and soil analysis, allow the identification of the species involved and the formation of future control strategies. No curative control is yet available to check damage to an already infected crop.

The major pests are the root-attacking cyst nematodes. These should be controlled by the integration of rotation, resistant varieties (if they exist) and pesticides, which are certainly more costly when compared to many other pest control measures.

Another agriculturally important nematode is the stem nematode. It has a wide host range but on each host it causes varying forms of necrosis and deformity of the shoot systems. Attacked crops will invariably have seriously impaired yields. Suppression is achieved by selecting the most suitable control measures, i.e. use of clean seed or planting material, resistant varieties, use of either hot water treatment and/or pesticide, adjustment of the rotation and crop and farm hygiene.

METHODS OF PEST CONTROL

Indirect control methods

These aim more at the prevention of the pest attack.

(i) *Rotations*
As a means of control, until recently it has not been considered so important because of more effective chemical control. However, with an increased emphasis on efforts to reduce the use of pesticides, rotations can be said to be assuming more importance again.

The principle behind control by rotation is that if the host plant (the crop) is continuously grown in the same field for too many successive years, the parasite will increase in large numbers, e.g. the nematode.

(ii) *Time of sowing*
(a) A crop may sometimes be sown early enough so that it can develop sufficiently to withstand an insect attack, e.g. wheat bulb fly in the wheat crop.
(b) A crop can be sown late enough to avoid the peak emergence of a pest, e.g. aphid transmitted barley yellow dwarf virus (BYDV) in winter cereals.

(iii) *Cultivations*
Ploughing exposes pests such as wireworms, leatherjackets and caterpillars, which are then eaten by birds. Well-prepared seedbeds encourage rapid germination and growth which will often enable a crop to grow away from pest attack.

(iv) *Encouragement of growth*
Good quality seed should be used which will germinate quickly and evenly. It is also important that the crop is not checked to any extent, say by lack of a plant food. A poor growing crop is far more vulnerable to pest attack than a quick growing crop. A top-dressing of nitrogen, just as a crop is being attacked, may sometimes save it.

(v) *Clean farming*
Weeds are alternate hosts to a great variety of insects and, as far as possible, these sources of infestation should be eradicated.

Biological control

A parasite or predator is used to control the pest. As yet the method has little application in farming in this country, but research and development is continuing. Two established examples of this type of control are used in horticulture:

(a) Where the red spider mite is successfully controlled by predatory mites, in cucumber production under glass.
(b) *Bacillus thuringiensis* for bacterial control of caterpillars. The crop is sprayed and the bacteria invades the target organism, e.g. used to control cabbage white butterfly.

Direct control methods

This means chiefly chemical control using a pesticide. These can be used in a number of ways:

(a) Sprays and dusts.
(b) A granular form for controlling aphids.
(c) Baits for controlling soil pests such as leatherjackets, slugs and snails.
(d) Seed dressing—mainly for the protection of cereals against wireworm, and brassica crops against flea beetle. Usually the insecticides are combined with a fungicide.

Gases, smokes and fumigants are commonly used in greenhouses against aphids chiefly, and in granaries for the control of beetles and weevils.

Basically, there are two ways in which pesticides kill pests:

(i) *By contact*

The pest is killed when it comes in contact with the chemical, either when:

(a) it is directly hit by the spray or dust;
(b) it picks up the pesticide as it moves over foliage which has been treated;
(c) it absorbs vapour;
(d) it passes through soil which has also been treated.

(ii) *By ingestion*

As a stomach poison, the pest eats the foliage treated with the pesticide, or the chemical is used in a bait.

As a systemic compound it is applied to the foliage or to the soil around the base of the plant. It gets into the sap stream of the plant and thus the pest is poisoned when it subsequently sucks the sap.

Most pesticides kill by more than one method, which makes them very effective. But many of them are extremely toxic to animals and humans and, by law, certain precautions must be observed by the persons using them.

Integrated pest management

With increasing awareness of the side effects of pesticides, integrated pest management (IPM) is becoming a more favoured method of pest control. Whilst it involves the use of pesticides, their use is minimized in an attempt to protect and enhance the activities of beneficial insects (natural enemies and pollinating insects). The resistance of the environment is augmented in various ways to limit pest problems by non-chemical means. This may include the introduction of natural enemies of pests, host plants for beneficials, and generally increasing the health of the crop, and thus its resistance to pest attack, e.g. by rotation, planting date, fertilizers, etc. Economic thresholds and forecasts of pests, where available, are consulted to reduce insurance spraying, and when pesticides must be applied they are used in a selective way. In the selective use of pesticides the following parameters may be considered:

Selectivity in space—	The area treated with pesticide may be limited to protect beneficial insects, e.g. part of a crop may be left untreated, or attractive baits may be used to separate the pest from non-target organisms.
Selectivity in time—	The timing of pesticide application may coincide with the time when beneficials are less likely to come into contact with the chemical, e.g. to protect bees, sprays should be applied to oilseed rape in flower at times when bees are inactive, i.e. in early morning, late evening or on cloudy days.
Selectivity of pesticide—	More selective pesticide applications may be made by exploiting various features of the pesticide itself. These include the biochemical activity of the pesticide, e.g. in IPM, selected pesticides are used in preference to broad spectrum chemicals. The formulation of the pesticide may also render it more specific, e.g. slug pellets incorporating an attractive bait. Finally, the dose rate of the pesticide may influence the survival of natural enemies, e.g. by reducing the dose rate of the pesticide used, better control of the pest can be achieved as the natural enemy:aphid ratio is increased.

CLASSIFICATION OF PESTICIDES

*Insecticides

(1) *Organochlorines*

These insecticides are all stomach and contact poisons and the main one is:

HCH. As a spray and dust it is used extensively on fruit crops. It controls, amongst other insects, aphids, caterpillars and weevils and is also a useful soil insecticide for control of wireworm and leatherjackets, but it can taint some crops such as potatoes. HCH is also used as a seed dressing on cereals and brassica crops.

(2) *The organophosphorus compounds*

As a group, the organophosphorus compounds and the carbamate compounds are dangerous to use. They should be handled strictly in accordance with the manufacturer's instructions.

Examples of some of the organophosphorus insecticides in common use are:

(a) *Non-systemic*

Chlorfenvinphos. Controls cabbage root fly, carrot fly and, in the maize crop, frit fly. As a seed dressing it helps to reduce wheat bulb fly.

Pirimiphos-methyl. For control of stored grain pests. Some formulations control wheat bulb fly and frit fly in cereals.

Chlorpyrifos. For control of cabbage root fly on brassicae, leatherjacket on cereals, frit fly and fruit pests.

Fenitrothion. For control of aphids and, in a bran bait, leatherjackets and cutworms. It also controls stored grain pests.

Fonofos. For control of cabbage root fly, wheat bulb fly, wireworm and yellow cereal fly.

(b) *Systemic*

Dimethoate. This is used for the control of aphids on many agricultural crops. As an emergency treatment it can be used as a spray against wheat bulb fly. This insecticide is not very selective.

Disulfoton. This insecticide controls aphids on brassicae, beans, potatoes, sugar beet and carrot fly. It is used in the granular form.

Phorate. This is used in a granular form chiefly for the control of aphids in beans, potatoes and sugar beet, and frit fly in maize.

(3) *The carbamate compounds*

Bendiocarb. This controls pests of sugar beet and frit fly in maize.

Carbofuran. For control of cabbage root fly, flea beetle and turnip root fly. It is also a nematicide.

Pirimicarb. For control of aphids. This insecticide is very specific and does not affect beneficial insects.

Carbosulfan. This is a systemic insecticide for sugar beet and brassica pests.

(4) *Synthetic pyrethoids*

These chemicals have a particularly high insecticidal power whilst generally being safe to humans and farm animals. They are efficient contact insecticides with a rapid knockdown action and some are thought to act as antifeedants. These properties are being seen as valuable in the chemical control of plant viruses by controlling the insect vectors, e.g. cypermethrin and deltamethrin.

Persistency is a constant cause of concern for food producers, and research is continually directed towards finding safer, less persistent compounds. In this context, as a chemical group, the carbamates, and synthetic pyrethoids, can be considered an important development.

Molluscicides

Metaldehyde. This is used as a mini-pellet for the control of slugs and snails. It works by dehydrating the molluscs.

Methiocarb. This is used as a mini-pellet for the

control of slugs and snails. It is a stomach-acting carbamate.

Nematicides (including soil sterilants)

Nematicides have been developed for the control of nematode pests of crops, such as

Aldicarb. This also controls aphids and docking disorder in sugar beet.

Oxamyl. This also controls docking disorder in sugar beet.

Certain chemicals in this group are classified as soil sterilants, e.g. *Dazomet* and *Dichloropropene*.

It is important to remember that a certain interval must be observed between the last application of the pesticide and the harvesting of edible crops as well as the access of animals and poultry to treated areas.

With some pesticides this interval is longer than others. This is another reason for very careful reading of the manufacturer's instructions.

*An asterisk marks those chemicals included in the Health and Safety (Agriculture) (Poisonous Substances) Regulations. Certain precautions (including the use of protective clothing) must by law be taken when using these chemicals. It is advisable to read the joint MAFF/HSE publication "Pesticides: Code of Practice for the Safe Use of Pesticides on Farms and Holdings" 1990 from HMSO.

For up-to-date information on pesticides available, and the regulations and advice on the use of these chemicals, reference should be made to the UK Pesticide Guide, published by CAB/BCPC; the Crop Chemicals Guide, published by ACP; in addition, on-line computer information is available from CABPESTCD, CAB International.

Table 71 indicates the major pests attacking farm crops, and their control.

OTHER PESTS OF CROPS

Birds

Generally, birds are more helpful than harmful, although this will depend on the district and type of farming carried out. To the grassland farmer in the west of England, birds are not such a problem as they are to the arable farmer in the Midlands and East Anglia. Although birds will eat some cereal seed, most of them help the farmer by eating many insect pests and weed seeds; the diet of some also includes mice, young rats and other rodents.

The wood pigeon certainly does far more harm than good. Not only will it eat cereal seed and grain of lodged crops, it also causes considerable damage to young and mature crops of peas and brassicae. The only effective ways of keeping this pest down are by properly organized pigeon shoots and nest destruction, and the use of various scarers.

Mammals

Wild animals which cause most damage to crops are:

Rabbits and hares which can be very serious pests; they eat many growing crops, particularly young cereals. Clearance of scrub and gassing are helpful in controlling rabbits. Organized shoots can control hares.

The worst pest of all is the brown rat which eats and damages growing and stored crops.

The mouse is another serious pest; it damages many stored crops.

The local rodent officer will give advice on methods of extermination.

The harmless mammals, as far as crops are concerned, are:

Badgers and hedgehogs which eat insects, slugs, mice, etc. Foxes which kill rats and rabbits. Squirrels which eat pigeons' eggs.

PLANT DISEASES

Although there are many causes of unhealthy crops, such as poor fertility and adverse weather conditions, the chief cause is disease.

Diseases, like pests, annually cause millions of pounds worth of damage and loss to the agricultural industry.

TABLE 71 *Major pests and their control*

Crop attacked	Pest	Description	Life-cycle	Symptoms of attack	Control	Notes
Cereals	*Adult:* Clickbeetle *Larva:* Wireworm	*Adult:* Brown, 6–12 mm long *Larva:* Growing to 25 mm long, yellow colour	Larvæ hatch out during summer from eggs laid just below soil surface, mainly in grassland. They take 4–5 years to mature, and after pupation in the soil, the adult appears in early autumn	Yellowing of foliage followed by the disappearance of successive plants in a row. This is caused by wireworm moving down the row. Larvæ eat into the plants just below soil surface. They are usually found in soil around the plants	Good growing conditions to help the crop grow away from an attack. Wheat and oats more susceptible than barley; they should not be grown where the wireworm count is over 2 million/ hectare. All seed should be dressed with gamma-HCH	Do not confuse wireworm attack with other pests such as nematode
	Adult: Cranefly *Larva:* Leather-jacket	*Adult:* Is the "Daddy longlegs" *Larva:* Leaden in colour, 30 mm long	Eggs laid on grassland or weedy stubble in the autumn from which the larvæ soon emerge. They feed on the crop the following spring, pupating in the soil during the summer	Crop dies away in patches, root and stem below ground having been eaten. Larvæ found in soil	If possible, plough the field before August to prevent the eggs being laid. HCH can be applied as a low volume spray, or a bait such as fenitrothion broadcast late in the day. Rolling and nitrogen top-dressing help crop to recover	
	Wheat bulb fly	*Larva:* Whitish-grey, 12 mm long	Eggs laid on bare soil in July, August. Eggs start hatching from early January. Larvæ feed on the crop until following May. Pupation then follows either in the soil or plant	Central shoot of plant turns yellow and dies in early spring. Larvæ found in base of tiller	If possible, avoid sowing wheat where the field has lain bare from late summer. Protection includes selection from one of following: (i) Granules or sprays at or about sowing time. (ii) Seed treatment, e.g. fonofos for late-drilled crops. (iii) Spray at egg hatch, e.g. chlorpyrifos. (iv) Spray at first sign of damage, e.g. use omethoate	Winter wheat and barley are attacked especially if drilled deep and late after a susceptible crop. Wheat bulb fly is a problem of central and eastern counties

TABLE 71 cont'd

Crop attacked	Pest	Description	Life-cycle	Symptoms of attack	Control	Notes
	Frit fly	*Larva:* Whitish, 3 mm long	Three generations in the year; the first when in spring eggs are laid on spring oats and larvæ feed on crop in May and June. The second generation damages the oat grain, whilst the third generation over-winters on grass, but when the latter is ploughed for autumn cereals the larvæ move on to the cereals	In early summer, the central shoot of the oat plant turns yellow and dies, but the outer leaves remain green; blind spikelets and shrivelled grains are caused by second generation; autumn cereals can have a kink above the coleoptile at single-shoot stage, shoot then turns yellow	Sow spring oats early, and try and get them past the 4-leaf stage as quickly as possible. With *late sown spring oats* can spray at 2-leaf stage at first sign of attack, but advice should be sought. Allow a 6-week interval between ploughing grass and sowing the *winter cereal*. Spray only at first sign of attack if more than 10% of crop is affected; can use chlorpyrifos	Spring oats and maize are particularly susceptible
	Yellow cereal fly	*Larva:* Yellowish; slender, about 8 mm long, pointed at both ends	Eggs laid near wheat plant in October and November, hatching early in new year; larvæ move down between outer leaves to feed on main tiller. Pupation in early summer, adults appear in June and a month later they migrate to hedgerows before returning to wheat field in late autumn	Circular or short spiral band at base of tiller producing brownish scar and then deaths of shoots – "deadhearts"	Early sown wheat in the eastern counties at greatest risk. Chemical treatment using a pyrethroid within 3–4 weeks of egg hatch is a possibility, but more trials are necessary	Winter barley rarely attacked; spring sown cereals virtually immune
	Gout fly	*Larva:* Legless, yellowish-white, 6 mm long	Two generations in the year, the most important being the first. Larva hatch and feed in plant	Leaf sheath surrounding the ear is swollen and twisted. Poorly developed grain emerges	Good growing conditions will help to keep it growing	Barley is chiefly affected
	Cereal aphids	Various species of green fly, 1.6–3.3 mm long	Winged females found feeding on cereal crops in May and June. Wingless generations produced which continue feeding	Depending on the species, the damage caused to the cereal varies from stunted withered growth, occasionally with	The grain aphid causes most concern. Spray in summer when infestation reaches threshold level of 5 aphids/ear or 66%	Aphids carry virus diseases from infected to clean plants. See barley yellow dwarf

TABLE 71 cont'd

Crop attacked	Pest	Description	Life-cycle	Symptoms of attack	Control	Notes
			during summer. Most species move back to winter quarters (woody hosts and some grasses and cereals, depending on aphid species) in autumn, although some may be found on young cereal crops at the end of the year, especially in the milder parts of the country	reddish-brown to purple spots on the leaves. The grain aphid causes empty and/or small grain; by puncturing the grain in the milk-ripe stage, the grain contents seep out. This also reduces the weight of the grain	of ears infested. Organophosphorus or carbamate compounds are used	virus disease (Table 72). Treatment only worth while in the wheat crop
	Stem and bulb nematode	Too small to be seen without magnification	Live and breed in the plant. If the plant dies, nematodes become dormant in dead tissue or soil, becoming active again when conditions are suitable	Twisting and swelling, and this normally prevents plants from elongating and producing an ear	Resistant varieties. Rotation to starve out the nematodes. Clean seed	Attacks oats
	Cereal cyst nematode	Dark-brown lemon-shaped cysts about 1 mm long	Live and breed in the roots. White-looking cysts (female cysts containing large numbers of eggs) are found on roots. Later these cysts (now dark brown) become free in the soil to infect the host plant again	Crop shows patches of stunted yellowish-green plants. Root system very bushy. Cysts visible on roots from June onwards	Avoid growing oats too often in the field. Grow resistant varieties when necessary	Oats chiefly infected
	Slugs and snails	Field slug lightish-brown in colour, about 40 mm long	Wheat grain damaged by being eaten in the ground before it germinates. Young cereals can be completely grazed off by a severe autumn attack. Most active in moist and humid conditions. An attack can be more serious when the seed is direct-drilled if the slit has not been properly covered or if the seedbed is cloddy and trashy		Baits containing metaldehyde or methiocarb spread evenly over the field prior to drilling. Extra cultivations in preparing the seedbed help to check the pests. Rolling and early nitrogen top-dressing can help a damaged crop to recover	Winter wheat chiefly attacked. Often worse after oilseed rape

TABLE 71 cont'd

Crop attacked	Pest	Description	Life-cycle	Symptoms of attack	Control	Notes
Stored grain	Saw-toothed grain beetle	*Adult:* Dark brown, 3 mm long *Larva:* White and flattened	Eggs are laid on the stored grain; larvæ feed on the damaged grain. Pupation takes place in the grain or store	The grain heats up rapidly; it becomes caked and mouldy. This is seen with the appearance of the beetles	Pirimiphos-methyl applied as the grain is fed into the store	
	Grain weevil	*Adult:* Reddish-brown, about 3 mm long with an elongated snout	During autumn the weevils bore into the stored grain to lay their eggs. The larvæ feed inside the grain where they also pupate	Hollow grains. Sudden heating of the grain. Weevils found a few feet below the surface of stored grain	See the saw-toothed grain beetle	
Maize	Frit fly	*Larva:* Whitish, 3 mm long	As for frit fly on oats	Twisting of leaves surrounding the growing point. In severe attacks this is killed, leading to secondary tillers	As a preventive, an insecticide such as bendiocarb in granular form applied at time of sowing. Seed treatment applied by seeds merchant using bendiocarb	
	Wireworm	See wireworm on cereals			Seed treatment using bendiocarb	
	Cyst nematode	Maize reduces the number of nematodes because they do not reproduce and form cysts		Bronze coloration; severe stunting of plant	Avoid planting maize on fields where nematode population is relatively high	Maize is susceptible to quite low populations of nematodes, especially on chalk soils following cereals
Beans	Bean aphid (black fly)	Very small oval body, black to green colour	There are many generations in the year. In summer winged females feed on the crop; wingless generations are then produced which	On all summer host plants, colonies of black aphids are seen on the stem leaves (especially the underside) and on the flowers. The plant	Apply an insecticide, e.g. pirimicarb, when 5% of plants infected on headland	It also attacks sugar beet and mangels. Mainly a pest of spring beans

TABLE 71 cont'd

Crop attacked	Pest	Description	Life-cycle	Symptoms of attack	Control	Notes
			continue to feed. Eventually a winged generation flies to the spindle tree on which eggs are laid for over-wintering	wilts; it can become stunted and with a heavy infestation it may be killed		
Peas, beans and other legumes	Pea and bean weevil, and striped pea weevil	*Adult:* Yellowish-brown with lighter stripes, 6 mm long long *Larva:* Legless, white with brown head	During early spring eggs laid in the soil near plants. Larvae feed on roots, whilst adults feed on leaves. Pupation takes place in the soil in mid-summer	Seedling crops checked. U-shaped notches at the leaf margins caused by adults. Larvae eat root nodules	Apply organo-phosphorus insecticide as spray or granules when attack is first noticed. Some pyrethroid sprays (e.g. deltamethrin) are approved for use when symptoms first appear	
	Pea moth	*Adult:* Dull greyish-brown, about 6 mm long *Larva:* Yellowish-white with darker head; legless, about 8 mm long	Eggs laid June–mid-August, hatch in a week. Larvae enter pods and feed on peas until fully grown. Larvae leave pod and make way to soil; pupate in spring and adult emerges in early summer	Holes in peas caused by larvae	One or more sprays using synthetic pyrethroid insecticide. The timing is important. A pea moth monitoring service operates in some regions and this can be used as a general guide. Can use pheromone trap	(1) Early and very late sown crops suffer less damage. (2) Dried peas for harvesting most vulnerable but only worth treating those for human consumption or seed
	Pea aphid	Large green aphid	Appear on crops from mid-May. Numbers peak in June/July	Direct feeding causes distortion of plant and pods. Can also transmit virus	Treatment worth while when 1 plant in 5 infested. Treat with aphicide	
	Pea midge	Small midge	Adult lays its eggs on the buds from June onwards. Maggots hatch and feed for 10 days	Larvae cause severe damage to the flower buds. Can also contaminate produce	Treat in vining peas when first midges are seen in high risk areas	Vining peas are more at risk than dry harvested peas because their flowering period is longer
Brassicae (cabbage, kale, oil-seed rape,	Flea beetle	A minute black beetle with a yellow stripe down each	Adults emerge from hibernation during late spring to feed on crops. Eggs are laid,	Very small round holes are eaten in the seed leaves of the plants	Sow the crop either early or late, i.e. avoid April and May. Good growing	Sugar beet, mangels and cereal crops can be attacked on

TABLE 71 cont'd

Crop attacked	Pest	Description	Life-cycle	Symptoms of attack	Control	Notes
swedes, turnips)		wing case	but larvæ do little damage. Pupation takes place in the soil during the summer		conditions to get the crop quickly past the seed leaf stage. Seed dressing containing gamma-HCH should be carried out. A dust or spray containing gamma-HCH can be applied as soon as the attack is noticed	occasions
	Cabbage stem flea beetle	*Adult:* 4 mm, metallic green/black	Adult moves into rape crop in September and lays eggs. Larvæ invade plant from October	Larvæ burrow into leaf petioles and cause stunting and leaf loss	Treat when 5 or more larvæ per plant across field. Use pyrethroid or gamma-HCH	Important pest in oilseed rape
	Cutworm	*Adult:* Turnip moth; garden dart moth; heart and dart moth; large yellow underwing moth *Larva:* Caterpillars – up to 40 mm long; vary in colour according to adult – dull/greyish brown to green; one species has black marks along the back	All similar except the garden dart moth; eggs laid and hatched in 10–14 days. After early feeding on leaves, caterpillars go into soil and feed (mostly at night) on stem above and below ground level Most, but not all, are fully fed in the autumn, and over-winter to pupate in the spring in the soil. Earlier fed caterpillars will pupate in the autumn to produce the adult and a second brood of caterpillars in the autumn. The garden dart lays eggs in late summer which hatch out the following spring; caterpillars fully fed in mid-summer then pupate, moths emerging in August	Young plants cut off at base of stem; poor top growth, and plant wilts. Turnip and swede roots can also be damaged by feeding caterpillars	Chlorpyrifos sprays used as soon as attack is confirmed	Also attacks potatoes – particularly in dry summers. Late sown crops of sugar beet can also be affected

TABLE 71 cont'd

Crop attacked	Pest	Description	Life-cycle	Symptoms of attack	Control	Notes
Brassica seed crops (N.B. oil-seed rape)	Pollen beetle (Blossom beetle)	Metallic-greenish black in colour, 3 mm long	Adults emerge from hibernation during spring to feed on buds and flower parts. Eggs laid and similar damage caused when larvæ emerge	Damaged buds wither and die, and the number of pods set is lessened	Treat with approved pyrethroid if more than 15/plant. Best results obtained by treatment at green/yellow bud stage	Extreme caution should be taken to ensure that no serious damage occurs to pollinating insects. Spring oilseed rape is most susceptible
	Pod midge	*Larva:* Whitish-cream	Females can insert eggs in pods through holes in pods left by emerging seed weevil larva. Larvæ feed on developing seed and walls of pod; pupate in soil after 4 weeks. Adults emerge 2 weeks later except for last generation which over-winters in cocoon and pupates in spring	Adults lay eggs in pods; larvæ cause pods to ripen prematurely and seed is shed early	As for seed weevil	Winter oilseed rape most affected
	Seed weevils	Lead-grey in colour, about 2.5 mm long	From hibernation near previous year's seed crops adults lay eggs in young pods. Larvæ feed on seeds in developing pods; they leave the pods and fall to the ground where they pupate in the soil	Seeds destroyed in pods by larvæ. Damage also caused by adult which makes holes for the pod midge to enter	Spray with an approved insecticide, e.g. alpha-cypermethria, when one or more weevils per plant is seen at end of flowering for autumn-sown crops, and late yellow bud stage for spring crops	As for pollen beetles
Sugar beet, fodder beet, mangels	Flea beetle	See flea beetle on brassicae			Seed dressing is not possible	
	Mangel fly	*Larva:* Yellow-white, legless, 20 mm long	White oval-shaped eggs are laid in the underside of leaves in May. Larvæ bore into the leaf tissue and after about 14 days they drop into the soil where they pupate	Blistering of leaf which can become withered. Retarded growth and in extreme cases death of the plant	Good growing conditions to help the crop pass an attack. Spraying carried out using an approved organo-phosphorus or carbamate insecticide	

TABLE 71 cont'd

Crop attacked	Pest	Description	Life-cycle	Symptoms of attack	Control	Notes
					when more than 25 hatched larvæ or eggs are counted per plant in the 6–8-leaf stage	
	Aphids (black and green fly)	The green-fly (peach potato aphid) has a very small oval-shaped body of various shades of green to yellow	During spring winged aphids migrate to the summer host crops. They move from one plant to another, thus transmitting the virus from an unhealthy to a healthy plant	A severe infestation can cause the death of the plant, but chiefly it will mean a bad attack of virus yellows, as both aphids are responsible for carrying the virus causing this disease	As for the bean aphid	
	Cyst nematode	See cyst nematode on cereals		Crop failing in patches. Plants which do survive are very stunted in growth	Soil sterilants or carbamate nematicides can be used on infected soil	If necessary the soil can be tested for a nematode count
	Wireworm	See wireworm on cereals		The roots of seedling plants are bitten off	A seed dressing, e.g. gamma-HCH	
	Docking disorder	A complex problem. Causes *irregularly* stunted plants with fangy root growth. Often caused by nematodes, but soil structure giving poor growing conditions can be a causal factor. The disease is only found on sandy soils, normally alkaline with a low organic matter content. Losses can be minimized by good growing conditions. Success has been achieved by combining measures to improve soil fertility and the incorporation of pesticides such as aldicarb			Soil incorporation of systemic carbamate-type insecticide and nematicide will help	
Carrots	Carrot fly	*Adult:* About 8 mm long; shining black reddish/brown head and yellowish wings *Larva:* When fully	Usually two generations a year. Eggs laid in soil surface near carrot. Hatch in 7 days and eventually burrow into root, forming "mines"; after third moult pupate in soil close to tap root. Some	Brown and rusty tunnels (mines often with larvæ protruding) becoming progressively worse as season proceeds. Foliage of badly infected plants turns red, wilts and dies	(1) Rotation and avoid sowing susceptible crops close together (2) Hygiene round edge of field to cut down shelter for adult (3) Chlorfenvinophos or diazinon incorporated into	Celery and parsnips also attacked

TABLE 71 cont'd

Crop attacked	Pest	Description	Life-cycle	Symptoms of attack	Control	Notes
		grown creamy-white, 8–10 mm long	of the second, and possibly a third, generation (especially in eastern counties) over-winter in the roots, emerging the following spring		soil as granules. Triazaphos can be foliar applied. Use pheromone traps to aid control	
Potatoes	Peach potato aphid (green fly)	See aphids on sugar beet, fodder beet and mangels		A bad infestation (5 aphids per compound leaf) will check the growth of the plant	Apply a granular organo-phosphorus insecticide. This will help to check the spread of disease by killing the aphids. An aphicide spray could also be used. There is resistance to some materials	Potato virus diseases are spread in seed crops by aphids
	Wireworm	See wireworm on cereals		Maincrop tubers are riddled with tunnel-like holes	Use ethoprophos granules applied to the soil before planting. Lift the crop in early September if possible	
	Potato cyst nematode (PCN)	White and golden types found in UK	Cysts in soil hatch and eelworms invade roots. New cysts formed from fertilized females	Stunted plants with restricted "hairy" root system	Where soil analysis indicates, use either a resistant variety or a nematicide in the seedbed preparation. Wide rotations help keep cyst levels low	
	Slugs	See slugs on cereals		Maincrop potatoes damaged by pests eating holes in the tubers	Very difficult. Some varieties resistant	
Grass	Leatherjacket	See leatherjacket on cereals		Grass dying off in patches, the roots having been eaten away. Larvæ found in the soil	Spray with gamma-HCH or use a bait	

TABLE 71 cont'd

Crop attacked	Pest	Description	Life-cycle	Symptoms of attack	Control	Notes
	Frit fly	See frit fly on cereals		The third generation larvæ can reduce the chances of a successful establishment of some autumn-sown grass seed mixtures	When reseeding, allow at least 6 weeks from the destruction of the old sward to the sowing of the new ley. Chemical control with chlorpyrifos spray	
	Slugs	See slugs on cereals	Seed destruction below the ground is particularly serious	Slug pellets – methiocarb and metaldehyde broadcast at least 3 days before drilling		
	Wireworm	See wireworm on cereals	Base of plant chewed just below ground level; plant wilts and turns yellow	Risk factor highest in second and third year of a new sward when previous crop was grass. If necessary, incorporate gamma-HCH into seedbed prior to sowing		
Red clover	Stem nematode	*Adult:* Slender and colourless, difficult to see, about 1.2 mm long	Lives and breeds continuously in plant; passes into soil to infect other plants. Can remain dormant in hay made from an infected crop, becoming active again when conditions are suitable	Thickening at base of stem; some distortion of leaves, petioles and stems. Plants are stunted; infested patches increase in size each year	(1) Some varieties of red clover show resistance. (2) Fumigated seed. (3) Rotation – several years' break from red clover	White clover can be affected by a different race, but it is not considered important. Lucerne also affected by different race and the seed is fumigated

The main agencies of disease are:

Fungi

Fungi cannot manufacture their own carbohydrate and so they obtain it from living or dead plants. Thus it is convenient to divide fungi into two main classes:

(1) *Parasites*
  (a) *Obligate parasites*. These are dependent on the living host; they are responsible for causing many plant diseases.
  (b) *Facultative parasites or semi parasites*. These kill the host tissues and live on the dead cells.

(2) *Saprophytes*
  These live on dead organic matter and are often present in plants attacked by parasites or plants that have reached maturity and died. They are also found on leaves coated in aphid honeydew. Saprophytes play an important part in helping to break down plant remains into organic matter.

There are many thousands of different species of fungus, the majority of which are invisible to the naked eye.

A typical fungus is composed of long thin filaments (made up of single cells) termed hyphæ. Collectively, these are known as mycelium. It is through the haustoria of the mycelium that the fungus absorbs nutrients from its host. With most parasitic fungi, the mycelium is enclosed within the host (only the reproductive parts protruding), although some fungi are only attached to the surface of the host.

Reproduction

Fungi can reproduce simply by fragments of the hyphæ dropping off, but usually reproduction is by **spores**. Spores can be compared to the seeds in ordinary plants, but they are microscopic and occur in vast numbers. The mycelium produces spores which when ripe are dispersed by elaborate discharge mechanisms.

The dispersal of spores

It is important to understand how the spores are dispersed, causing infection to be spread from one plant to another. Being aware of the particular form of dispersal will help in deciding on disease prevention and control methods.

Spores can be dispersed by:
(i)   *The seed*. The fungus is carried from one generation to the next by surviving on the seed coat or inside the seed itself, e.g. smut diseases of cereals.
(ii) *The soil*. The spores drop off the host plant and remain in the soil until another susceptible host crop is grown in the field. A suitable rotation will go a long way to check diseases caused in this manner, e.g. clubroot of brassicæ.
(iii) *The air*. Spores of many foliar diseases as well as loose smut are dispersed on dry air currents and also sometimes by rain splash, e.g. cereal rust diseases.

Many fungi, especially the obligate parasites, show great specialization and attack only one host or even a restricted range of varieties of one host.

The extent of the disease caused by the fungi does depend upon soil and weather conditions and also upon the state of the host crop. A healthy crop which is growing well will withstand an attack far more successfully than a stunted slow-growing crop.

Viruses

The virus was discovered in the last century. It is a very small organism. Something like one million viruses could be contained on an average bacterium. Only by using electron microscopes can it be seen that plant viruses have a sort of crystalline form.

All viruses are parasitic. They are not known to exist as saprophytes. In many, but not all, virus diseases the actual disease is not transmitted through the seed. Examples of exceptions to this are the bean and pea mosaic, and the lettuce mosaic virus disease.

The virus is generally present in every part of the infected plant (apical meristem tissues are often not infected) except the seed. Therefore, if part of that plant, other than the seed, is propagated, then the new plant is itself infected, e.g. the potato. The tuber is attached to the stem of the infected plant, and infection is carried forward when the tuber is planted as "seed".

With most plant virus diseases, the infection is transmitted from a diseased to a healthy plant by aphids and other insect vectors; even nematodes can transmit some virus diseases. A fungus can also be a vector, e.g. barley yellow mosaic virus (BYMV) transmitted by the soil-borne polymyxa graminis; rhizomania in sugar beet transmitted by polymyxa betæ.

Bacteria

Bacteria are very small organisms, only visible under a microscope. They are of a variety of shapes, but those that cause plant diseases are all rod-shaped. Like fungi, bacteria feed on both live and dead material. Although they are responsible for many diseases of humans and livestock, in the United Kingdom they are of minor importance compared with fungi and viruses as causal agents of crop diseases.

Bacteria reproduce themselves simply by the process of splitting into two. Under favourable conditions this division can take place about every 30 minutes. Thus bacterial disease can spread very rapidly indeed, once established.

Lack of essential plant foods (mineral deficiency)

When essential plant foods become unavailable to particular crops, deficiency diseases will appear. In "marginal" situations, where intensive systems of cropping are practised, mineral deficiencies are likely to be more apparent simply because there are insufficient trace elements present to cope with a large crop output. Most of the diseases are associated with a lack of trace elements, but shortage of any essential plant food will certainly reduce the yield, cause stunted growth, and make the crop more vulnerable to pest and disease attack.

Physiological diseases (stress)

These are often triggered by adverse environmental conditions which can upset the normal physiological processes of the plant. Normally this is only temporary, but there may be occasions when the effect is more permanent.

Temporary, e.g. a high water table in the early spring. This will cause yellowing of the cereal plant as its root activity is restricted, considerably reducing its oxygen and plant food requirements. When the water table falls, the plant is able to grow normally once more, assuming a healthy green colour.

Permanent, e.g. where the soil has become compacted the root activity of the plant can become restricted. This will result in poor stunted growth with the plant far more vulnerable to pest and disease attack. The yield will certainly be reduced.

THE CONTROL OF PLANT DISEASES

Before deciding on control measures it is important to know what is causing the disease. Having ascertained the cause as far as possible, the appropriate preventive or control measure can then be applied.

Crop rotation

A good crop rotation can help to avoid an accumulation of the parasite. In many cases the organism cannot exist except when living on the host. If the host plant is not present in the field, in a sense the parasite will be starved to death, but it should be remembered that:

(i) Some parasites take years to die, and they may have resting spores in the soil waiting for the susceptible crop to be planted again, e.g. club root of the brassicæ family.

(ii) Some parasites have alternative hosts, e.g. the fungus causing take-all of wheat is a parasite on some grasses.

Soil fertility

A crop under stress from low nutrient status can be more prone to disease attack. Too lush growth can also encourage disease. Early application of nitrogen in winter cereals can cause increased infection from foliar diseases.

Seedbed

Puffy seedbeds can increase the risk of take-all. Over-consolidation or compaction can lead to poor growth and more disease.

Crop hygiene

Diseases should be discouraged by avoiding sources of infection on the farm, e.g. blight from old potato clamps and virus yellows where sugar beet was stored prior to being sent to the factory. Good stubble cleaning will minimize certain cereal diseases being carried over from one cereal crop to the next year's cereal crop.

Some parasites use weeds as alternative hosts. By controlling the weeds, the parasite can be reduced, e.g. cruciferous weeds such as charlock are hosts to the fungus responsible for club root.

Clean seed

The seed must be free from disease. This applies particularly to wheat and barley which can carry the fungus causing loose smut deeply embedded in the grain. Seed should only be used from a disease-free crop.

With potatoes it is essential to obtain clean "seed", free from virus. In some districts where the aphid is very prevalent, potato seed may have to be bought each year. Under the Seed Certification Scheme it is mandatory to have field-inspected and virus-tested seed.

Resistant varieties

In plant breeding, although the breeding of resistant varieties is better understood, it is not by any means simple.

For some years plant breeders concentrated on what is called single or major gene resistance. However, with few exceptions, this resistance is overcome by the development of new races of the fungus to which the gene is not resistant.

Breeding programmes are now concentrating on multigene or "field resistance", which means that a variety has the characteristics to **tolerate** infection from a wide range of races with little lowering of yield. Emphasis is now on tolerance rather than resistance.

Varietal diversification

By choosing cereal varieties with different disease-resistance ratings for growing in adjacent fields, or in the same field in successive years, the risk in any year of a serious infection of yellow rust and mildew in winter wheat and mildew in spring barley can be considerably reduced (NIAB Leaflet No. 8—Cereals).

Cereal mixtures (blends)

This is a natural extension of diversification in that varieties from different diversification groups are grown together in the same field. In this way a disease-carrying spore from a susceptible variety, but within a blend of varieties (two or three varieties) making up the crop, has less chance of successfully infecting a neighbouring plant than in a pure crop. Yields of blended crops can be more reliable than pure crops, and this may be achieved with the use of fewer fungicides.

Time of sowing

Early-drilled winter cereals are more likely to be infected by eyespot, take-all and some foliar diseases.

Early-drilled spring barley is more susceptible

to rhynchosporium, but late-drilled barley is more susceptible to mildew.

The control of insects

Some insects are carriers of parasites, causing serious plant diseases, e.g. control of the green fly (aphid) in sugar beet will reduce the incidence of virus yellows. Furthermore, fungi can very often enter through plant wounds made by insects.

Table 74 indicates the main diseases affecting farm crops, and their control.

The use of chemicals

Broadly speaking, chemical control of plant disease means the use of a fungus killer—a fungicide. A fungicide may be applied to the soil, the seed or the growing plant. It can be used in the form of a spray, dust or gas. To be effective, it must in no way be harmful to the crop nor, after suitable precautions have been taken, to the operator or others, and it must certainly repay its cost.

Because of the extreme difficulties, if not impossibility, of breeding varieties resistant to cereal diseases, fungicides are now playing a very important part in the growing of cereals, although tolerance against certain diseases should not be ignored.

It should be understood that, apart from drought, disease can significantly influence yield. Some cereal diseases are more common in certain parts of the United Kingdom (Fig. 90). Wet weather diseases, such as rhynchosporium and septoria, are more common in the south and west of the country, and on the coastal fringes, and yellow rust in wheat in the drier east. Virus yellows is more of a problem in the major sugar beet areas in the east, rather than the west of the country.

Fungicides are used in the following way:

(1) *Soil use of fungicide*. This has little application in agriculture, more in horticulture. Dazomet is a soil sterilant used for control-ling club root on brassicae either by dusting the seedling roots or dipping them in a paste.

(2) *The dressing of seed with a fungicide—seed disinfection*. This is carried out to prevent certain soil- and seed-borne diseases. In many cases an insecticide is added to help prevent attacks by soil-borne pests. Various fungicides can be used, depending upon the disease to be controlled and the crop.

(3) *Application to the plant*. Different treatment systems involving the use of fungicides can now be considered as an essential part of crop production programmes. There are three main systems:

(a) Assessment of the disease risk:
Fungicides are used only when it is considered that a specific disease has developed to a point (the threshold) which will actually cause a loss of yield, e.g. mildew on spring barley. It is recommended that the crop should be sprayed when 3% of the second or third leaf is affected by mildew. This can be described as specific control.

A potato crop may have to be treated for blight when weather conditions are forecast to favour the disease.

(b) Prophylactic control:
This system involves the use of a fungicide or fungicides before the disease or diseases are present. It can be described as a programmed approach whereby factors such as minimum varietal resistance and other predisposing circumstances could bring about a disease situation. It is based on experience and trial work showing that standard timings for fungicide treatment normally give a yield response.

(c) Managed disease control:
This treatment involves both prophylactic and specific control according to the assessment of the disease risk. It aims to produce maximum economic response rather than maximum yields.

TABLE 72 *Main plant diseases and their control*

Crop attacked	Disease	Causal agent	Symptoms of attack	Life-cycle	Methods of control
Cereals	(1) Bunt, covered or stinking smut of wheat (2) Covered smut of barley. Leaf spot of oats (3) Covered and loose smut of oats	Fungus	Brown or black spore bodies with distinct smell replace the grain contents	Infected grain is planted; seed and fungus germinate together and thus young shoots become infected. The spores are released when the skin breaks, and so combining contaminates healthy grain	Seed dressing, e.g. products containing guazatine and carboxin
	(1) Leaf stripe of barley (2) Leaf stripe of oats	Fungus	The first leaves have narrow brown streaks. Subsequently, brown spots appear on the upper leaves	Infected grain is planted; seed and fungi germinate together and thus young shoots are infected. From the secondary infection, spores are carried to developing grain	Seed dressing containing imazalil
	(1) Loose smut of wheat (2) Loose smut of barley	Fungus	Infected ears a mass of black spores. They do not remain enclosed within the grain as with the covered smuts	Similar to the covered smuts, but the fungus develops within the grain. The spores are dispersed by the wind to affect healthy ears	(1) Resistant varieties (2) Clean seed (3) The seed can be dressed with carboxin or broad-spectrum fungicides
	Net blotch of barley	Fungus	Short brown stripes on older leaves; on younger leaves and adjacent plants, in addition to the striping, irregular-shaped dark brown blotches which can run together. Can spread to ears	Seed-borne disease and it can spread from previously infected crop or volunteers. Disease encouraged by cool, wet weather	(1) Seed dressing (2) Crop hygiene to clear stubble of previously infected crop (3) Fungicides now available, e.g. prochloraz, propiconazole
	Yellow rust	Fungus	Yellow-coloured pustules in parallel lines on the leaves, spreading in some cases to the stems and ears. In a severe attack the foliage withers and shrivelled grain results	The fungus mainly attacks wheat. Infection appears on the plant from May onwards. From the pustules, spores are carried by the wind to infect healthy plants. During winter, spores are normally dormant on autumn-sown crops. Cool, humid conditions favour disease	(1) Resistant varieties, although new races appear against which these varieties soon have no resistance (2) Fungicides, e.g. triazoles such as triadimenol (3) Diversification – *NIAB Cereal Leaflet*, No. 8
	Mildew	Fungus	On winter cereals, grey-white and brown mycelium on lower leaves in February.	From self-sown cereals in stubble, winter and spring cereals can be infected.	(1) Clean-up old stubbles (2) Resistant varieties, although new races of

TABLE 72 cont'd

Crop attacked	Disease	Causal agent	Symptoms of attack	Life-cycle	Methods of control
			Infection spreads to other leaves and plants. Disease common between May and August. Early infected leaves go yellow and shrivelled. Towards the end of season black spore cases formed among brown fungi	Warm, humid (but not wet) weather favours disease	the fungus may appear to nullify previous resistance in a variety (3) Barley seed can be dressed with ethirimol (4) Fungicides available, although field resistance now to the triazoles. Use tridemorph or ferpropimorph (5) Diversification – NIAB Cereal Leaflet No. 8
	Leaf blotch (rhynchosporium)	Fungus	When fully-formed lentil-shaped blotches (light grey with dark brown margins) up to 19 mm long are seen on the leaves. As disease progresses, blotches coalesce	Fungus over-winters on self-sown barley plants and on winter barley crops. From here spores are carried to planted barley crops. Disease is spread by rain splash	(1) Clean stubbles of all self-sown barley plants (2) Systemic fungicides, e.g. triadimefon Note: Attacks only rye and barley, six-row barley is more resistant (3) Seed dressing, e.g. triadimenol
	Brown rust of barley and wheat (less important in wheat)	Fungus	Numerous, very small and scattered, orange-brown pustules on the leaves in June. These gradually develop late in the season. A severe attack causes shrivelled grain	The resting spores over-winter on volunteers. From here the pustules are airborne to infect healthy plants. Encouraged by warm, humid weather; spores can be spread by wind	(1) Crop hygiene to clear stubble of volunteer plants (2) Resistant varieties (3) Systemic fungicides, e.g. propiconazole, triadimenol
	Septoria diseases (leaf and glume blotch, i.e. S. tritici and S. nodorum)	Fungus	Leaf: bleached or discoloured blotches of varying sizes and shapes (on which appear rows of minute black dots) seen from late autumn to early summer. Glume blotch: becomes prominent in July and August, especially in wet seasons. Irregular, chocolate spots or blotches on glumes, beginning at the	Leaf spores are liberated in wet weather, and they winter on volunteer crops; they transfer to winter cereals and then move on to spring crops. Glume blotch has a similar life cycle, but the fungus can also be carried on the seed. Weather conditions in late May which favour Septoria are	(1) Crop hygiene to clear stubble of volunteer plants (2) Clean seed (3) Tolerant varieties (4) Systemic fungicides, e.g. flutriafol, propiconazole. Treat when weather conditions favour disease. Too late once symptoms present

TABLE 72 cont'd

Crop attacked	Disease	Causal agent	Symptoms of attack	Life-cycle	Methods of control
			tips; later ears become blackened with secondary infection. Leaves also show yellowish areas and fungus can affect stems and leaf sheaths as well. Shrivelled grain results	3 or 4 days of consecutive rain. Totalling at least 10 mm, or one day with more than 5 mm	
	Barley yellow dwarf virus (BYDV)	Virus	Stunted plants in patches or scattered as single plants. Poor root development. Mostly red and yellow colour changes in leaves. Late heading and reduced yield	The virus is carried by cereal aphid vectors, e.g. bird cherry aphid and grain aphid. Volunteers and grasses act as source of infection. Early drilled crops most susceptible	Use a pyrethroid insecticide end October/early November. High risk areas may require earlier treatment. Oats most susceptible, wheat the least
	Barley yellow mosaic virus (BYMV) and barley mild mosaic virus	Virus	Appears in patches in field. Pale green streaks later turning brown, particularly at leaf tip; leaves tend to roll inwards, remain erect to give plant a spikey appearance. Plants stunted and late to mature	Soil-borne fungus. Polymyxa graminis is the vector. Disease encouraged by hard winter	On susceptible fields grow resistant varieties
	Eyespot	Fungus	Eye-like lesions on stem about 75 mm above ground. Grey "mould" inside stem; straws lodged in all directions. No darkening at base of stem. White heads at harvest	The fungus can remain in the soil, on old stubble and some species of grasses for several years. It usually attacks susceptible crops in the young stages. Early drilled crops most susceptible. Disease spread by rain splash	(1) Spring-sown barley and wheat are more resistant than autumn-sown crops. Arrange a break of at least 2 years from cereals (2) Systemic fungicides, e.g. prochloraz, if 20% of tillers affected
	Sharp eyespot	Fungus	Lesions more sharply defined than true eyespot. Brown/ purple border followed by cream-coloured area and brownish centre. Lesions more numerous and occur further up stem than true eyespot. Can cause lodging, "whiteheads" and shrivelled grain	The fungus is soil-borne but is also found on plant debris. Tends to be more severe on light soils and on early drilled winter crops. Often more serious after grass	Oats and rye most susceptible, and wheat more than barley. Not a serious disease, but can be accentuated when crop is sprayed against true eyespot. No chemical control

TABLE 72 cont'd

Crop attacked	Disease	Causal agent	Symptoms of attack	Life-cycle	Methods of control
	Fusarium – brown foot rot and ear blight	Fungus	Undefined brown discoloration at base of tillers and lower leaf sheaths. Interior of stem shows pink fungal growth. Premature ripening and "whiteheads" or "blind" ears	A number of fusarium spp. but infection is either seed-borne or from previously infected stubble and crop remains	(1) Stubble hygiene (2) Seed treatment using triadimenol with fuberidazole will help (3) Fungicides for control of eyespot and septoria will also help
	Crown rust of oats	Fungus	Orange-coloured pustules spread mainly on leaf blade. Later in season black pustules are produced. Severe attack prior to, and including, milk-ripe stage, causes shrivelled grain	Spores are air-borne from over-wintering volunteer plants, and winter oats, to the spring crop. Disease encouraged by warm, humid conditions	(1) Crop hygiene to clear stubble of volunteer plants (2) Keep winter and spring oat crops as far apart as possible (3) Some varieties show reasonable resistance (4) Some triazole fungicides recommended, e.g. triadimenol
	Take-all or "whiteheads"	Fungus	Black discoloration at base of stem. Grey colour of roots. Ease with which plant can be pulled from the soil. Infected plants ripen prematurely, and produce bleached ears containing little or no grain	Wheat is most affected. The fungus survives in the soil in root and stubble residues and the host plant is infected when it is grown in the field	Rotation to starve the fungus, but after some years of continuous wheat growing, infection appears to lessen. Extra nitrogen helps the growth of new roots. Bad drainage reduces plant vigour and it is more easily damaged. Direct-drilling appears to lessen the intensity of the disease
	Snow mould (mainly of wheat and rye)	Fungus	Patchy crop in autumn. Stunted seedlings occasionally with white-pink mycelium at base. After snow has thawed – withered plants in patches temporarily covered by white-pink mould. Thereafter, infected plants	A seed-borne fungus, but contamination of crop is also possible from the old stubble and other plant debris. Fungus is particularly favoured by low temperatures	(1) Good stubble hygiene (2) Clean seed (3) Seed treatment using triadimenol with fuberidazole

TABLE 72 cont'd

Crop attacked	Disease	Causal agent	Symptoms of attack	Life-cycle	Methods of control
			stunted, weak root system and shrivelled grain		
	Snow rot (Tuphyla rot) (mainly of barley)	Fungus	Thin, poorly-tillered crop; plants yellowing and withered in patches. Old leaves often covered by white mycelium, young leaves standing erect but eventually yellowed. Weak root system	Soil-borne fungus which can remain dormant for years; when active, infects emerging cereal plants, developing rapidly in dark and humid conditions, i.e. under snow	(1) Sow early in the autumn (2) Some varietal resistance (3) Seed treatment using triadimenol with fuberidazole
	Ergot	Fungus	Hard black curved bodies up to 20 mm long replacing the grain, and protruding from the affected spikelet	Ergots fall to ground and remain until next summer when they germinate and produce short stems with globular heads containing the spores which are then air-borne to the cereal flowers, and certain grasses, depending on the species. Open-flowering cereals most affected, e.g. rye	Although disease has little effect on yield, ergot is poisonous to mammals (but it does contain medicinal properties). Crop rotation and control of grass weeds (especially blackgrass) in the crop will help. Not considered important enough for special control measures. No ergot is permitted in grain for milling, and one piece only is permitted in Basic seed
	Manganese deficiency of cereals	Manganese deficiency	Yellowing on leaf veins followed by development of brown lesions. With oats the spots enlarge and can extend across leaf. Thus the leaf can bend right over in the middle. Older leaves wither and die. Can lead to shrivelled grains		9 kg/ha manganese sulphate in 340 litres/ha water. With a bad attack this may have to be repeated after 3 weeks
Linseed	Alternaria	Fungus	Damage affects seed-lings as they emerge. Brick red lesions are found on stems and roots. Can also affect mature plant	A seed-borne disease. Can spread up mature plant during periods of wet weather	Most effective method of control is seed treatment, e.g. with iprodione. Some evidence of varietal resistance

TABLE 72 cont'd

Crop attacked	Disease	Causal agent	Symptoms of attack	Life-cycle	Methods of control
	Grey mould	Fungus	Attacks leaves, stems and seed capsules. Reddish browning on stem base. Plants then become covered in grey mould	A seed-borne disease encouraged by warm, moist conditions	Sow certified seed. Seed dressings containing prochloraz or MBC can be effective. Low nitrogen rates and plant populations tend to reduce problem
Maize	Stalk rot	Fungus	Base of plant attacked in August/September; foliage grey/green colour and wilting. Pith brown/pink at base of stem; leads to premature senescence	Soil-borne fungus	Distinct differences in varietal susceptibility. Grain maize more likely to be affected because disease develops most rapidly in mature crops
	Smut	Fungus	Large black galls on any of the above ground parts of the plant, including the cob	Spores from galls can reinfect other maize plants, or they can remain dormant in the soil, surviving for several years	Would appear to be more serious when maize is cropped frequently. Not a seed-borne disease to the same extent as the other cereal smuts
Peas	Downy mildew	Fungus	Greyish-white to a grey-brown mycelium on underside of leaf on young seedlings which usually die. A secondary infection shows as isolated yellow-green spots on the upper surfaces of the leaves. Considerably reduced yield with fewer or no seeds per pod	The disease is mainly soil-borne (but also suggested to be seed-borne) from where spores infect the growing point of seedlings which usually die in about 3 weeks. A secondary infection can follow, with spores spreading by air currents and rain splash to the developing foliage of other plants. From plant remains pathogen returns to soil	Seed treatment using metalaxy or fosetyl-aluminium. Some varieties show more resistance
	Ascochyta (leaf and pod spot)	Fungus	Brown spotting on stems, leaves and pods can cause seedling loss. Peas can become stained	Disease is mainly seed-borne. Infected seed germinates and lesions develop on first leaves from where spores spread to rest of foliage, including the pods, and to other plants by air currents and rain splash	Sow healthy seed. Seed dressed with thiabendazole
	Botrytis (grey mould)	Fungus	Grey mould develops where petals have	Disease spread by damp and humid	Treat with a suitable fungicide,

TABLE 72 cont'd

Crop attacked	Disease	Causal agent	Symptoms of attack	Life-cycle	Methods of control
			fallen on to pods and leaves	conditions at end of flowering	e.g. chlorothalonil
	Mycosphaerella	Fungus	Small brown spots on leaves and pods. Can cause whole plants to die	Spread on seed. Wet weather favours disease	Use healthy seed. Chlorothalonil gives some control
	Bacterial blight	Bacteria	Dark brown lesions on leaves, stems and pods	Spread on seed. Wet windy weather favours disease	A notifiable disease. Only sow healthy seed
	Pea wilt (Fusarium wilt)	Fungus	In late May/June, lower leaves tend to turn grey before rolling downwards. All leaves eventually affected. Death of plant either before podding or before pods have swollen. White mycelium appears on stem after death	The pathogen is soil-borne. It invades the plant and only returns to the soil from the infected dead plant material	Rotation and use of resistant varieties Note: There are other Fusarium species – notably root rot
	Pre-emergence damping-off	Fungus Complex of different species	Poor germination and seedling establishment. The seed rots, or if not the stem is soft and dark brown in colour	Soil-borne pathogens invade seeds and/or plant stems at or just below soil level. Worst when peas sown in cold, wet soil	Seed protected by seed dressing with thiram
	Marsh spot	Manganese deficiency	Yellowing of leaves between veins which remain green; with severe deficiency growth is restricted and yield reduced. Brown discoloration (marsh spot) in centre of pea		Manganese sulphate at 5 kg plus wetter in 250–500 litres water/ha as soon as in full flower. Repeat 7 days later
Beans	Chocolate spot	Fungus	Small circular chocolate coloured discolorations on leaves and stems; with bad attack symptoms move to flowers and pods. In wet weather spots coalesce	Fungus carried over from previous year on debris of old bean haulm and on self-sown plants. But infection can start from almost any dead vegetation with this widespread fungus. Disease favoured by warm, wet weather	(1) Clean up stubbles containing remains of old crop of bean (2) Fungicide– some field resistance to MBCs; use chlorothalonil + carbendazim. First treatment in winter beans should be at early flowering stage

TABLE 72 cont'd

Crop attacked	Disease	Causal agent	Symptoms of attack	Life-cycle	Methods of control
					(3) Autumn-sown beans, especially early sown, are more liable to attack than spring-sown and they suffer more severely
	Ascochyta (leaf spot)	Fungus	Leaves affected by regular brown to black spots, some up to 2 cm in diameter; spots have slightly sunken grey centres (in which can be seen small black spots with brown margins). Pods and seed also affected, the latter covered with brownish-black lesions	Infected seeds when sown may produce seedlings with characteristic disease symptoms on stem at soil level or on lowest leaves. In cold moist conditions the disease will move up the plant and on to other bean plants	(1) Healthy seed – seed can be tested (2) Hygiene – kill any volunteer beans in other fields (3) Seed treatment, e.g. benomyl. Foliar treatment with chlorothalonil will give some control
	Sclerotinia (stem rot)	Fungus	Rotting of shoots and roots. Can cause death in severe attacks on winter beans	Soil-borne. The strain that attacks spring beans also attacks peas, red clover and oilseed rape. It can persist in the soil for 10 years	No fungicide recommendations. Use wide rotations
	Bean rust	Fungus	Red/brown rust pustules appear on leaves	Can be spread on seed, trash or volunteers. More of a problem on spring beans	(1) Hygiene – kill any volunteer beans (2) Fungicides – fenpropimorph is very effective
	Downy mildew	Fungus	Pale green water-soaked lesions. Grey fungal growth on underside of leaf	Seed and soil-borne. Spring beans most affected. Disease encouraged by wet weather	(1) Rotation (2) Hygiene – destroy debris (3) Fungicide – treat with metalaxyl and chlorothalonil
Brassicae (Brussels sprouts, oil-seed rape, cabbage, kale, swedes and turnips)	Club root or finger and toe	Fungus	Swelling and distortion of the roots. Stunted growth. Leaves pale green in colour	A soil-borne fungus. The fungus grows in the plant roots and causes the typical swellings. Resting spores can pass into the soil, especially if diseased roots are not removed. They can remain alive for several years, becoming active when the	(1) Rotation. With a bad attack advisable not to grow the crop for at least 5 years in the field (2) Liming and drainage. The spores are more active in acid and wet conditions (3) Resistant crops. Kale is

TABLE 72 cont'd

Crop attacked	Disease	Causal agent	Symptoms of attack	Life-cycle	Methods of control
				host crop is again grown in the field	more resistant than swedes or turnips. Some varieties of swedes and turnips are more resistant than others
	Stem canker (leaf spot) in oilseed rape	Fungus	Beige coloured circular spots with distinct brown margin (0.5–1.0 cm diam.) on leaves; spores spread to produce brownish-black canker at base. Stem splits and rots causing lodging; rapid stem elongation and premature ripening	Air-borne spores produced on the stubble carry infection to young crops in the vicinity. Disease can also be spread from infected seed, but this is less significant	(1) Destroy stubble debris as soon as possible after harvest (2) Rotation; avoid growing rape crops in the same field more often than one year in six (3) Tolerant varieties (4) Use a seed treatment, e.g. thiabendazole foliar spray with prochloraz
	Alternaria (dark leaf and pod spot) on oilseed rape	Fungus	Circular small brown-to-black leaf spots sometimes coalescing on leaves and later on pods. Premature ripening and loss of seed	Seed-borne disease, although spores can be carried through the air from other infected brassica crops. Favoured by warm, humid conditions	No resistant varieties. Foliar treatment from late flowering with a fungicide, e.g. iprodione
	Light leaf spot on oilseed rape	Fungus	From January onwards, light, almost white spots found on leaf about 1 cm diameter	A sexual reproduction. Disease is spread from plant to plant, particularly during wet spells. Flower buds can become infected and killed at early extension stage	(1) Dispose of infected crop residues (2) Spray with carbendazim or prochloraz in spring at stem extension (3) Some varieties less susceptible than others
	Stem rot (Sclerotinia)	Fungus	Bleached skin lesions which contain black resting bodies (sclerotia)	Sclerotia left in soil after harvest and can survive for 8 years. They germinate in spring and produce spores which infect susceptible crops. Encouraged by warm, wet weather	(1) Wide rotation (2) Fungicides – at early flowering, e.g. iprodione, prochloraz

TABLE 72 cont'd

Crop attacked	Disease	Causal agent	Symptoms of attack	Life-cycle	Methods of control
	Powdery mildew (particularly Brussels sprouts, swedes and turnips)	Fungus	Upper surface of leaves show blue/black discoloration; sprouts turn black	Spores over-winter on infected plants and are carried by air currents to infect the following year's crop	(1) Varieties differ in their susceptibility to this disease (2) Fungicide, e.g. triadimenol
	Brown heart of swedes and turnips (Raan)	Boron deficiency	No external symptoms, but when the root is cut open a browning or mottling of the flesh is seen. Affected roots are unpalatable		See Heart rot of sugar beet
	Stem rot in kale	Boron deficiency	Cavitation in the pith followed by a brown rot and stem collapse		See Heart rot of sugar beet
Sugar beet, fodder beet, mangels	Virus yellows	Virus	First seen in June/early July on single plants scattered throughout the crop – a yellowing of the tips of the plant leaves. This gradually spreads over all but the youngest leaves. Infected leaves thicken and become brittle. The yield is seriously reduced by an early attack	The crop is infected by aphids which have over-wintered in mangel clamps and steckling beds. Several aphids carry the virus, particularly peach potato aphid	(1) Good growing conditions to keep the crop growing vigorously (2) All mangel clamps should be cleared by the end of March. If not, they should be sprayed to kill any aphids (3) Seed crop stecklings should not be raised in the main sugar beet area (4) Aphid warnings issued by British Sugar aid timing of aphicide (e.g. pirimicarb). There is now resistance to some of the foliar insecticides
	Powdery mildew	Fungus	Powdery greyish-white mycelium on foliage seen in dry weather in late summer, early autumn	The spores are air-borne and move from diseased plants found at loading sites and from roots left in the field and also from weed beet to infect the new crop. Disease favoured by dry, warm weather	(1) Spray with wettable sulphur (10 kg/ha) as soon as disease is seen but before mid-September (2) Some varieties show more resistance than others

TABLE 72 cont'd

Crop attacked	Disease	Causal agent	Symptoms of attack	Life-cycle	Methods of control
	Rhizomania (root madness)	Virus	Wilted plants (sometimes in patches in field) showing pale yellowing of veins; development of elongated, strap-like leaves often protruding above surrounding plants. Infected roots smaller than healthy roots, constricted below soil level, usually fanging with a proliferation of small lateral roots (bearding). Inside of root shows brown-streaked tissue from tip of tap root upwards	The virus is trans-ported by soil fungus – *Polymyxa betæ* which is now present in UK soils	(1) Widespread on Continent and now spreading in this country. Therefore precautions to prevent spread of the disease. New clause in Sugar Beet Contract: "Sugar beet and other betae not to be grown on same land more than one year in three years". Eradicate weed beet which can act as host to polymyxa fungus (2) Tolerant varieties, when available, may be the only answer
	Speckled yellows	Manganese deficiency	Small yellowish areas between leaf veins, later turning to buff-coloured angular, sunken spots (speckled yellows) which eventually coalesce	Symptoms often disappear as root system develops. But disease can be a problem on near-alkaline soils or those soils with high organic matter content	Foliar spray with manganese sulphate in May and June which may have to be repeated. Pelleted seed incorporated with manganous oxide can prevent early symptoms
	Heart rot	Boron deficiency	In young plants the youngest leaves turn a blackish-brown colour and die off. A dry rot attacks the root and spreads from the crown downwards. The growing point is killed, being replaced by a mass of small deformed leaves		This deficiency is more apparent on dry and light soils and can be made worse by heavy liming. Apply borax at 22 kg/ha as soon as the disease is seen. Use a boronated compound fertilizer on suspected soils. Deficiency diag-nosed by soil analysis
Potatoes	Blight	Fungus	Brown areas on leaves. Whitish mould on the underside of leaves. Leaves and stems become brown and die off	Infected tubers (either planted, ground keepers, or throw-outs from clamps) produce blighted shoots. From these shoots the fungus spores are	Blight spreads rapidly in warm, high humidity, weather and warnings of such conditions are given by ADAS. Early

TABLE 72 cont'd

Crop attacked	Disease	Causal agent	Symptoms of attack	Life-cycle	Methods of control
				carried by the wind or rain to infect the haulms. From the haulms the spores are washed into the silo to infect the tubers. Infection can also take place at harvest. The fungus cannot live on dead haulm. Risk of blight encouraged by certain weather conditions, e.g. two days with temp. above 10°C and relative humidity at 90% for at least 11 hours/day. Disease more likely to develop once crop meets between rows	preventative spraying, followed by repeated sprays every 10–14 days, is advisable – especially in areas where blight is a problem; a wide range of fungicides are available, both protectants (e.g. dithiocarbamates or organotins) and systemics, e.g. phenylamide group. Due to resistance problems, systemic fungicide mixtures used. The haulms should be destroyed chemically before harvest to prevent the tubers becoming infected whilst being lifted
	Leaf roll	Virus	Lower leaves are rolled upwards and inwards; they feel brittle and crackle when handled. The other leaves are lighter green and more erect than normal. Yield is lowered	The virus is transmitted by aphids from plant to plant. Infected tubers (which show no signs of the disease) are planted and thus the disease is carried forward from year to year	(1) Use certified "seed" which is grown in areas such as Scotland and Northern Ireland where aphids are not prevalent due to the colder climate. Thus the "seed" is usually free from virus infection (2) Systemic sprays or granules will control the aphids, and thus reduce the spread of the virus
	Mosaics	Virus	May range from a faint yellow mottling on leaves to a severe distortion of the leaves and distinct yellow mottling. Yield can be seriously reduced by the severe forms	Some of the viruses responsible are spread by aphids but some, e.g. Virus X, are spread by contact between leaves and roots and on machinery and clothing	As for leaf roll. Some varieties resistant

TABLE 72 cont'd

Crop attacked	Disease	Causal agent	Symptoms of attack	Life-cycle	Methods of control
	Common scab	Actinomycete	Skin-deep irregular-shaped scabs on tuber: these can occur singly or in masses. With a severe attack cracking and pitting takes place with secondary infection by insect larvæ and millipedes	The soil-borne organism attempts to invade the growing tuber which responds by development of corky tissue to restrict the parasite to the surface layers. Organism re-enters soil when infected seed is planted	(1) Avoid liming just prior to planting potatoes. Disease is particularly prevalent on light sandy, alkaline soils. (2) Irrigation of dry, light soils an advantage. Irrigate when soil moisture deficit reaches 18 mm. Dry conditions favour the spread of the disease. (3) Some varieties are more resistant than others
	Powdery scab	Fungus	Appearance can be similar to common scab, but the spots are rounder and formed as raised pimples under the skin which burst; sometimes cankers and tumours develop	Spore balls can remain in the soil for many years or may be planted on infected tubers and the zoospores attack the new tubers by way of the lenticles, eyes or wounds, resulting in scab development	Usually more troublesome in wet seasons and lime-rich soils. Potatoes should not be planted after a severe attack for at least 5 years. Avoid using infected seed or contaminated FYM; some varieties are very susceptible
	Dry rot	Fungus	Infected tubers are usually first noticed in January and February. The tuber shrinks and the skin wrinkles in concentric circles. Blue-pink or white pustules appear on the surface	The soil-borne fungus enters the tuber from adhering soil. Infection can only enter through wounds and bruises caused by rough handling at harvest. The disease can be easily spread during storage	(1) If the potatoes are handled carefully, infection is considerably reduced. Some varieties are more susceptible than others to the disease. (2) Use of thiabendazole will give some control at lifting
	Spraing	Tobacco rattle virus (TRV) or potato mop-top virus (PMTV)	Foliage – very variable; TRV – stem mottling; PMTV – yellow blotches and bunching of leaves on short stems – like a mop. Tubers – primary (after soil infection): wavy or arc-like brown, corky streaks in flesh of cut tuber. Secondary – from infected	TRV spread by free-living nematodes in soil, especially in light sandy soils. PMTV – spread by powdery scab fungus and can remain in the soil for years in fungal resting bodies	Plant only resistant varieties or infected soils. Control free-living nematodes with a nematicide, e.g. oxamyl

TABLE 72 cont'd

Crop attacked	Disease	Causal agent	Symptoms of attack	Life-cycle	Methods of control
			tubers: PMTV – badly formed and cracked tubers. TRV – brown spots in flesh		Rogue or reject seed crops where the disease shows on foliage. Do not plant infected tubers. No resistant varieties, although some are less susceptible
	Blackleg	Bacteria	Plants stunted and pale green or yellow foliage; easily pulled out of the ground and stem base is black and rotted. Infected and neighbouring tubers develop a wet rot in the field or in store, especially in damp and badly ventilated (warm) conditions	The bacteria move to tubers via the stolons and in wet soil to healthy tubers to enter via lenticles or damaged areas. Carried on seed tubers	
	Gangrene	Group of phoma fungi	A serious tuber rot which develops in storage, usually late; it shows as grey "thumb-mark" depressions on the tubers and the flesh beneath is rotted; also, pin-head black spore cases	The fungus remains alive in the soil and on trash, and can infect tubers in the soil and from tuber to tuber when handling	Do not plant diseased seed. Assist tuber wounds to heal by keeping them warm (13–16°C) and humid up to 10 days after any handling operation. Reduce mechanical damage. Treatment with a fungicide reduces incidence in store, e.g. thiabendazole. Cool storage conditions favour the disease
	Skin spot	Fungus	Tuber symptoms develop during late storage and appear as pimple-like, dark brown, shrunken spots with raised centres. The worst damage is the destruction of the buds in the eyes of seed tubers	Mainly spread by infected tubers. Tuber infection occurs at lifting and is worse in cold, wet seasons	As for gangrene

TABLE 72 cont'd

Crop attacked	Disease	Causal agent	Symptoms of attack	Life-cycle	Methods of control
Grass	Barley yellow dwarf virus (BYDV)	Virus	Leaves turn yellow, red or brown, discoloration starting at tips and going down leaves. Plants generally stunted, but can produce more tillers. Disease more conspicuous in single plants than in whole sward	Spread by aphid	Very difficult; not economic to spray. Some ryegrass varieties more resistant than others
	Ryegrass mosaic	Virus	Yellowish/green mottling or streaking of leaves. Severe infection can show more general browning of leaf	Spread by mites which are favoured by hot, dry weather	Some ryegrass varieties are more resistant than others
	Crown rust of perennial ryegrass	Fungus	Usually seen in late summer. Pale yellow leaf flecking, followed by bright orange-yellow oval spots. Badly affected crops have an overall yellow colour	Spores are air-borne and can quickly infect a clean sward	(1) Frequent grazing, but top over if badly infected sward is rejected by stock (2) Some grass varieties are more resistant than others (3) Propiconazole or triadimefon can be used
	Rhynchosporium	Fungus	Irregular scald-like blotches on leaves. Favoured by cool wet weather and is most apparent in spring and early summer	Spores are air-borne and move from a diseased to a clean sward in the spring	(1) Choose resistant varieties in the south and south-west where disease is always more apparent (2) Propiconazole can be used
	Mildew	Fungus	Disease found throughout the country from early spring onwards, particularly in dense swards of short-duration ryegrass. Greyish-white mycelium on leaves	Spores are air-borne, and move from an infected to a clean crop	(1) Resistant varieties (2) Triadimefon or propiconazole can be used

TABLE 72 cont'd

Crop attacked	Disease	Causal agent	Symptoms of attack	Life-cycle	Methods of control
Red and white clover; lucerne	Clover rot (Sclerotinia)	Fungus	Foliage turns olive-green and then black and eventually dies. The root can also die	Resting bodies of fungus produced on affected plants in winter and spring. They are small (size of clover seed), white at first and then turning black. Bodies remain dormant in summer but in autumn produce seed-borne spores which affect other plants	(1) Clean seed (2) Use resistant varieties where possible (at present fewer resistant varieties with white clover) (3) Rotation; may have to be an interval of at least 6 years *Note*: Also affects lucerne and sainfoin
Lucerne	Verticillium wilt	Fungus	Usually seen in fairly isolated patches in first harvest year; in next 2 years spreads to many parts of the field. Normally after first cut, lower leaves turn pale-yellow colour and then white, and eventually shrivel from base upwards. Whole plant finally dies	Disease can be introduced by contaminated seed; spores can also be transported by air, as well as being spread by contaminated fragments of the crop moving from plant to plant and then from field to field by machine	(1) Where suspected use tolerant varieties (2) Use clean seed (3) Harvest healthy crops first before moving onto older infected crops

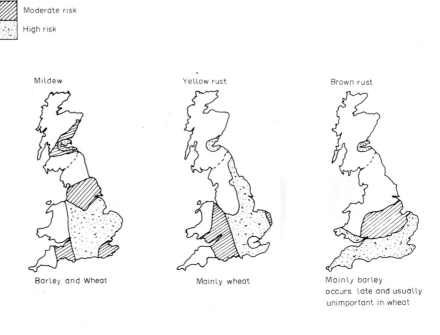

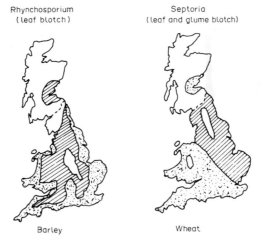

FIG. 90. Generalized maps of regions where leaf diseases of cereals are most likely to occur.

Fungicide resistance

When a fungal disease is controlled effectively by a fungicide, the fungus is "sensitive" to the chemical. However, other strains of the fungus can and do occur over a period of time, and some of these may be resistant ("insensitive" or "tolerant") to the fungicide which means that the disease is then not controlled adequately.

The more often the same chemical, or chemical in the same group, is used, the greater are the chances of resistant strains developing. There is also an increased risk of this happening with fungicides which act at one specific site in the fungus compared with multi-site fungicides, i.e. those fungicides which act at many sites in the fungus.

There are other ways of avoiding a build-up of resistance by a fungus:

(1) A reduction of disease levels by good husbandry.
(2) Use of resistant varieties; diversifying varieties in adjacent fields and in successive years; cereal mixtures.
(3) Where possible, fungicides with different modes of action (i.e. from different groups) should be used when more than one has to be used on the same crop. Some manufacturers are now formulating broad-spectrum sprays with combinations of chemicals from different groups.
(4) A late application (after flowering) of a fungicide should, as far as possible, be avoided.
(5) Where feasible, by the use of appropriate fungicide mixtures. These mixtures must have clearance under the Control of Pesticides Regulations 1986.

Unsatisfactory disease control following the use of fungicides is, at present, not always due to fungicide resistance. There are several other reasons, the main ones being wrong timing, the use of too low a dose and poor application.

FURTHER READING

Cereal Diseases, NIAB.
Cereal Fungicides Supplements, Farmers' Weekly.
Cereal Pests, Bulletin No. 186, HMSO.
Control of Crop Diseases, W. R. Carlile.
Crop Protection Handbook: Cereals, Grassland, Potatoes, British Crop Protection Council.
Insecticide and Fungicide Handbook, Blackwell.
Potato Diseases, HMSO.
Potato Diseases, NIAB/PMB.
Schering Guides: *Cash Crops Diseases; Cereal Diseases; Field Crops Nutrient Disorders*.
Sugar Beet Pests, Bulletin No. 162, HMSO.
UK Pesticides for Farmers and Growers, The Royal Society of Chemistry.
The UK Pesticide Guide 1993, Ed. G. W. Ivens, British Crop Protection Council.

LAND USE IN THE UNITED KINGDOM

Total area: 24 million hectares, consisting of approximately:

4.8 million ha (20%)—mountains, forests, urban areas, motorways, etc.
7.2 million ha (30%)—rough grazing, including deer forest.
5.1 million ha (20%)—permanent grassland (5 years +).
1.7 million ha (10%)—temporary grass (leys).
5.2 million ha (20%)—arable crops.

The areas for the various arable crops are approximately:

4.03 million ha (77%)—cereals (2.0 wheat, 1.9 barley, 0.1 others).
0.30 million ha (5.7%)—oilseed rape.
0.21 million ha (4.0%)—sugar beet.
0.18 million ha (3.5%)—potatoes.
0.12 million ha (2.3%)—forage crops (kale, swedes, turnips, etc.).
0.09 million ha (1.7%)—peas harvested dry.
0.06 million ha (1.1%)—field beans.
0.21 million ha (4.0%)—horticultural crops.
0.07 million ha (1.3%)—other crops and fallow.

AGRICULTURAL LAND CLASSIFICATION IN ENGLAND AND WALES

MAFF LAND CLASSIFICATION MAPS AND REPORTS

THE 12 million hectares of agricultural land in England and Wales have been classified into five grades by the Ministry of Agriculture. These are used by planners when considering requests for planning permission which would take land out of agricultural use.

This survey work is published on coloured Ordnance Survey maps on two scales:

 (i) 1:63,360, i.e. 1 inch to 1 mile.

(ii) 1:250,000, i.e. 1 inch to 4 miles. There are seven of these, each covering approximately one MAFF region.

The details shown in (ii) are more generalized than in (i) and the classification is based mainly on:

Climate—rainfall, transpiration, temperature and exposure.
Relief—altitude, slope, surface irregularities.
Soil—wetness, depth, texture, structure, stoniness and available water capacity.
Chemical composition—toxicity, deficiency, acidity and alkalinity.

These characteristics can affect:

The range of crops which can be grown, the level of yield, the consistency of and the cost of obtaining the yield.

Land is graded into five grades under this classification with the third grade divided into two sub-grades.

Grade 1—Excellent. Land with no, or very minor, limitations to agricultural use. A very wide range of agricultural and horticultural crops can be grown. Yields are high and consistent. This grade occupies 3% of agricultural land in England and Wales and is coloured dark blue on the maps.

Grade 2—Very good. Land with minor limitations which affect crop yield, cultivations or harvesting. A wide range of crops can be grown, but there may be difficulties with very demanding crops. The level of yield is generally high, but it can be lower or more variable than Grade 1. This grade occupies 14% of agricultural land and is coloured light blue.

Grade 3—Good to moderate. Land with moderate limitations which affects choice of crop, timing and type of cultivations, harvesting or level of yield. Yields are generally lower or more variable than on land in Grades 1 and 2.

Sub-grade 3a—Moderate/high yields of a narrow range of crops (especially cereals) or moderate yields of a wide range of crops. Some good grassland.

Sub-grade 3b—Moderate yields of a narrow range of crops and lower yields of a wide range of crops.

Grade 3 is the largest group of soils, occupying 50% of agricultural land and is coloured green.

Grade 4—Poor. Land with severe limitations restricting range and/or yield levels. Suited to grass plus some low and variable yields of cereals. It also includes droughty arable land. It occupies 20% of agricultural land and is coloured yellow.

Grade 5—Very poor. Land with very severe limitations. Permanent pasture or rough grazing. It occupies 14% of agricultural land and is coloured light brown.

Urban areas are coloured red and non-agricultural areas are coloured orange on the maps.

Additionally, the MAFF is undertaking a physical classification of the hill and upland areas of Grades 4 and 5 which will be divided into two main categories:

The Uplands—The enclosed and wholly or partially improved land; there will be five sub-grades.

The Hills—Unimproved areas of natural vegetation; there will be six sub-grades.

The physical factors to be taken into account will be vegetation, gradient, irregularity and wetness.

The object of this classification is to help the farmer plan possible improvement schemes when conditions permit. It is also intended to be a guide to those concerned with conservation and amenity who are trying to preserve scenery and ecology.

Booklets giving full details of all the classifications and grades and maps can be obtained from the MAFF.

Soil Series and Soil Survey Maps

In the grading of land for the Land Classification Maps and Reports, use was made of the maps which show the Soil Series Classification of England and Wales. A "soil series" is a group of soils formed from the same, or similar, parent materials, and having similar horizons (layers) in their profiles. Each soil series is given a name, usually where it was first recognized and described. The name is used for the soil group, however widespread, throughout the country; it is also used on the maps which are produced on 1:63,000 and/or 1:250,000 scales. A few examples of the named soil series are:

Romney series. These soils are deep, very fine, sandy loams found in the silt areas of Romney Marsh, and also in parts of Cambridgeshire, Lincolnshire and Norfolk. They are potentially very fertile soils.

Bromyard series. Red-coloured, silt loam soils (red marls) found in Herefordshire and other parts of the West Midlands and in the south-west. They should not be worked when wet. Suitable crops are cereals, grass, fruit and hops.

Evesham series. Lime-rich soils formed from Lias (or similar) clays, found in parts of Warwickshire, Gloucestershire, Somerset, East and West Midlands. They are normally very heavy soils and are best suited to grass and cereals.

Sherborne series. Shallow (less than 25 cm deep), reddish-brown, loam-textured soils of variable depth and stoniness (and so subject to drought). They are mainly found on the soft oolitic limestones of the Cotswolds, and in parts of Northamptonshire, and the Cliff region of Lincolnshire. The soils of this series are moderately fertile, easy to manage and mainly grow cereals and grass, but some root crops and potatoes, e.g. the good quality King Edward in Lincolnshire.

Worcester series. Red silt loam (or silty clay loam) formed from Keuper Marl and found in the East and West Midlands and in the south-West. They are slow-draining, require subsoiling regularly and are best suited to grass and cereals.

Newport series. Free-draining, deep, easy-working, sandy loams over loamy sands with varying amounts of stones. They are formed from sands or gravels of sandstone or glacial origin and are found in many areas of the Midlands and north. They are well suited to arable cropping.

Several different soil series may be found on the same farm and sometimes in the same field.

The Land Use Capability Classification in the United Kingdom has been modelled on the United States Department of Agriculture Classification Scheme. It uses the soil series groups in conjunction with limitations imposed by wetness (w), soils (s), gradient and soil pattern (g), erosion (e) and climate (c). The Classification divides all land into seven classes (1–4 classes are very similar to grades 1–4 of the Land Classification Maps prepared by MAFF; Classes 6 and 7 are virtually useless for agriculture). Each class has sub-divisions and limitations as indicated by the letters w, s, g, e, or c, after the class number, e.g. Class 2w. This classification is being carried out by the Soil Survey unit which is based at Rothamsted Experimental Station with the assistance of ADAS.

Maps and booklets are published as survey work is completed.

SOIL TEXTURE ASSESSMENT IN THE FIELD

THE texture of a soil (i.e. the amount of sand, silt, clay and organic matter present) can be measured by a mechanical analysis of a representative sample of the soil. It is therefore important for the farmer and his adviser to be able to assess the texture of the soil in the field, not only as a guide for cultivations and general management, but also because soil-acting pesticides (mainly herbicides) are becoming increasingly important. Many of these chemicals are absorbed by the clay and/or organic matter in the soil and so higher dose rates may be required on soils which are rich in these materials. On sandy soils, surface-applied residual herbicides may be washed into the root zone of the crop too easily and so cause damage.

With practice, it is possible to become reasonably skilled at assessing soil texture by feeling the soil in the following way:

Carefully moisten a handful of stone-free soil until the particles cling together (avoid excess water). Work it well in the hand until the structure breaks down; rub a small amount between the thumb and fingers to assess the texture according to how gritty, silky or sticky the sample feels. The handful of moist soil can also be assessed by the amount of polish it will take, and the ease or difficulty of moulding it into a ball and other shapes.

Sands feel gritty, but are not sticky when wet (loose when dry) and do not stain the fingers. Four grades may be distinguished from the coarseness or fineness of the gritty material in the sample.

Clays (at the other extreme of particle size) take a high polish when rubbed, are very sticky, bind together very firmly, and need some pressure to mould into shapes. Three grades may be distinguished, depending on the amounts of sand (grittiness) and/or silt present.

Silty soils have a smooth silky feel and the more obvious this is, the greater is the amount of silt present. The amount of polish the sample takes, and its grittiness, are guides to the amount of clay and sand present.

Loams have a fairly even mixture of sand, clay and silt and, because these tend to balance each other, loam soils are not obviously gritty, silky or sticky and they take only a slight polish. A ball of moist loam soil is easily formed, and the particles bind together well.

These are the main texture grades, but a wide range of intermediate grades exist, each having different amounts of sand (of various sizes), silt and clay particles. All this can be complicated by the amount of organic matter present which has a soft silky feel and is usually dark brown or black in colour.

SOIL TEXTURE
KEY 1 (MINERAL SOILS)

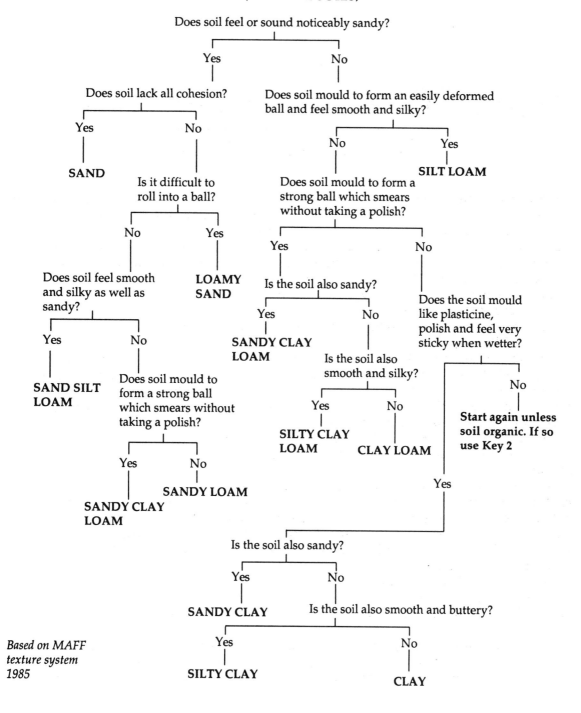

Based on MAFF
texture system
1985

SOIL TEXTURE
KEY 2 (ORGANIC SOILS)

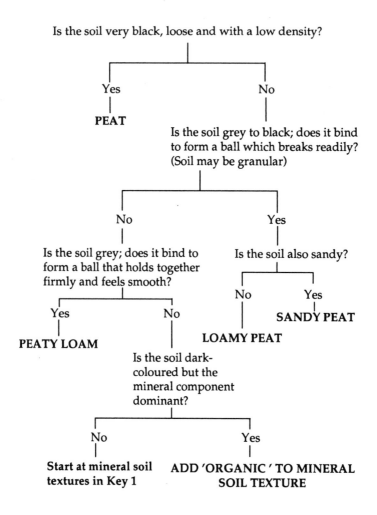

Is the soil very black, loose and with a low density?

Yes — **PEAT**

No — Is the soil grey to black; does it bind to form a ball which breaks readily? (Soil may be granular)

No — Is the soil grey; does it bind to form a ball that holds together firmly and feels smooth?

Yes — **PEATY LOAM**

No — Is the soil dark-coloured but the mineral component dominant?

No — **Start at mineral soil textures in Key 1**

Yes — **ADD 'ORGANIC' TO MINERAL SOIL TEXTURE**

Yes — Is the soil also sandy?

No — **LOAMY PEAT**

Yes — **SANDY PEAT**

Once the particular qualities described are recognized, it should be possible to use the following two Texture Keys to arrive at a single texture (pages 285 and 286):

ADAS use the following soil textural groups in their advisory work and set them out fully in *MAFF Pamphlet 3001—Soil Texture System and Pesticide Use*:

Sands—coarse sand (CS), sand (S), fine sand (FS), loamy coarse sand (LCS).
Very light soils—loamy sand (LS), loamy fine sand (LFS), coarse sandy loam (CSL).
Light soils—sandy loam (SL), fine sandy loam (FSL), sandy silt loam (SZL), silt loam (ZL).
Medium soils—sandy clay loam (SCL), clay loam (CL), silty clay loam (ZCL).
Heavy soils—sandy clay (SC), clay (C), silty clay (ZC).

The prefix *organic* can be applied to the above classes if organic matter levels are relatively high, i.e. about 6% in sands or loamy sands, and about 10% in clays; prefix *peaty* when 20–35% organic matter; *peat soils* when more than 35% organic matter (peat has over 50% organic matter). The prefix *calcareous* (Calc) can be applied if more than 5% calcium carbonate is present.

The above textural groups are used in advisory work for making recommendations on dose rates for soil-acting herbicides and for assessing available-water capacity, suitability for mole drainage, workability and stability of soils.

NOMENCLATURE OF CROPS

Crop names	Botanical names
Cereals—wheat	*Triticum aestivum (T. vulgare)*
durum	*T. durum*
barley	*Hordeum sativum*
oats	*Avena sativa*
rye	*Secale cereale*
triticale	*Triticale hexaploide*
maize	*Zea mays*
Potato	*Solanum tuberosum*
Sugar beet, mangel, fodder beet	*Beta vulgaris*
Cabbage group	*Brassica oleracea*
savoy	*B. oleracea* var. *bullata*
drumhead	*B. oleracea* var. *capitata*
red (white)	*B. oleracea* var. *rubra (alba)*
Brussels sprouts	*B. oleracea* var. *bullata* s. var. *gemmifera*
Cauliflower	*B. oleracea* var. *botrytis*
Sprouting broccoli, calabrese	*B. oleracea* var. *botrytis* s. var. *italica*
Kohlrabi	*B. oleracea* var. *gongyloides*
Kale—marrow stem	*B. oleracea* var. *acephala* s. var. *medullosa*
thousand head	*B. oleracea* var. *acephala* s. var. *millicapitata*
curly	*B. oleracea* var. *acephala* s. var. *laciniata*
Turnip group	*B. rapa* var. *rapa*
turnip oilseed rapes	*B. campestris* var. *oleifera*
Swede group	*B. napus*
swedes	*B. napus* var. *napobrassica*
swede forage rape	*B. napus* var. *oleifera*
swede oilseed rapes	*B. napus* var. *oleifera*
hungry gap and rape kale	*B. napus*
Mustard—brown	*B. juncea*
white	*Sinapsis alba (Brassica alba)*
Fodder radish	*Raphanus sativus* var. *campestris*
Buckwheat	*Fagopyrum esculentum*
Carrot	*Daucus carota*
Parsnip	*Pastinaca sativa*
Celery	*Apium graveolens*
Onion	*Allium cepa*
Pea	*Pisum sativum*
Beans—field and broad	*Vicia faba*
green, dwarf, French	*Phaseolus vulgaris*
runner	*P. coccineeus*
soya	*Glycine max*

Crop names	Botanical names
Vetch (tares)	*Vicia sativa*
Lupins—yellow (white)	*Lupinus luteus (L. albus)*
pearl (blue)	*L. mutabilis (L. augustifolius)*
Lucerne (alfalfa)	*Medicago sativa*
Sainfoin	*Onobrychis viciifolia*
Linseed and flax	*Linum usitatissimum*
Sunflower	*Helianthus* spp.
Grasses: ryegrass—Italian	*Lolium multiflorum*
hybrid	*L. (multiflorum × perenne)*
perennial	*L. perenne*
Grasses: cocksfoot	*Dactylis glomerata*
Timothy	*Phleum pratense*
meadow fescue	*Festuca pratense*
tall	*F. arundinacea*
red	*F. rubra*
Clovers: red	*Trifolium pratense*
white	*T. repens*
alsike	*T. hybridum*
crimson	*T. incarnatum*
Borage	*Borage officinalis*
Evening primrose	*Oenothera biennis*
Fenugreek	*Trigonella fenumgraecum*
Quinoa	*Chenopodium quinoa*

NOMENCLATURE OF WEEDS

(A) annual, (B) biennial, (P) perennial

Common names	Botanical names
Allseed—goosefoot (A)	*Chenopodium polysper—um*
Amphibious bistort (P)	*Polygonum amphibium*
Autumn crocus (P)	*Crocus nudiflorus*
Barley—meadow (P)	*Hordeum secalinum*
—wall	*H. murinum*
Bent—black (P)	*Agrostis gigantea*
—common (P)	*A. tenuis (A. capillaris)*
—creeping (watergrass) (P)	*A. stolonifera*
Bindweed—black (A)	*Polygonum* or *Bilderdykia convolvulus*
—field (P)	*Convolvulus arvensis*
Birdsfoot trefoil—common (P)	*Lotus corniculatus*
Blackgrass (A)	*Alopecurus myosuroides*
Borage	*Borage officinalis*
Bracken (P)	*Pteridium aquilinum*
Bristly oxtongue (A) or (B)	*Picris echioides*
Brome—soft (A) or (B)	*Bromus mollis (B. hordeaceus)*
—barren (A) or (B)	*B. sterilis*
—rye (A) or (B)	*B. secalinus*
—erect (upright) (P)	*B. erectus*
—meadow (A) or (B)	*B. commutatus*
Broomrape—common	*Orobanche minor*
Burdock—greater (B)	*Arctium lappa*
Burnet—salad (P)	*Poterium sanguisorba*
—fodder (P)	*P. polygamum*
Buttercup—bulbous (P)	*Ranunculus bulbosus*
—corn (A)	*R. arvensis*
—creeping (P)	*R. repens*
—meadow (crowfoot) (P)	*R. acris*
Campion—white (B)	*Silene alba*
—red (B)	
—bladder (P)	
Campion (A)	*S. noctiflora*
Canary grass—awned (bristle-spiked) (A)	*Phalaris paradoxa*
Carrot—wild (A)	*Daucus carota*
Cat's ear (P)	*Hypochaeris radicata*
Chamomile—corn (A) or (B)	*Anthemis arvensis*
Charlock (yellow) (A)	*Sinapsis arvensis*

Common names	Botanical names
Chervil—rough (B)	*Chaerophyllum temulentum*
Chickweed—common (A)	*Stellaria media*
—mouse-ear (A)	*Cerastium holosteoides*
Cleavers (A)	*Galium aparine*
Coltsfoot (P)	*Tussilago farfara*
Corncockle (A)	*Agrostemma githago*
Cornflower (A)	*Centaurea cyanus*
Corn mint (P)	*Mentha arvensis*
Cornsalad—common (A)	*Valerianella locusta*
Couch—common (P)	*Elymus repens*
—onion (false oat grass) (P)	*Arrhenatherum elatius* (var. *bulbosum*)
Cowbane (water hemlock) (P)	*Cicuta virosa*
Cow parsley (wild chervil) (P)	*Anthriscus sylvestris*
Cow wheat—field (A)	*Melampyrum arvense*
Crane's–bill (A)	*Geranium* spp.
Cress—hoary (P)	*Cardaria draba*
Cuckoo flower (P)	*Cardamine pratensis*
Daisy—common (P)	*Bellis perennis* (*Leucanthemum vulgare*)
Dandelion (P)	*Taraxacum officinale*
Darnel (A)	*Lolium temulentum*
Dead nettle—red (A)	*Lamium purpureum*
—Henbit (A)	*L. amplexicaule*
Docks—broadleaved (P)	*Rumex obtusifolius*
—curled (P)	*R. crispus*
Duckweed—common	*Lemna minor*
—ivy-leaved	*L. triscula*
Fat hen (A)	*Chenopodium album*
Fescue—red (P)	*Festuca rubra*
—sheep's (P)	*F. ovina*
Fiddleneck (A)	*Amsinckia intermedia*
Fleabane—Canadian (A)	*Conyza canadensis*
Flixweed (A)	*Descurainia sophia*
Fool's parsley	*Aethusa cynapium*
Forget-me-not (field) (A)	*Myosotis arvensis*
Foxglove (B) or (P)	*Digitalis purpurea*
Foxtail—meadow (P)	*Alopecurus pratensis*
Fritillary (P)	*Fritillaria meleagris*
Fumitory—common (A)	*Fumaria officinalis*
Gallant soldier (A)	*Galinsoga parviflora*
Garlic—field (P)	*Allium oleraceum*
Gorse (whin, furze) (P)	*Ulex europaeus*
Gromwell—corn (A)	*Lithospermum arvense*
Ground elder (goutweed) (P)	*Aegopodium podagraria*
Ground ivy (P)	*Glechoma hederacea*
Groundsel (A)	*Senecio vulgaris*
Hawkbit—autumn (P)	*Leontodon autumnalis*
—rough (P)	*L. hispidus*
Hawk's-beard—rough (A) or (B)	*Crepis biennis*
—smooth (A) or (B)	*C. capillaris*
Heather (P)	*Calluna vulgaris*
—bell (P)	*Erica cinerea*
Hedge mustard (A)	*Sisymbrium officinale*
Hedge parsley (A)	*Torilis japonica*
Hemlock (A) or (B)	*Conium maculatum*
Hemp nettle—common (A)	*Galeopsis tetrahit*
Henbane (A)	*Hyoscyamus niger*

Common names	Botanical names
Hogweed (B)	*Heracleum sphondylium*
—giant (P)	*H. mantegazzianum*
Horsetail—field (P)	*Equisetum arvense*
—marsh (P)	*E. palustre*
Knapweed (P)	*Centaurea nigra*
Knawel (A) or (P)	*Scleranthus annuus*
Knotgrass (A)	*Polygonum aviculare*
—Japanese (P)	*P. cuspidatum*
Loose silky bent (A)	*Apera spica-venti*
Marigold—corn (A)	*Chrysanthemum segetum*
Mayweed—scented (A)	*Matricaria* or *Chamomilla recutita*
—scentless (A)	*Tripleurospermium maritimum* spp. *inodorum*
—stinking (A)	*Anthemis cotula*
Meadow grass—annual (A)	*Poa annua*
—rough-stalked (P)	*P. trivialis*
—smooth-stalked (P)	*P. pratensis*
Medick—black (A)	*Medicago lupulina*
Mercury—annual (A)	*Mercurialis annua*
—dog's (P)	*M. perennis*
Mignonette—wild, cut-leaved (A) or (P)	*Reseda luta*
Mugwort (P)	*Artemisia vulgaris*
Mustard—black (A)	*Brassica nigra*
—white (A)	*Sinapsis alba*
—treacle (A)	*Erysimum cheiranchoides*
Nettle—common (P)	*Urtica dioica*
—small, annual (A)	*U. urens*
Nightshade—black (A) or (B)	*Solanum nigrum*
—deadly (P)	*Atropa belladonna*
Nipplewort (A)	*Lapsana communis*
Nutsedge—yellow	*Cyperus rotundus*
Oat—bristle (greys) (A)	*Avena strigosa*
—spring, wild (A)	*A. fatua*
—winter, wild (A)	*A. sterilis* spp. *ludoviciana*
Oat grass—downy (P)	*Helictotrichon pubescens*
—false (onion couch) (P)	*Arrhenatherum elatius* (var. *bulbosum*)
Onion—wild (crow garlic) (P)	*Allium vineale*
Orache—common (A)	*Atriplex patula*
Pansy—field (A)	*Viola arvensis*
—wild (A)	*V. tricolor*
Parsley—cow (B) or (P)	*Anthriscus sylvestris*
—fool's (A)	*Aethusa cynapium*
Parsley-piert (A)	*Aphanes arvensis*
Pearlwort (A) or (P)	*Sagina procumbens*
Pennycress—field (A)	*Thlaspi arvense*
Persicaria—pale (A)	*Polygonum lapathifolium* spp. *pallidum*
Pineapple-weed (A)	*Chamomilla suaveolens* (*Matricaria matricarioides*)
Plantain—greater (P)	*Plantago major*
—ribwort (narrow-leaved) (P)	*P. lanceolata*
Poppy—corn or common field (A)	*Papaver rhoeas*
—opium (A)	*P. somniferum*
—Californian (A)	*Eschscholzia californica*
Primrose (P)	*Primula vulgaris*
Radish—wild (white charlock) (A) or (B)	*Raphanus raphanistrum*

Common names	Botanical names
Ragwort—common (B) or (P)	*Senecio jacobia*
—Oxford (B)	*Squalidus*
Ramsons (P)	*Allium ursinum*
Red bartsia (P)	*Odontites verna*
Redshank (A)	*Polygonum persicaria*
Reed—common (P)	*Phragmites australis (P. communis)*
Restharrow—common (P)	*Ononis repens*
—spiny (P)	*O. spinosa*
Rush—jointed (P)	*Juncus articulatus*
—soft (common) (P)	*J. effusus* and *conglomeratus*
—hard (P)	*J. inflexus*
—heath (P)	*J. squarrous*
Saffron—meadow (P)	*Colchicum autumnale*
St. John's wort (P)	*Hypericum perforatum*
Scabious—field (P)	*Knautia arvensis*
Scarlet pimpernel (A)	*Anagallis arvensis*
Sedges (P)	*Carex* spp.
Selfheal (P)	*Prunella vulgaris*
Shepherd's-needle (A)	*Scandix pecten-veneris*
Shepherd's-purse (A)	*Capsella bursa-pastoris*
Silverweed (P)	*Potentilla anserina*
Soft grass—creeping (P)	*Holcus mollis*
Sorrel—common (P)	*Rumex acetosa*
—sheep's (P)	*R. acetosella*
Sow-thistle—corn or perennial (P)	*Sonchus arvensis*
—smooth, milk (A)	*S. oleraceus*
Speedwell—common, field (Buxbaum's) (A)	*Veronica persica*
—germander (P)	*V. chamaedrys*
—procumbent (A)	*V. agrestis*
—ivy-leaved (A)	*V. hederifolia*
Spurge—sun (A)	*Euphorbia helioscopia*
—dwarf (A)	*E. exigua*
Spurrey—corn (A)	*Spergula arvensis*
Stork's-bill—common (A) or (B)	*Erodium cicutarium*
Thistle—creeping (field) (P)	*Cirsium arvense*
—spear (Scotch) (B)	*C. vulgare*
Toadflax—common (P)	*Linaria vulgaris*
Traveller's-joy (old man's beard)	*Clematis vitalba*
Trefoil hop (A)	*Trifolium campestre*
Tussock grass (P)	*Deschampsia cespitosa*
Venus's-looking-glass (A)	*Legousia hybrida*
Vetch—common (tares) (A) or (B)	*Vicia sativa*
—kidney (P)	*Anthyllis vulneraria*
Viper's bugloss (B)	*Echium vulgare*
Water-dropwort (hemlock) (P)	*Oenanthe crocata*
Watergrass (P)	*Agrostis stolonifera*
Willow-herb (rosebay or fireweed) (P)	*Epilobium angustifolium (Chamerion angustifolium)*
Woodrush—field (P)	*Luzula campestris*
Yarrow (P)	*Achillea millefolium*
Yellow iris or flag (P)	*Iris pseudacorus*
Yellow rattle (A)	*Rhinanthus minor*
Yorkshire fog (P)	*Holcus lanatus*

FURTHER READING

The Dictionary of Weeds of Western Europe, Ed. G. H. Williams, Elsevier.

APPENDIX 6

CROP DISEASES

Crop names	Common names	Botanical names
Cereals		
All cereals	mildew	*Erysiphe graminis*
Wheat (barley)	yellow rust	*Puccinia striiformis*
Barley	brown rust	*P. hordei*
Wheat	brown rust	*P. recondita*
Oats	crown rust	*P. coronata*
Barley	leaf blotch (rhynchosporium)	*Rhynchosporium secalis*
Wheat (barley)	glume blotch	*Septoria nodorum*
Wheat (barley)	leaf spot	*S. tritici*
Oats	dark leaf spot, speckle blotch	*Leptosphaeria avenaria*
Oats	stripe, leaf spot, seedling blight	*Pyrenophora avenae*
Barley	leaf stripe	*P. graminae*
Barley	net blotch	*P. teres*
Oats	halo blight	*Pseudomonas coronafaciens*
Rye, wheat, barley	ergot	*Claviceps purpurea*
Wheat	bunt, covered, stinking smut	*Tilletia caries*
Barley (oats)	bunt, covered, smut	*Ustilago hordei*
Oats	bunt, covered, smut	*U. avenae*
Maize	bunt, covered, smut	*U. maydis*
Wheat, barley	loose smut (different races on wheat and barley)	*U. nuda*
Wheat, barley, oats	eyespot	*Pseudocercosporella hervotrichoides*
Wheat, barley, oats	sharp eyespot	*Rhizoctonia cerealis*
Wheat, barley	take-all	*Gaeumannomyces graminis* var. *graminis*
Oats	take-all	*G. graminis* var. *avenae*
Wheat, barley, oats	brown foot rots (and ear blight)	*Fusarium* spp.
Wheat, barley	black (sooty) mould	*Cladosporium herbarum*
Wheat (barley), oats, maize	scab	*Gibberella zeae*
Maize	stem rot	*Fusarium* spp.
Winter cereals (esp. barley)	snow rot	*Typhula incarnata*
Potatoes	blight	*Phytophthora infestans*
	pink rot	*P. erythroseptica*
	common scab	*Streptomyces scabies*
	powdery scab	*Spongospora subterranea*
	gangrene	*Phoma exigua* var. *foveata*
	watery wound rot	*Pythium ultimum*
	wart disease	*Synchytrium endobioticum*
	sclerotinia	*Sclerotinia sclerotiorum*
	blackleg	*Erwinia (carotovora)* var. *atroseptica*
	skin spot	*Oospora pustulens*
	black scurf and stem canker	*Corticum (Rhizoctonia) solani*
	dry rot	*Fusarium caeruleum*

Crop names	Common names	Botanical names
	silver scurf	*Helminthosporium solani*
	ring rot	*Corynebacterium sepedonicum*
Sugar beet	blackleg	*Pleospora bjoerlingii*
	downy mildew	*Peronospora farinosa*
	powdery mildew	*Erysiphe* species
	leaf spot	*Ramularia beticola*
	rust	*Uromyces betae*
	violet root rot	*Helicobasidium purpureum*
	rhizomania	
Legumes		
Peas, beans	downy mildew	*Peronospora vicia*
Peas	leaf and pod spot	*Ascochyta pisi*
Beans	leaf and pod spot	*A. fabae*
Beans	chocolate spot	*Botrytis cinerea. B. fabae*
Beans,	stem rot (bean sickness)	*Sclerotinia trifolium*
Red clover	clover rot (sickness)	*S. trifolium*
White clover	phyllody virus	
All legumes	damping-off of seedlings	*Pythium* spp.
Peas	pea wilt	*Fusarium oxysporium*
Lucerne	vertieillimm wilt	*Verticillium albo atrum*
Lucerne	bacterial wilt	*Corynebacterium insidiosum*
Peas, beans	grey mould	*Botrytis cinerea*
Peas	powdery mildew	*Erysiphe pisi*
Peas	halo blight	*Pseudomonas phaseolicola*
Peas	anthracnose	*Colletotrichum lindemuthianum*
Peas	downy mildew	*Peronospora viciae*
Brassicae	club root (finger and toe)	*Plasmodiophora brassicae*
	powdery mildew	*Erysiphe cruciferarum*
	downy mildew	*Peronospora parasitica*
	white blister	*Cystopus candidus*
	light leaf spot	*Cylindrosporium concentricum*
	dark leaf and pod spot	*Alternaria* spp.
	stem canker and leaf spot	*Phoma lingam*
	ring spot	*Mycosphaerella brassicicola*
	soft rot	*Pectobacterium cartovorum*
	stem rot	*Sclerotinia sclerotiorum*
	wilt	*Fusarium oxysporum*
	white leaf spot	*Pseudocercosporella capsellae*
Onions	white rot	*Sclerotium cepivorum*
	downy mildew	*Peronospora destructor*
	neck rot	*Botrytis allii*
	leaf spot	*Botrytis* spp.
Grasses	choke	*Epichloe typhina*
	blind seed disease	*Gloeitinia temulenta*
	ergot	*Claviceps purpurea*
	mildew	*Erysiphe graminis*
	rusts	*Puccinia* species
	leaf fleck	*Mastigosporium rubricosum*
	leaf blotch	*Rhynchosporium secalis*
	crown rust	*Puccinia coronata*

INSECT PESTS

Crop	Common names	Latin names
Cereals		
Wheat, barley, oats	cyst nematode (root eelworm)	*Heterodera avenae*
Oats	stem and bulb eelworm	*Ditylenchis dipsaci*
Wheat	bird cherry aphid	*Rhopalosiphum padi*
	grass aphid	*Metopolophium festucae*
	rose grain aphid	*Metopolophium dirhodum*
	grain aphid	*Macrosiphum avenae*
All	leatherjackets	*Tipula* spp. and *Nephrotoma maculata*
All	wireworms	*Agriotes* spp.
All	slugs	*Agriolimax reticulatus*
Wheat	wheat bulb fly	*Leptohylemyia coarctata*
Wheat	opomyza flies	*Opomyza florum*
Barley	gout fly	*Chlorops pumilionis*
Oats, barley, wheat	frit fly	*Oscinella frit*
	saddle-gall midge	*Haplodiplosis equestris*
	grain weevils	*Sitophilus* spp.
	saw-toothed grain beetle	*Oryzephilus surinamensis*
	mites	*Acarus siro*
Potatoes	root eelworm (pathotype A)	*Heterodera rostochiensis*
	yellow cyst nematode (pathotype Ro 1)	*Globodera rostochiensis*
	white cyst nematode	*Globodera pallida*
	aphid—peach potato	*Myzus persicae*
	aphid—"false top roll"	*Macrosiphum euphorbiae*
	Colorado beetle	*Leptinotarsa decemlineata*
	cutworms (noctuid moths)	
	e.g. turnip moth	*Agrotis segetum*
	heart and dart moth	*A. exclamationis*
	large yellow underwing	*Noctura pronuba*
	garden dart moth	*Euxoa nigricans*
	slugs, leatherjackets, wireworms, see Cereals	
Sugar beet	cyst nematode (root eelworm)	*Heterodera schachtii*
	aphid—peach potato	*Myzus persicae*
	aphid—black bean (blackfly)	*Aphis fabae*
	mangels fly	*Pegomyia hyoscyami* var. *betae*
	millipedes	*Blaniulus guttulatus*
	millipedes	*Brachydesmus superus*
	docking disorder	*Longidorus* and *Trihodorus* spp.
	cutworms, slugs, wireworms, leatherjackets, see Potatoes and Cereals	

Crop	Common names	Latin names
Peas and Beans	pea and bean weevil	*Sitona* spp.
	pea cyst nematode	
	(pea root eelworm)	*Heterodera gottingiana*
	stem and bulb eelworm	*Ditylenchus dipsaci*
	black bean aphid (blackfly)	*Aphis fabae*
Brassicae	flea beetle	*Phyllotreta* spp.
	cabbage root fly	*Erioischia brassicae*
	blossom (pollen) beetle	*Meligethes aeneus*
	seed weevil	*Ceuthorhynchus assimilis*
	mealy cabbage aphid	*Brevicoryne brassicae*
	cabbage stem flea beetle	*Psylliodes chrysocephala*
	bladder pod midge	*Dasyneura brassica*
	cabbage white butterflies	*Pieris* spp.
Onions	onion fly	*Hylemyia antiqua*
Carrots	carrot fly	*Psilia rosae*
	carrot willow aphid	*Cavariella aegopodii*

CROP SEEDS

THE FOLLOWING are average figures and are only intended as a general guide and for comparisons. Precision drilling of crops requires seed counts per kilogram to be known for the stock of seed being sown and merchants will usually supply these figures.

Crop	1000 seeds weight (g)	Seeds per kilogram (000's)	Seeds per m² for every 10 kg/ha sown	kg/hl
Cereals—wheat	48	21	21	75
barley	37	27	27	68
oats	32	31	31	52
maize	285	3.5	3.5	75
Grasses—ryegrass—Italian	2.2	455	455	29
ryegrass—Italian tetraploid	4	250	250	
ryegrass—hybrid	2.15	465	465	
ryegrass—perennial S 24	2	500	500	35
ryegrass—perennial tetraploid	3.3	303	303	
cocksfoot	1	1000	1000	33
Timothy	0.3	3333	3333	62
meadow fescue	2	500	500	37
tall fescue	2.5	400	400	30
Clovers—red	1.75	571	571	80
white—cultivated	0.62	1613	1613	82
wild	0.58	1724	1724	82
Lucerne (alfalfa)	2.35	425	425	77
Sainfoin—milled	21	48	48	
Peas—marrowfats	330	3	3	
large blues and whites	250	4	4	78
small blues	200	5	5	
Beans—broad	980	1	1	
winter and horse types	670	1.5	1.5	
tick	410	2.4	2.4	80
dwarf, green, French	500	2	2	
Vetches (tares)	53	19	19	77
Linseed and flax	10	100	100	68
Carrots	1.5	660	660	
Onions—natural seed	4	246	246	
mini-pellets	18	55	55	
Sugar beet—pelleted	62	16	16	
Cabbages	4.1	240	240	
Kale	4.5	220	220	
Swedes	3.6	280	280	

Crop	1000 seeds weight (g)	Seeds per kilogram (000's)	Seeds per m² for every 10 kg/ha sown	kg/hl
Turnips	3	330	330	
Oilseed rape	4.5	220	220	
Brussels sprouts	4.7	210	210	
Cauliflower	4.1	240	240	

FACTORS AND LEGISLATION AFFECTING THE APPLICATION (AND MIXING) OF CROP PROTECTION CHEMICALS

Weather conditions

Normally, dry settled conditions are preferable for most spray chemicals. The effects of rain after spraying vary according to the chemical involved; some, such as paraquat, are not affected by rain shortly after application, whereas others, such as glyphosate and contact herbicides, require at least six to eight hours of dry weather if they are to act efficiently. A number of wild oat herbicides may benefit from light rain after application to concentrate the herbicide on the lower part of the leaf blade where it acts more effectively. Frost on leaves at time of spraying can affect the intake of the chemical, and cold poor growing conditions can reduce the effectiveness of growth-regulating herbicides such as MCPA and 2,4-D.

Windy conditions prevent uniform application and may cause problems with drift on to neighbouring crops. Following an application of some ester-formulated herbicides, a short period of warm weather may cause these to become volatile, damaging susceptible neighbouring crops such as oilseed rape by drift.

Crop conditions

Certain herbicides, e.g. the wild oat herbicide flamprop-M-isopropyl, rely on good crop competition to work satisfactorily. Poor results may be obtained if the crop is under stress. Some crops are more susceptible to crop damage by pesticides if they are under stress. Size of target weed or stage of disease development can affect the effectiveness of pesticide control.

Soil conditions

Soil-acting residual herbicides work better when the soil is in a damp, finely divided condition. Depth of drilling can be important in reducing crop damage.

Volatile chemicals, e.g. tri-allate, EPTC and trifluralin, must be incorporated into the soil which should be in a free-working condition and as free of stones and clods as possible.

Most residual herbicides are not recommended for use on soils with more than 10% organic matter. This is because there is poor activity of the chemical due to its being adsorbed on to the organic matter. On some clay soils, where minimum cultivations have been undertaken for several years, residual herbicide activity may be reduced. Again, high soil adsorption of the residual herbicide may be to blame. Rotational ploughing relieves the problem. Application rates of residual herbicides on some light soils are often reduced to avoid crop damage. Some residual herbicides are not recommended on sands.

Wet conditions may prevent the use of ordinary ground sprayers, and so aircraft (for approved chemicals) or special low ground pressure vehicles may have to be used.

Formulation

Most spray chemicals are formulated in water solutions to be diluted with water, but the less soluble ones are often formulated as wettable powders which have to be mixed carefully with some water before adding to the tank. These materials are now being marketed ready mixed with water to make them easier for the operator to use. A number of chemicals which are only soluble in oil are formulated as emulsions.

A few chemicals, e.g. tri-allate, may also be formulated as very small granules to be applied on the soil surface by a special applicator. There are others formulated as very fine powders which are made to stick to the leaves by electrostatic charges.

There are many factors which influence the formulation of a product such as the properties of the active ingredient, transport and storage stability, ease of application, activity against weeds, pests or diseases, crop tolerance and all aspects of safety.

Mixing two or more chemicals

The application of a single product for the control of weeds, pests or diseases is a fairly simple operation provided the weather and soil conditions are suitable and the sprayer is in good working order. Problems arise when several chemicals are required at the same time—such as one or more herbicides, fungicides, insecticides, trace elements and, possibly, chlormequat. Some of these can be mixed without reducing the desired effects (sometimes, an improved effect may be obtained). However, in many cases it is very risky to mix chemicals either because of reduced effect or because of damaging effects on the crop, such as scorch, yield reduction and toxicity to consumers. Information is given with each product concerning its compatibility with some other products, but this is of limited value.

Method of application

The ordinary agricultural sprayers are fitted with hydraulic nozzles to control the flow and disperse the liquid into drops. There are two types of nozzle—the fan (the most common) and the hollow cone (or swirl). In both types the liquid is forced through the orifice as a sheet which then breaks into drops of various sizes. However, some of these drops are so small that they drift away, whilst others are so large that they bounce off the plants and are useless with those chemicals which should remain on the foliage. The farm sprayer is used for applying many different chemicals for many different purposes; the results are reasonably acceptable because, normally, there will be some drops of the correct size for the application of the chemical being used. This is a waste of chemical in situations where only a part of the spray is retained on the foliage, although the recommended dose rates allow for this loss. The need for very accurate, uniform coverage is not so important with systemic chemicals which can move through the plant, or with soil-acting residual chemicals. However, it is important with contact chemicals, such as some herbicides, insecticides and fungicides, which only act at the point of contact.

To maximize contact with the concentrated pesticide, many spray manufacturers are developing closed spraying transfer systems.

The research and development work with tooth-edged spinning discs, starting with the Micron Herbi and Ulva hand held battery operated applicators and the types from the AFRC Silsoe Research Institute, showed that this method of application could produce droplets of very uniform size. Furthermore, the size of droplet could be varied, as required, by changing the speed of the discs. This is known as CDA (controlled droplet application). Development work has concentrated on producing sophisticated tractor-mounted or light self-propelled machines.

Chemicals such as glyphosate, sometimes specially formulated, can be applied selectively by a machine, such as the weed-wiper, to tall weeds above the level of the crop leaves, e.g. weed beet in sugar beet, tussock grass and thistles in grassland; there are no drift problems and the chemical is only taken up by the weed.

The volume rates normally used for spraying farm chemicals are:

high volume	650–1100 litres/hectare
medium volume	220–650 litres/hectare
low volume	55–220 litres/hectare
very low volume	up to 55 litres/hectare
ultra low volume and/or CDA	0–50 litres/hectare

Farm sprayers are available in many sizes and forms. Those with very wide spray booms are difficult to use on undulating ground. Many ingenious devices have been developed to prevent the booms swinging about and bouncing excessively. The tramline system for spraying cereal crops has solved many of the problems of covering a field quickly and accurately (avoiding overlaps and misses).

Safety and approval of spray chemicals

It is important that the instructions on the leaflets which accompany spray chemicals should be followed as carefully as possible, especially when dealing with the more poisonous types. An operator who has been working with pesticides and who feels unwell should seek medical attention immediately, drawing attention to the spray leaflets so that proper treatment can be given.

Regulations concerning the control of pesticides

The Food and Environment Act (FEPA) 1985 covers:

(1) the continuous development of means to
     (i) protect the health of human beings, creatures and plants;

(ii) safeguard the environment;

(iii) secure safe, efficient and humane methods of controlling pests;

(2) making information about pesticides available to the public.

This Act provides Ministers with power—by regulation or order—to control the import, sale, supply, storage, use and advertisement of pesticides, and to approve pesticides. The Control of Pesticides Regulations 1986, made under the Act, replaces the former voluntary Pesticides Safety Precautions Scheme with statutory powers to approve pesticides, making it an offence to sell, supply, store, use or advertise an unapproved pesticide.

From October 1986 it has been mandatory that only approved pesticides may be supplied, stored or used, and only provisionally or fully approved pesticides may be sold. Only pesticides specifically approved for aerial application may be applied from the air; very detailed rules are imposed on aerial application. General obligations are imposed on the marketing, storage and use of pesticides. All reasonable precautions must be taken to protect the health of people, animals and plants; to safeguard the environment and avoid pollution of water. All those who handle pesticides must be competent in their duties, and employers must ensure that commercial users have received adequate instruction and guidance.

Only provisionally or fully approved pesticides may be advertised and only in relation to their approved use; the active ingredient must be stated. A certificate of competence is required for any person who stores approved pesticides for sale or supply.

A certificate of competence is also required for a person who sells or supplies pesticides.

More specific control on the use of pesticides, such as maximum application rate and minimum harvest interval following application, must also be followed.

The Agricultural Training Board publishes a list of approved pesticide training programmes. The recognized tests of competence are those of the National Proficiency Tests Council (MTPC).

A certificate of competence is also required by contractors and those under 25 years old who are using pesticides for agriculture, horticulture and forestry unless working under the direct and personal supervision of a certificate holder. Tank mixing of pesticides will be controlled. Only MAFF listed adjuvants may be used.

The use of pesticides is legal, provided the conditions of approval and any other controls over their use in the Act and the Regulations are observed.

The Control of Substances Hazardous to Health Regulations (COSHH) 1988 came into force in October 1989. The aim of COSHH is to reduce or eliminate exposure to substances classified as "hazardous to health". The risks associated with the use of any substance hazardous to health (e.g. pesticides, harmful micro-organisms, dusts) must be assessed before it is used and the necessary action taken to control the risk (e.g. a less toxic product should be used if possible). Staff must be adequately trained and informed. All operations involving pesticides must be recorded and retained for at least three years.

A list of pesticides published annually by HMSO shows the approved and provisionally approved products under the 1986 Regulations. The brand name, active ingredient, marketing company and approval number (registration number) are listed alphabetically for each product and grouped as herbicides, fungicides, insecticides, etc.

FURTHER READING

Code of Practice for Suppliers of Pesticides for Agricultural, Horticultural and Forestry Use, MAFF.
COSHH Assessments, HSE Publications.

COSHH in Agriculture (1989), HSE Publications.
Crop Spraying—Agricultural Safety (1988), HSE Publications.
Pesticide-related Law, British Crop Protection Council.
Pesticides: Code of Practice for the Safe Use of Pesticides on Farms and Holdings (1990), HMSO.
Pesticides Register (current edition), MAFF/HSE.
The UK Pesticide Guide 1993, Ed. G. W. Ivens, British Crop Protection Council.
Training in the Use of Pesticides (1990), HSE Publications.
UK Pesticides for Farmers and Growers, The Royal Society of Chemistry.
Working with Pesticides Guide: The Regulations and Your Responsibilities (1990), Schering Agriculture.

APPENDIX 10

METRICATION

IT HAS been assumed that, when this edition is published, metrication will be well established in agriculture and so only metric units are used throughout the text. There is still considerable doubt about the use of certain units; for example, when metrication was introduced it was clearly understood that centimetres should only be used in very exceptional circumstances, but they are now being commonly used in many cases, e.g. spacing of crop plants. In the text, centimetres are used where it would seem they are more sensible than millimetres (note, 1 centimetre = 10 millimetres).

Many farmers and other agriculturalists are likely to use Imperial units for many years to come and so the following simplified conversion factors may be helpful. They are correct to within 2% error, which is acceptable in agriculture!

The following factors are approximate conversions for easy mental calculations; for accuracy, reference should be made to conversion tables. To convert metric to Imperial the multiplying factors should be inverted.

Imperial	Metric	Multiply by	Example
inches	to millimetres	$^{100}/_4$	12 in = 300 mm
feet	to metres	$^{3}/_{10}$	30 ft = 9 m
yards	to metres	$^{9}/_{10}$	100 yd = 90 m
square feet	to square metres (m^2)	$^{1}/_{11}$	55 sq ft = 5 m^2
square yards	to square metres (m^2)	$^{10}/_{12}$	60 sq yd = 50 m^2
cubic feet	to cubic metres (m^3)	$^{11}/_{400}$	800 cu ft = 22 m^3
cubic yards	to cubic metres (m^3)	$^{3}/_4$	12 cu yd = 9 m^3
nos. per foot	to nos. per metre	$^{10}/_3$	30 per ft = 100 per metre
nos. per sq ft	to nos. per sq metre	11	20 per sq ft = 220 per sq metre
lb per cu ft	to kg per cu metre	16(4 × 4)	50 lb/ft^3 = 800 kg/m^3
lb per bushel	to kg per hectolitre	$^{10}/_8$	60 lb bush = 75 kg/hl
pounds	to kilograms	$^{9}/_{20}$	100 lb = 45 kg
chain (22 yards)	= chain (20 metres)		
acres	to hectares	$^{4}/_{10}$	16 acres = 6.4 ha
nos. per acre	to nos. per hectare	$^{10}/_4$	40 per acre = 100 per ha
fert. units per acre	to kg per hectare	$^{10}/_8$	40 units/ac = 50 kg/ha
cost per unit	to cost per kilogram	2	9p/unit = 18p/kg
tons	to tonnes	1	for accuracy, × by 1.016
cwt	to kilograms	50	for accuracy, × by 50.8
tonnes per acre	to tonnes per hectare	$^{10}/_4$	2 tons/ac = 5 tonnes/ha
cwt per acre	to tonnes per hectare	$^{1}/_8$	32 cwt/ac = 4 tonnes/ha
cwt per acre	to kg per hectare	$^{1000}/_8$	2 cwt/ac = 250 kg/ha
pounds per acre	to kg per hectare	$^{11}/_{10}$	10 lb/ac = 11 kg/ha
pints per acre	to litres per hectare	$^{7}/_5$ or $^{11}/_8$	5 pints/ac = 7 litres/ha
gallons per acre	to litres per hectare	11	20 gal/ac = 220 litres/ha

Imperial	Metric	Multiply by	Example
tons per cubic yard	to tonnes per cubic metre	$\frac{4}{3}$	1 ton/yd^3 = 1.3 tonnes/m^3
pints	to litres	$\frac{4}{7}$	7 pints = 4 litres
gallons	to litres	$\frac{9}{2}$	1100 gal = 5000 litres
lb per gallon	to kg per litre	$\frac{1}{10}$	4 lb/gal = 0.4 kg/litre
pence per lb	to pence per kg	$\frac{20}{9}$	18p/lb = 40p/kg
horsepower	to kilowatts	$\frac{3}{4}$	100 h.p. = 75 kW
acres per hour	to hectares per hour	$\frac{4}{10}$	5 ac/hour = 2 ha/hour
lb ft in^2 (psi)	to kilopascals (kPa)	7	28 psi = 196 kPa (2 bars)
lb ft in^2 (psi)	to bars	$\frac{7}{100}$	200 psi = 14 bars
irrigation inch/acre	= cm/ha = 100 m^3		

Index